Bizarre Bioethics

BIZARRE BIOETHICS

Ghosts, Monsters, and Pilgrims

Henk A.M.J. ten Have, MD, PhD

Johns Hopkins University Press

Baltimore

© 2022 Johns Hopkins University Press
All rights reserved. Published 2022
Printed in the United States of America on acid-free paper
2   4   6   8   9   7   5   3   1

Johns Hopkins University Press
2715 North Charles Street
Baltimore, Maryland 21218-4363
www.press.jhu.edu

Library of Congress Cataloging-in-Publication Data
Names: ten Have, H., author.
Title: Bizarre bioethics : ghosts, monsters, and pilgrims /
Henk A.M.J. ten Have, MD, PhD.
Description: Baltimore : Johns Hopkins University Press, 2022. |
Includes bibliographical references and index.
Identifiers: LCCN 2021018586 | ISBN 9781421443027 (hardcover ; alk. paper) |
ISBN 9781421443034 (paperback ; alk. paper) | ISBN 9781421443041 (ebook)
Subjects: MESH: Bioethical Issues | Metaphor | Decision Making—ethics |
Internationality | Socioeconomic Factors
Classification: LCC R724 | NLM WB 60 | DDC 174.2—dc23
LC record available at https://lccn.loc.gov/2021018586

A catalog record for this book is available from the British Library.

Special discounts are available for bulk purchases of this book. For more
information, please contact Special Sales at specialsales@jh.edu.

For Felix, miracle and hope

I have seen no more evident monster or miracle in the world than myself (Je n'ay veu monstre et miracle au monde plus expres que moy-mesme)

    —Michel de Montaigne, *Essais,* Livre III, Chapitre XI, 1029

CONTENTS

Bizarre Bioethics

CHAPTER ONE

Questioning the Paradigm of Bioethics

"Bioethics" is an academic discipline that emerged after the 1970s and is now firmly embedded in medical and nursing schools and many health care facilities across the world. It is a broad field of discussion around ethical concerns about illness, disability, death, suffering, and care as reflected in popular discourse. Not merely a domain of specific scholarly expertise, bioethics concerns all people because health and disease, aging and dying, are universal human experiences. Expanding medical knowledge and technologies confront health professionals, scientists, and policy makers as well as patients and the general public with questions about what should be done given increasing opportunities for prevention, diagnosis, and intervention. Such questions arise at the level of general debate, triggering controversies concerning, for example, experimental drugs, gene-edited babies, euthanasia, and life extension. They also appear in specific cases of patient care, posing difficult challenges for patients, families, and caregivers in determining the best care for an individual patient.

As an academic approach, bioethics has developed methods and procedures to deal with such questions. It usually examines scientific and technological possibilities, balancing their harms and benefits, in order to guide their application. This approach, however, is too simplistic. Science and technology are not value-free endeavors that provide facts and evidence. Many times, the values incorporated in medical knowledge and technological practices are not critically examined in ethical analyses. Bioethics, furthermore, is focused on individual cases rather than the larger context in which ethical problems emerge. This is true for patients who are usually surrounded by families and communities, but it is equally so for institutions that are part of larger health and insurance systems and embedded in social, political, and economic contexts. Care for individual patients is often related to larger issues of justice, solidarity, discrimination, or racism. Finally, bioethics proceeds using rational explanation and justification. It assumes that individuals are rational decision makers, even when they are diseased, disabled, or dying, and the role of feelings and emotions is frequently downgraded in the analysis of ethical problems. To counteract these criticisms, various ethical approaches have developed, such as casuistry, narrative ethics, and care ethics.

In this book, I explore a different viewpoint. I start with the question of how the agenda of ethical debate is determined. It seems that frequently in the public realm, the ethical issues that capture attention are extraordinary and rare—what I call "bizarre bioethics." Although these cases are undeniably real, they present a limited and skewed view of everyday moral reality. That they draw enormous public and media attention points to their power to elicit value perspectives that are often not taken into account in expert bioethical analyses; that is, perspectives that go beyond the level of individual patient care, that engage emotions and feelings, and that invoke notions such as hope, faith, miracles, and vulnerability.

The thesis of this book is that current bioethical debate is bizarre because it concentrates on exceptional cases but does not pay sufficient attention to underlying value perspectives that in fact influence the agenda of the debate. These perspectives are often based on metaphors and worldviews from philosophical and theological traditions. Metaphors provide ways of thinking and shape the imagination. They help to make sense of phenomena in the world.[1] Ghosts, monsters, pilgrims, prophets, and relics, as examples, can illustrate how science and medicine are animated by imaginations that fuel the search for hope, salvation, healing, and a predictable future. Bioethics is usually not critically addressing these value perspectives because philosophical and theological traditions that used to articulate them have been exorcised and marginalized. The assumption in this book is that bioethics can instead learn from these traditions to develop a broader approach that critically explores the interpretive frameworks and imaginative views that determine our understanding of the world and human existence. This broader approach is especially needed now that academic bioethics is increasingly confronted with social media and popular protest movements.

Charlie Gard

The case of Charlie Gard dominated the news media during the summer of 2017. Apparently healthy during his first few weeks of life, Charlie ultimately demonstrated signs of muscle weakness and increasing difficulties with feeding and breathing. By two months old, he was admitted to intensive care in Great Ormond Street Hospital in London where he was diagnosed with an extremely rare genetic disease, infantile-onset encephalomyopathic mitochondrial DNA depletion syndrome (MDDS). The disease was rapidly progressing. It was discovered, too, that Charlie was deaf, and he eventually became paralyzed. He needed continuous respiratory support. His heart, liver, and kidneys were affected and he developed persistent seizures. The hospital team emphasized that his prognosis was extremely poor and recommended only palliative care.[2]

In early January 2017, Charlie's parents searched the internet and found an experimental treatment, called nucleoside therapy, used for patients with other mitochondrial diseases. They contacted the researcher, Dr. Michio Hirano, who worked as a neurologist specializing in neuromuscular diseases at Columbia University Medical Center in New York. Dr. Hirano explained that the nucleoside therapy had never been tested on animals or on patients with Charlie's disorder. However, it was a noninvasive oral therapy and had not caused significant side effects when administered to other patients. Hirano offered to treat Charlie at his hospital in the United States. The parents discussed this potential therapy with the physicians in London. The doctors argued that they had explored this possibility in an earlier stage and had prepared an application to an ethics committee for experimental use. However, that application was withdrawn after Charlie developed persistent seizures. The team concluded that he now had irreversible brain damage and any chance of benefit from the experimental treatment was gone.

Soon after the parents started a public campaign. They believed that Charlie deserved a chance to try the new medication. The parents set up a crowdfunding page to raise money for treatment and received an overwhelming response. Within a few months they raised almost $1.7 million from more than 80,000 donors.[3] The campaign spread worldwide and catapulted the case into the news media. More than 350,000 people signed a petition to allow Charlie to travel to the United States. Followers of social media created a group known as "Charlie's Army" dedicated to saving Charlie. The hospital, however, went to court. It demanded three legal orders: permission to withdraw artificial ventilation, provide palliative care only, and prevent nucleoside therapy. All three requests were justified using the best-interest standard of decision-making for Charlie's treatment. In April 2017, the judge ruled in favor of the hospital. Charlie's parents appealed, first to the Court of Appeals

(May 2017), then to the UK Supreme Court (June 2017), and finally to the European Court of Human Rights (June 2017). All appeals were dismissed.

These legal decisions created a firestorm on the internet. Pope Francis and US President Donald J. Trump intervened and offered support for Charlie. A group of international experts forwarded and published a letter to Great Ormond Street Hospital with new evidence about the experimental treatment and offered the treatment in the United States and Italy. The hospital team decided to reopen the case and submit the new evidence for legal assessment in a court hearing while at the same time maintaining their position. They argued that Charlie's brain was irreversibly damaged and that the experimental treatment would be futile. Furthermore, they reiterated that Charlie was suffering every day, and being alive was not in his best interest. The hospital also blamed Dr. Hirano. He had not examined Charlie or read his medical records—which suggests that he gave false hope to the parents and prolonged Charlie's suffering.[4] In fact, the judge indicated that the new evidence was not compelling and criticized Hirano for not seeing the patient. He repeated his previous conclusion that it was in the best interest of Charlie "to be allowed to slip away peacefully."[5]

A week earlier, a multidisciplinary meeting of experts had recommended further MRI scans of Charlie's body. The scans showed that muscle tissue was disappearing. The physicians, including the foreign experts, had to conclude that Charlie was beyond help. This time, the parents agreed. They admitted that the battle to save Charlie was lost. The parents expressed their final wishes to bring Charlie home and spend some time with him to say good-bye. The hospital resisted. It argued that Charlie could not spend any significant time outside of an intensive care environment, as it would only prolong his suffering. The judge ordered that he be transferred to a children's hospice. He gave the parents twenty-four hours to find a specialized team to provide intensive care in the hospice. When the parents were unable to find such a team, even when several nurses of Great Ormond Street Hospital volunteered to help, the judge ruled that Charlie should be transferred to the hospice and disconnected from artificial ventilation "shortly thereafter." Charlie died the next day, July 28, 2017.

Bizarre Cases

Why did the case of Charlie Gard attract so much attention across the globe? At first it seems bizarre to focus on this case. MDDS is an extremely rare disease. In the medical literature, only fifteen infants have ever been diagnosed with the condition, which means that scientific expertise with the disease is very limited. It was peculiar to focus on such a unique, controversial issue as opposed to more common pressing ones. For example, a few days after Charlie died, it was Earth Overshoot Day. On August 2, 2017, humanity had reached its demand for more planetary resources for that year than Earth could regenerate. In other words, we had depleted the yearly biocapacity of the planet, and this overconsumption essentially jeopardizes the survival of humankind in the long run.[6] While the death of Charlie was widely discussed in the media, almost nothing was reported on the observation that human beings are rapidly exhausting the resources of our planet—the only common home we have. For neutral observers, the exhaustion of planetary resources would seem a much more fundamental bioethical challenge than Charlie's individual calvary.

A similar argument is made when comparing the frequency of problems. In the same year that Charlie died of his rare disorder, many more children were suffering and dying from common diseases. UNICEF, for example, announced in February 2017 that over 1,400 children under five years old die each day from diarrhea. This equates to the death of about 527,000 children annually. This is a serious ethical problem because these diseases are easily preventable, and simple and effective treatments that do not require sophisticated technologies of intensive care are available.[7] When we are concerned about the death and suffering of children, our attention would be more effectively focused on the frequent and preventable diseases.

What explains the public concern with Charlie's case if it was not related to the severity or frequency of bioethical problems? What determines the agenda of bioethical debate? One answer rests on the fact that bioethics discourse has developed based on cases of either renowned legal struggles (for example, the Karen Ann Quinlan case in 1975) or scandals in research and medical practice (such as the Tuskegee syphilis research revealed in 1972). In this regard, the case of Charlie follows the typical pattern of evolution in the history of bioethics. But this emphasis on individual cases has limitations. The attractiveness of centering on a real person illustrates the relevancy and concrete concerns of ethics but at the same time it decenters broader concerns related to the social, economic, and cultural context of the case. These concerns seem too abstract, theoretical, and distanced when compared to the real worries and complexities of the person suffering in the case. The focus on cases is, therefore, a way to manage the tension between individual and collective tragedy.

It is also assumed that ethics is initially a matter of personal conscience and choice. In Charlie's case, responsible actors such as parents and health professionals were considering Charlie's well-being and making decisions regarding his personal fate. Even if the social, economic, and environmental context of such decisions is important, it is not fair in this perspective to compare the death of Charlie with that of a category of children in different circumstances. Both situations are evil but not within the reign of bioethics to address. The extraordinary attention to Charlie's case can thus be explained because the dominant approach of bioethics is individualistic. It is concerned with individual persons facing difficult decisions regarding medical interventions and health care. It often emphasizes respect for autonomy as the most fundamental principle of bioethics. Mainstream bioethics is less concerned with efforts to enhance the flourishing of humankind as a whole. I argue that a broader approach to ethical issues related to health care and health technologies is needed, an approach that goes beyond the individual perspective and represents a social ethic.

Power Struggle

Something more seemed to be at work in Charlie's case than the predicament of one baby. What was striking was the implosion of standard bioethical debate. Hundreds of thousands of people showed interest in his case because they sympathized with his parents. They sensed that what was going on with Charlie was, in fact, a power struggle. It is reminiscent of the paternalism that gave rise to bioethics decades ago. This time, however, it was a more refined and benevolent paternalism that was enhanced by bioethics itself. Great Ormond Street Hospital continuously praised Charlie's parents as brave individuals. At the same time, as the parents desired more time to spend with their baby before he died, the hospital affirmed that "Charlie's needs have taken priority," suggesting that the parents did not focus on his needs.[8] Recalling that the parents disputed the expert claims that Charlie's brain was irreversibly damaged, and that experimental treatment was futile, the hospital statement moralistically mentioned that the hospital "treats patients and not scans."[9] While the ruling judge referred to the love and care of the parents and called them "fine parents," the hospital emphasized that if Charlie had a relationship with the world around him, it was one of suffering, not love.[10] For most followers and sympathizers, the parents were pitted in their love for Charlie against the establishment of experts that wanted him to die as soon as possible.

Questioning the Paradigm of Bioethical Case Discussion

The standard debate on bioethical issues, as advocated by medical, legal, and ethical experts, commonly proceeds with two assumptions. First, rational debate is based on scientific evidence. Without facts, there cannot be an ethical analysis. Second, ethical examination demands application of moral principles. However, both assumptions became questionable in the case of Charlie Gard.

Great Ormond Street Hospital is considered the most prestigious children's hospital in the United Kingdom. It is also held around the world to be, as the July 2017 court ruling states, in "exceptionally high regard . . . for its excellence and expertise in the treatment of sick children."[11] The judge commented in April 2017 that Charlie was served by "the most experienced and sophisticated team that our excellent hospitals can offer."[12] Who would be in a better position to provide facts and evidence? The input of Dr. Hirano concerning experimental treatment was disregarded in the same ruling as "a lone voice in the USA."[13] Several months later, the judge complained that many opinions in the case were based on feelings rather than facts. Numerous people voiced their opinions while knowing "almost nothing" about the case.[14]

Experts used a set of ethical principles to guide the debate, especially best-interest standards, futility, and quality-of-life standards. The judge also referred to "Charlie's dignity" and permitting Charlie to "die with dignity."[15] In none of these cases were the principles defined or explained. On the contrary, it was simply assumed that they formed the core considerations of the ethical debate. They were also narrowly interpreted, concluding that Charlie's quality of life was "not worth sustaining."[16]

While similar cases of disagreement between parents and experts are often addressed in a private environment, within a process of mediation, and assisted by bioethics consultants and ethics committees, Charlie's parents decided to go public. At that moment, the bioethical paradigm broke down and the ideal of bioethics discourse imploded. Charlie became a symbol for a fight against an establishment that determines that it is not reasonable to prolong the life of a seriously ill and disabled baby, or to even ask for an experimental treatment with uncertain benefits and harms. This sort of normative elitism disregards the love and dedication of parents. Emotions, relationships, hope, faith, and imagination—important for most people in everyday life—are considered irrelevant.

The Charlie Gard case demonstrates the problem with assuming that ethical analysis should be based purely on objective facts. In clinical settings, facts are not self-evident. Ethical deliberation devotes much time to determine what is actually and objectively known about the condition and prognosis of a patient. In legal proceedings, opinions and judgments are often framed as objective facts. In Charlie's case, there was serious controversy over whether there was structural brain damage or brain dysfunction. The hospital statement that Charlie's brain was irreversibly damaged was contested by the parents. In their experience, the baby was responsive and showed signs of recognition. Similar disputes emerged over the hospital's assessment that Charlie's existence was painful. The parents' experience was different. Also, if there was pain and suffering, it was possible to manage the symptoms with analgesia and sedation, rather than conclude that death was in Charlie's interest.[17] Further controversies emerged over the experimental treatment. The hospital's conclusion that this treatment was futile (which was subsequently reiterated by the judge) was rejected by international experts. Academic centers in Rome and New York offered the treatment although they had no experience with it. Money was not a problem since the parents had raised funds for treatment. Nonetheless, it was ruled that treatment was not in Charlie's best interest.

If this decision was based on facts, it is obvious that the "facts" represented only the evidence endorsed by the experts from the hospital. The scientific validity of the evidence was disputed.[18] Even if there was a very small chance of benefit, the conclusion that treatment was futile was not based on facts but instead implied a value judgment. There was also disagreement on what constituted benefit. The hospital argued that treatment would not affect the existing brain damage, while the parents wanted Charlie to continue to live. In April 2017, the judge introduced new facts and ultimately issued a ruling. He stated that the pioneering treatment had not been tested in mice (suggesting that the impact of the treatment is completely unknown) and that there was no evidence that nucleoside therapy could cross the blood-brain barrier.[19] Foreign experts contested this last statement. References to "facts" were operating as rhetorical devices to swing the opinion in a specific direction.

More fundamental is the view that Charlie's life was doomed. He had a progressive, deteriorating, and debilitating disease that made his daily existence one of continuous suffering. But this is not a statement of fact. It is ironic that access to new drugs before they are approved for widespread use is called "compassionate use." Although benefits are unknown and potential risks might be involved, preapproved experimental drugs have become more easily available recently, especially for patients with serious and life-threatening conditions without any other options. In the United States, the Food and Drug Administration established regulations for compassionate use in 2009. In Charlie's case, however, the need for evidence was more often expressed than the need for compassion.

The case furthermore illustrates that assuming ethical discussion be based on moral principles is questionable. For many people, it seems that bioethics presents a particular framework within which rational persons are supposed to operate. This framework narrows down options and perspectives. Emphasizing best interest, futility, and quality of life as relevant ethical considerations reduces the ethical debate to the commonly used principles of beneficence and nonmaleficence. At the same time, it seems that there already is a preconceived idea of what is the best interest. For Charlie, given his condition, it was to die as quickly as possible. His quality of life was not worth sustaining, and the judge explained why: "He is lying in bed, unable to move, fed through a tube, breathing through a machine."[20] In this view, continued existence was considered a form of abuse. Ruling that Charlie should die with dignity suggests that his current existence was not compatible with dignity. The case materials give the impression that from the beginning the experts concluded that Charlie's prognosis was bleak and his life span very limited, suggesting that death was perhaps the best solution. This was not the result of an analysis of best interest or dignity. It was the other way around. It was a perspective formed very early in the process of care that determined how subsequent stages were interpreted. It influenced how benefits and harms of experimental treatment were perceived, how disability was thought to impede human flourishing, and whether suffering was considered bearable and maybe didn't make life worth living.

In other words, it is questionable whether bioethical debate is neutral in the sense that it gathers objective facts and analyzes them in the light of a set of ethical principles. It is more likely, as this case illustrates, that bioethics starts with certain value presuppositions and preexisting perspectives about what is an acceptable human existence. The rationalist logic of bioethics does not seem to apply in this case. Moral reasoning seems to follow prior moral judgments concerning quality of life and dignity, rather than being a method to produce moral judgments.[21] This conclusion may be contested because the media presentation and public debate were primarily influenced by the court proceedings. Therefore, it is not fair to blame bioethics for the one-sidedness of legal reasoning. Nonetheless, contributions of bioethics experts to the debate mostly focused on the rational balancing of abstract principles. But, as the responses of the parents and their supporters in the social media illustrated, bioethical debate is not simply a matter of moral reasoning. Emotions and intuitions are involved, related to value perspectives that differ from the ones expressed by legal and bioethical experts.

The Global Context

Global responses to the Charlie Gard case and massive support for the parents indicate that Charlie's fate was the subject of a power struggle. Analyzing such ethical concepts as best interest and futility, as marks of the bioethical paradigm, did not solve anything. The bigger question was whose interpretation was determinative. Ultimately the ethical debate focused on the question of who had the right to decide. Was it the parents who are responsible for their child or the medical and legal establishment representing the state that needs to protect children against abuse? The day before Charlie died, his mother exclaimed, "We've had no control over our son's life and no control over our son's death."[22] Politicians in the United States suggested that Charlie was a prisoner of the British National Health Service. Many commentators, especially in the United States, sided with the parents. They questioned whether medical and legal experts were the right persons to determine what quality of life is and what constitutes death with dignity. They regarded the case as a threat to parents in general and warned about the growing intrusion of the state in family life.[23] They also pointed out that the case revealed biases against disability since the potential benefits of treatment were discussed in relation to brain damage.[24] Disabled patients themselves have testified that disability is not well tolerated in medicine.[25] They argued that even if the London physicians were right about the evidence, it does not follow that they are the best experts to make value judgments.[26] Why should the moral point of view of a team of physicians override other normative positions? Quality-of-life assessment is often differentiated between health care professionals and disabled people themselves. Why should elite institutions, lawyers, and bioethics experts define what is or is not morally reasonable and whether experimental treatment should be provided to a seriously ill and disabled baby? Saving Charlie became a rallying cry for a populist movement against the state and establishment.[27]

The decision of the parents to bring the case to social media brought massive support and many followers. This helped Charlie's parents raise a substantial amount of money for possible treatment. It also started widespread grassroots action. The parents found experts willing to support their case. Many commentators, however, regretted that the case went from the private to public sphere.[28] Rather than following the ideal of shared decision-making between parents and physicians, the case expanded and became a concern for the global public. Instead of celebrating this enormous upsurge of interest for bioethics, many experts were concerned that the two assumptions of the paradigm of bioethical debate were not observed. The judge complained that many opinions were voiced but were often based on ignorance. The public also did not focus on bioethical principles. Instead it was primarily motivated by emotions, including sympathy for the parents. Some commentators acknowledge that the case brought more attention to such issues as futility of treatment and the rights of children and parents. They admit that the mobilization of sympathy through social media can support families that are vulnerable when they oppose the medical establishment.[29]

In Defense of Bizarre Cases

As mentioned earlier, bizarre cases are not uncommon in bioethics. Focusing on high-profile individual cases illustrates quandaries that often arise in the context of health care. The case of Charlie Gard highlights the disagreement between parents and a medical team concerning the best treatment for a seriously ill child. This is an almost everyday problem in pediatric intensive care. Cases like this underline how the wishes of parents should be weighted, and there is no simple remedy, even if academic bioethics suggests a procedure based on balancing principles. Furthermore, exceptional cases with very rare conditions may engage a broader audience in what otherwise would be an academic discussion, just as happened in the debate on Charlie Gard. The point is that the scholarly bioethical debate abstracted from the concerns in the social media. The complexities of the case and the ambiguities of the ethical concerns were also not reflected in the legal decision that "resolved" the case.

Finally, it can be argued that Charlie Gard is a "stigmata case": the external manifestation of a fundamental value conflict.[30] Such cases are common and attract extraordinary attention because they are novel and ethically controversial and contested. They bring to the surface profound moral questions about, for example, the value of kinship, family, parenthood, and relatedness. They demonstrate the interconnection between private and public issues. What is at stake is not merely a private affair but a public interest that is significant for all of us. They require a moral vision focused on what is worthwhile. As stigmata, such cases indicate radical shifts in what contemporary medicine is able to do. However, in practice, the stigmata of bizarre cases are usually ignored. The dispute among legal and ethical experts asks what medicine should be allowed to do, but not why, or whether the achievable is acceptable. The emphasis is on regulation, not on critical engagement with value perspectives.

Symbolic Nature of the Case

So far I have argued that the case of Charlie is exceptionally rare and not a product of a general or frequent bioethical problem. The case is also not a good example of the paradigm of bioethical debate based on facts and ethical principles. So the question still remains as to why it was the centerpiece of so much public attention and concern.

The case indeed represented a specific individual problem, but at the same time it referred to something else that had a much wider meaning for human beings in general. Maybe everybody could identify with the personal fate of Charlie and his parents, but that was not the only reason why so many people became moved and fascinated by this case. It certainly was a power struggle that motivated many to side with the parents, but the issue was much broader than the concern about parental authority. The case was appealing to so many people because of its symbolic meaning. The individual predicament of Charlie was symbolic of value concerns that many people recognized as fundamental.

At least three such concerns were at play in the public debate. First, the issue of relatedness. The case articulated the concept of parenthood and family. Charlie was not on his own. He was embedded in a context with loving parents who sought the best care available and did not want him to die. The power struggle that developed poised these loving parents against medical and legal experts who argued from scientific evidence. Second, the significance of hope. The parents continuously sought possibilities of a cure while the experts were primarily emphasizing futility. Third, the concept of human life. The parents proceeded with respect for Charlie's life without distinctions or qualifications, with the assumption that disabled persons have the same rights as everybody else. None of these three concerns were advanced in the bioethical discussion. In fact, there were two different bioethical debates: the formal one—based on mainstream bioethics and proceeding on the basis of evidence and rationality, advocated by the medical and legal professionals, and disseminated in publications and in traditional media—and the one on social media, giving voice to grassroots bioethics, criticizing mainstream bio-ethical expertise, and sympathizing with the parents. The first debate was driven by scientific facts and normative principles, while the second debate was motivated by concerns of relatedness, hope, and human life.

These debates clearly focus on different issues. Many people who identified and sympathized with the difficulties and, ultimately, the fate of Charlie and his parents emphasized the role of emotions in ethics. They were distressed about the impossible struggle between parents and establishment. They acted on the feeling that nobody deserves to die, especially when potential treatments are available. And the final decision of the court and the hospital not to accommodate the desperate desire of the parents to take Charlie home to die outraged them. The image of loving parents fighting for the best possible care for their baby against elite doctors who want him to die inspired the social movement to save Charlie.

Grassroots bioethics was also inspired by fundamental existential views on the meaning of life. It rejected the idea that ethical debate is founded on rational, scientific logic, proceeding from evidence and objective facts and applying abstract principles. For grassroots bioethics, it is important to engage the hearts of people and not just their minds. It also rejected the idea that ethical quandaries can be resolved through decision-making focused on the interest of the individual alone. What is at stake is more than an individual perspective. Charlie was part of a community—first of which was his family but also a community of concerned citizens empathizing with the family, and a global community of humankind. What happens to Charlie reflects how human beings treat each other and how they interpret disability, disease, and suffering. It illuminates what kind of care is provided to the most vulnerable members of the human community. Therefore, his case was more than a power struggle—it was a search for meaning.

This search for meaning in the grassroots response to the case is evidenced in the concerns mentioned earlier: relatedness, hope, and human life. Hope provides a perspective for many people confronted with disease and suffering that goes beyond the bare facts of biology and medicine. It motivates patients to travel to the other end of the planet for medical interventions or to use remedies that are rejected by scientific medicine as unproven and dangerous. Hope is not an issue considered in mainstream bioethics. It is regarded as a transcendental perspective, more associated with religions than with the rational discourse of bioethics.

The uneasiness with the bioethics discourse that became clear in the public response to the case of Charlie seems related to the discarding of such perspectives. Bioethics is about "argument, reasons, and rational reflection."[31] It is about facts not faith; it is argumentation, not emotional opinion. At the same time, bioethicists admit that decisions are value judgments.[32] However, the values involved are not elaborated and scrutinized. If values are fundamental, medical, ethical, and legal professionals cannot claim specific expertise. The debate then is open to everyone. But this also points to an important question. If bioethical debates revolve around values, what are the perspectives from which values arise and are articulated? Are such values solely based on science or also on philosophical and religious worldviews? If cases such as Charlie's are symbolic for broader value concerns, what can we say about such concerns?

Conclusion

"Bizarre" is defined as something very strange, odd, or unusual. I have labeled the case of Charlie as bizarre for two reasons: First, the case does not represent everyday reality. Of course, the case is real and concerns existing individuals, but the chance that it will happen again is extremely low, and most people will never confront such a case. In that sense, it is a rarity that will not be helpful for future experiences. Second, the case is unusual in that it shows the deficiencies and impotencies of the bioethical discourse. It is not so much the issue that ultimately the court must decide the case. More important is that mainstream bioethics and its emphasis on facts and rational arguments ignores the concerns and engagement of many people, like those who sympathized with Charlie and his parents. As a stigmata case, it could have raised critical examination of value perspectives, but value conflicts were actually obscured. The case attracted so much attention at a global scale because various values were at stake, but in the bioethical approach, only a limited perspective was addressed. The symbolic nature of the case, explaining why such a rarity became common knowledge in public and social media, was not discussed in mainstream bioethics. This makes bioethics itself bizarre in the sense that it does not address the sources and perspectives in which moral values and concerns emerge and propel cases to the rostrum of public debate. The next chapter will examine the characteristics of the current bioethics discourse.

CHAPTER TWO

The Establishment of Bioethics

The new discipline of bioethics has existed now for five decades. Its history is written, but there is not one dominant story. Several stories explain how ethical issues have emerged in the context of modern health care and why traditional medical ethics has been superseded with a new and broader discourse to elaborate, clarify, and analyze these issues. Others relate to advances in science and technology (with a long series of scandals and revelations), to criticism of medical paternalism and professional authority, to protest against the power of science and technology in the field of health care, to a crisis in values and an increase in cultural and moral pluralism, and to the articulation of human rights (especially civil rights and patient rights).[1] "Bioethics," therefore, is a term that covers a variety of phenomena.

The discipline's date of origin even differs across the literature. For Albert Jonsen, it is 1960, when the invention of the arteriovenous shunt made repeated dialysis treatment possible. The shortage of hemodialysis machines raised the question of how to select patients for this lifesaving intervention. In this instance, a special committee composed of mostly nonmedical members decided.[2] For David Rothman, bioethics was born in 1966. In that year, Henry Beecher published a famous article on ethics and clinical research, presenting examples of clearly unethical studies published in prestigious medical journals over the previous decade. Beecher's article showed that unethical conduct was common even though clear guidelines and rules for research existed, particularly the Nuremberg Code (1947) and the Helsinki Declaration (1964). His article pointed out that professional self-regulation was inadequate.[3]

Regardless of these different interpretations, scholars agree that the turning point was around 1970. At that time bioethics started to develop rapidly.

The first bioethics institute, the Hastings Center, was established in 1969. In 1970, the Society for Health and Human Values was founded, followed a year later by the creation of the Kennedy Institute at Georgetown University. The first professional journal appeared in 1971. In the same year, Van Rensselaer Potter proposed the new term "bioethics" for this emerging field.[4] Only a few years later similar developments of institutionalization took place in Western Europe and Latin America.

Whatever the specific conditions and reasons that gave rise to the new field, it is important to note that it was initially taken up and practiced by experts from theology and philosophy. The original core of bioethical literature, as aptly concluded by Tina Stevens, was produced by philosophers and theologians.[5] Both disciplines were associated with broad and critical perspectives on relevant moral challenges. However, these disciplinary characteristics quickly disappeared when bioethics evolved into a separate area of study with its own professional and disciplinary specifics. Theologians and philosophers rapidly morphed into bioethicists, distancing themselves from exclusively academic analysis and focusing on the practical issues in clinical medicine and research, especially presented in cases. Bioethics became a specific discourse: secular, rational, and analytic. The question, then, is what was lost when bioethics discourse was narrowed down into the professional expertise and discipline that it has become today? The focus on cases has reestablished the need for appealing to these broader perspectives that were ejected during the maturation of bioethics.

The Exorcism of Theology

In his reconstruction of the history of bioethics, sociologist John Evans argues that theology was initially heavily involved in ethical debates contesting the claims of science and medicine. However, it retreated when the new profession of bioethics took shape.[6] This thesis has been advanced earlier by Albert Jonson in his history of bioethics. Himself educated as a Jesuit, Jonsen narrates how he transmuted into a bioethicist. Many of the early scholars and founders of bioethics, in fact, came from theology and philosophy.[7]

The first professional organization, the Society for Health and Human Values, originated from the initiatives of religious ministers. Theologians advocated a broad approach to medical humanities, and advances in science and medicine generated questions of human values that were traditionally addressed by theology but also philosophy, art, and literature. Theology brought specific perspectives that went beyond scientific materialism. Theologians argued that in moral life, larger themes are at stake than relate to the ends of scientific progress and technological innovation, themes such as destiny, creation, grace, sin, blessing, responsibility, and personhood.

Nevertheless, this initial influence of theologians rapidly disappeared. The new area of bioethics offered expanding opportunities, and the professional identity of theologians, as Jonsen points out, "faded as they migrated into the bioethical world."[8] Similar observations were made by James Gustafson.[9] In his view, theology's contributions to medical ethics are not very important to most people involved in medical care and practice. Moral principles and values are justifiable without reference to God. Furthermore, practical problems in medical ethics can be solved without recourse to theology or religious discourse.[10]

A common thread in the history of bioethics is that the more it flourished the less the engaged theologians used their disciplinary backgrounds. Religious perspectives were increasingly translated into secular language. One common argument of how and why this happened is that the increased recognition of pluralism gave rise to the development of secular ethics as a means to apply a neutral common language.[11] Western countries have an increasing plurality of religious and nonreligious beliefs. This is even more obvious at the global level. Therefore, bioethics should transcend this plurality of value systems through developing a secular, neutral approach based on rational criteria.[12]

Another reason is the growing involvement of governments and the need for public policy. National committees identified principles and drafted guidelines that would govern the conduct of scientists and health care professionals. These principles were supposed to be acceptable to everyone, all citizens, so that trust in science and health care could be restored and maintained. The focus of such efforts was on common morality. It was unclear what contributions theologians could make. In policy areas, religious claims were translated into the language of principlism. The result is that religious discourse and imagery lost influence in bioethics. Daniel Callahan, the founder of the Hastings Center, pointed out that bioethics was accepted because it pushed religion aside.[13]

The Neutralization of Philosophy

A similar process of marginalization and growing irrelevance affected the discipline of philosophy. Although theologians came first, philosophers have also been "shapers of the field of bioethics."[14] For a long time, professional philosophers had ignored medical ethics. However, by the late 1980s, according to Warren Reich, the majority of bioethics scholars trained in ethics were philosophers,[15] and like theologians, philosophers were compelled to "move beyond their training."[16] Moral philosophy used to be a metaethical activity focused on the analysis of ethical language. Philosophers were concerned with logical clarification, moral reasoning, and the meaning of terms. They assisted bioethics in framing a discourse that went beyond specific religious perspectives and diverging value systems. In particular, they helped to articulate an ethics of universal principles, as proposed in the *Belmont Report* and the influential textbook of Beauchamp and Childress (*Principles of Biomedical Ethics*), promoting the secularization of bioethics and a conception of bioethics as applied ethics.[17]

Nonetheless, philosophers soon discovered that their contributions were limited. Moral theories like utilitarianism and Kantian deontology were not immediately relevant in clinical settings and medical classrooms. The focus was very much on actual problems, real cases, and urgent decisions. There was a need for conceptual analysis and clear argumentation but not so much for critical and substantive perspectives. Rather than theoretical views, practical solutions were needed. And since no theory of bioethics was generally accepted, bioethics as a practical endeavor was in need of principles to guide actions. Philosophers therefore concentrated on the articulation of these principles. They narrowed the realm of moral life into what was relevant in a medical context. As Jonsen observed, "Almost from its birth, bioethics was an ethics of principles, formulated as 'action-guides' and little else."[18]

Most philosophers adapted to this context. Rather than critical reflection and analysis, they formulated methods and procedures for medical decision-making. An example is the "ethical work-up" of David Thomasma, a six-step process of moral reasoning that brings the decision maker from facts to values, and from weighing values to ethically justified decisions.[19] Though a more streamlined example is David Seedhouse's Ethical Grid, a practical problem-solving instrument composed of four layers with boxes in different colors, going from principles to duties to outcomes. It is presented as an algorithm that leads to a final decision about the moral problem.[20]

What Was Lost?

What is lost if theology and philosophy as initial shapers of bioethics have been shaped by the new field itself? Is the articulation and specification of bioethical discourse an advantage or disadvantage? To begin with, some scholars will deny that something has been lost. They argue that the suggestion that theological discourse has adapted to and retreated from bioethics is false. Theologians have continued to contribute to bioethics, even if their contributions are no longer deemed relevant for mainstream bioethics. Lisa Cahill, for example, argues that theological bioethics emphasizes such issues as justice, the common good, and solidarity as particularly relevant in a global perspective.[21] Bioethics in this century must also be social ethics but this message is not palatable for the current bioethical paradigm. Evans's thesis about the retreat of theology does not recognize that the theological emphasis on justice (as well as on dignity, health, and care) has not been compatible with the efforts of bioethics toward institutionalization and acceptability within science and medicine.[22]

Similar arguments are used against the supposed retreat of philosophy. Its contributions were not lost but rather ignored when they could not be framed within the mold of mainstream bioethics. Julian Savulescu, the former editor of the *Journal of Medical Ethics,* recently complained that medical ethics today is "more like a religion." It is based on faith, not argument, and dominated by moralists that defend entrenched positions using bad arguments and empirical data to decide normative disputes. In his view, bioethics can only progress with more and better philosophy.[23]

Another opinion is that philosophical analysis cannot solve problems of ethical decision-making but can bring other perspectives. Philosophy may introduce a new grammar for normative analysis, for example, by introducing notions of vulnerability, embodiment, interdependence, and biopower. It may clarify ontological presuppositions that frame normative questions. It may furthermore examine and criticize the historical and political background of key debates. Doing so, it helps to overcome "conceptual stagnation."[24] Addressing philosophical issues beyond the usual scope of bioethics was exactly the purpose of the book series *Philosophy and Medicine* (starting in 1975) and the *Journal of Medicine and Philosophy* (since 1976). However, a closer look at the publications shows that many address specific bioethical issues.

"Philosophy" is often interpreted as moral philosophy while other areas of the domain—such as philosophy of science, epistemology, and philosophical anthropology—remain underdeveloped. The emergence of the philosophy of medicine as a separate field of study assisted in creating a special niche for philosophical reflection that was harmless to the development of mainstream bioethics. Philosophers now had a choice: either contribute to applied ethics and operate as "bioethicists" or examine broader, critical questions that transcend the domain of ethics and operate as "philosophers of medicine."

The Paradigm of Bioethics

To have a better appreciation of what was lost, it is helpful to consider what has become the dominant conception of bioethics. Beauchamp and Childress's well-known textbook defines biomedical ethics as "the application of general ethical theories, principles, and rules to problems of therapeutic practice, health care delivery, and medical and biological research."[25] This conception of applied ethics implies the following set of interdependent presuppositions:[26]

1. Bioethics is the application of ethical theory and ethical principles.
2. There is a body of available ethical theories, principles, and rules to be applied to a variety of practical biomedical problems. The idea of applied ethics supposes that there are ethics to be applied.
3. Professional ethicists have a special expertise in applying ethical theories and principles, whereas nonethicists (for example, physicians) provide moral problems for applied ethics.
4. Bioethics is general ethics applied to medicine. That is, the context in which these problems arise is not unique; it is not characterized by specific values that generate special problems. Indeed, the medical context is viewed as a practice ground for a new profession of biomedical ethicists.
5. Medical ethics aims to offer practical recommendations and prescriptions based on or deduced from ethical theories and principles.

This set of presuppositions to some degree clarifies why many perceive bioethics as an independent discipline. For example, some hold that ethics should perform four tasks: to clarify concepts, analyze and structure arguments, weigh alternatives, and advise a preferable course of action. The central contribution of bioethics is therefore restricted. It does not necessarily result in judgments regarding what should be done. The bioethicist only provides the topography of arguments and objectifies the options. He or she acts as a disinterested and neutral observer of medical practice who is in the best position to weigh moral alternatives.

Focus on Application and Principles

The consolidation of bioethics as applied ethics has been critically questioned. Over the last few decades, many innovative approaches to bioethics have been advocated: phenomenological ethics, hermeneutic ethics, narrative ethics, and care ethics, for example. Traditional conceptions have been revitalized, notably casuistry (drawing from the classical casuistic mode of moral reasoning) and the virtue approach (emphasizing qualities of character in both individuals and communities). But mainstream bioethics is still dominated by two major notions: application and principles.[27]

The notion of application is useful for two reasons. First, one can employ it in many areas where ethical problems arise, not merely in biomedicine but also in business, journalism, or the environment. Secondly, a wide range of topics falls into the purview of ethical analysis but the analytic method is the same. Perhaps the issues are new, but the ethics are not. Application also refers to practical purposes. Instead of the theoretical abstractions of traditional moral philosophy, applied ethics contributes to analyzing dilemmas, resolving complex cases, and clarifying practical problems arising in the health care setting. "Application" here has a double connotation: it indicates that ethics is available for what we usually do (it applies to our daily problems) but it also is helpful and practical in the sense that ethics is something to do (it works to resolve our problems).

If ethics is conceived as applied ethics, then subsequent reflection is needed on what is being applied. The emerging consensus is that principles should provide the answer to this query, which is coherent with the moralities of obligation that have dominated modern ethical discourse, especially since Kant. Behavior in accordance with moral obligations is considered morally right. The morality of behavior is a morality of duty. And morality is understood as a system of precepts or rules people are obliged to follow. Particularly in the early days of bioethics, medical power was criticized and the rights of patients were emphasized as requiring respect. In this context, the moralities of obligation presented themselves as a common set of normative principles and rules that we are obliged to follow in practice. As Diego Gracia has pointed out, the *Belmont Report* in 1978 was influential because it was the first official body to identify three basic ethical principles: autonomy, beneficence, and justice.[28] A basic principle was defined as a general judgment serving as a justification for prescriptions and evaluations of human actions. From these principles, ethical guidelines were derived that could be applied to the biomedical area.

About the same time, Beauchamp and Childress, in the first edition of their book, introduced the four-principles approach, adding nonmaleficence to the aforementioned three principles.[29] In their view, principles are normative generalizations that guide actions. However, as general guides, they leave considerable room for judgment in specific cases. Various types of rules are needed to specify the principles into precise action guides. Although Beauchamp and Childress have considerably nuanced their theoretical framework in later editions, their work contributed to the conception of bioethics that is dominating the practical context—in ethics committees, clinical case-discussions, ethics courses, compendia, and syllabi, for example.

The predominance of principles has been increasingly criticized over the past few decades. Nonetheless, almost all policy documents in bioethics articulate sets of principles. For example, the Oviedo Convention states such principles as the primacy of the human being (Article 2) and equitable access to health care (Article 3). (These are called "provisions" but, as explained later in the document (Article 31), are really "principles.")[30] The Universal Declaration on Bioethics and Human Rights proclaims a universal framework for bioethics listing fifteen principles, including human dignity, autonomy, consent, vulnerability, and solidarity.[31]

Problems with the New Paradigm

As soon as the paradigm of bioethics was consolidated, it rapidly developed into an established and recognized discipline. Working with a constellation of fundamental principles facilitated the search for answers and solutions in clinical practice, research assessment, and public policy. However, four types of criticism have been directed against this paradigm.

Oversimplification

The focus on solutions has led to a simplification that prioritizes the design of procedures to make practical decisions and resolve disputes over the development of substantive theories. The focus on application is not embedded in a broader philosophical approach. Arthur Caplan has labeled this approach the "engineering model" of applied ethics.[32] In daily practices, bioethics focuses on mid-level principles—respect for autonomy, beneficence, nonmaleficence, and justice. These principles are applied to dilemmas, cases, and problems encountered in the practice of health care. From a specific principle, guidelines or recommendations can be derived to resolve various problematic situations. Considering applied ethics as a deduction from principles is now regarded as too simplistic.

The emphasis on principles is less stringent. Edmund Pellegrino observed that the mentioning of principles disappeared from the definition of bioethics in the second edition of the *Encyclopedia of Bioethics* in 1995. It now refers to "a variety of ethical methodologies in an interdisciplinary setting."[33] There also is a reinterpretation of principles away from principlism. Rather than deducting moral judgments from principles, what is important is the dialectic process of interpretation, specification, and balancing of principles. Principles are not simply recipes or tools for producing clear-cut answers to moral dilemmas. In fact, they require continuous critical labor since there is always ambiguity and conflict.[34]

Principles are not the solid foundations of ethical decision-making but, as Dewey said, "hypotheses with which to experiment."[35] Principles therefore need revision, adaptation, expansion, and altering. The need for interpretation is advanced by the recognition that ethical decision-making is not abstract but takes place within concrete contexts and practices. Rational application of principles requires a particular moral experience of the agents involved. Ethical reasoning, then, often starts with moral perception. Moral principles are necessarily abstract and therefore not immediately relevant to the particular circumstances of actual cases, the concrete reality of clinical work, and the specific responsibilities of health care professionals. By appealing to principles, norms, or rules, applied ethics may fail to realize the importance of the concretely lived experience of health care professionals or patients.

Contemporary philosophers critically elaborated this point that moral agents are taken to have an abstract existence. Ethics, according to Bernard Williams, does not respect the concrete moral subject with personal integrity.[36] It requires that the subjects give up their personal point of view in exchange for a universal and impartial point of view. This is, Williams argues, an absurd requirement, because the moral subject is requested to abandon what is constitutive for his or her personal identity and integrity. The idea that knowledge of normative theories and principles can be applied to medical practice ignores the fact that moral concerns tend to emerge from experiences in medical settings themselves. A similar issue is raised by Charles Taylor in his book *Sources of the Self*, in which morality and identity are considered two sides of the same coin.[37] To know who we are is to know to which moral sources we should appeal. The community, the particular social group to which we belong, is usually at the center of our moral experience. Even the use of ethical language depends on a shared form of life.

The conclusion is that bioethics should become more appreciative of the actual experiences of practitioners and patients. It should also become more attentive to the context in which physicians, nurses, patients, and others experience their moral lives (e.g., the roles they play, the relationships in which they participate, the expectations they have, and the values they cherish). The physician-patient relationship is neither ahistorical, acultural, nor an abstract rational notion; persons are always persons-in-relation, members of communities, immersed in a tradition, and participants in a particular culture. As a domain of philosophy, ethics should, therefore, proceed from moral experience. The moral dimension of the world is not given but first and foremost experienced. Moral experience is humanity's way of understanding itself in moral terms. Ethics is, therefore, the interpretation and explanation of this primordial understanding. Before acting morally, we must already know, at least to some extent, what is morally desirable or right. Otherwise, we would not recognize what is appealing in a moral sense. On the other hand, what we recognize in our experience is typically unclear and in need of further elucidation and interpretation. In short, we approach the moral dimension of the world from a set of prior understandings; they form the basis of our interest in what at first seems odd and strange to us, requiring us to continuously reconstruct the moral meaning of our lives. Such an interpretive perspective will be helpful for integrating the experiences disclosed in clinical-ethical studies or debates on specific cases, as well as utilizing the insights gained from analyzing the value-contexts of health care practices.

In other words, a conception of bioethics as an application of principles is insufficient because it has lost the significance of moral experience and the need for interpretation. Ironically, it also is paradoxical. Moral issues have emerged, and have promoted the rise of bioethics, because of an almost exclusively technological orientation to the world and a predominant scientific conceptualization of human life. Human beings resisted the tendencies of medicine to focus primarily on their bodies and biological existence. They protested against the overwhelming power of health professionals and health care institutions, reducing patients to cases, numbers, and objects. They objected to the lack of involvement of individual beings within decision-making processes, as well as to the lack of respect for the authenticity and subjectivity of individual persons. Bioethics has emerged as a movement to reintroduce the subject of individual patients into the health care setting, emphasizing patients' rights, respect for individual autonomy, and the need to set limits to medical power. The paradox is that we try to address the moral issues of medicine with a conception of bioethics that is itself impregnated with scientific-technical rationality.

Nonneutrality

Another criticism is that bioethics is not the neutral common language that it pretends to be. Jonsen's historical study presents bioethics as a quintessential American enterprise. It is the expression of what Jonsen calls the "American ethos," a specific panorama of ideas and values that determines how morality is interpreted and applied. Its main characteristics are a focus on capitalism (market economy), progress and technological optimism regarding medicine as a source of miracles, and individual freedom.[38] This ideological context, however, is almost always implicitly assumed. There is hardly any reflection on this value context. On the contrary, the neutral, value-free, and apolitical nature of bioethical discourse is taken for granted.

With the advancement of globalization, it is currently recognized that the particular cultural context shapes the discourse of bioethics. The fundamental ethos of applied ethics, its analytical framework, methodology, and language, its concerns and emphasis, and its very institutionalization have been shaped by beliefs, values, and modes of thinking grounded in specific social and cultural traditions. Nowadays, bioethics literature and debates serve as powerful means to express and articulate these traditions. However, bioethics only rarely attends to or reflects upon the sociocultural value systems within and through which it operates.

Scholars usually assume that the principles, theories, and moral views of bioethics are transcultural. H. Tristram Engelhardt, for example, distinguishes between two levels of discourse: that of the secularized pluralistic society and that of the many particular moral communities with competing visions of the good life.[39] Bioethics, in his opinion, should focus on the societal level, speaking across gulfs of moral discourse; it is a common neutral language, a secular moral grammar, guaranteeing a peaceable society. The most interesting task of ethics is on the first level: promoting and defending, in the context of health care, the general secular moral language of mutual respect. Critics agree that this is an important task, but it flows from a rather thin or minimalist conception of ethics.[40] Ethics is conceptualized as procedural; it is the regulation of social relations through peaceable negotiation. To speak the language of mutual respect, all other moral languages must be pacified.

The question is why we should abstain from our particular moral languages in favor of a neutral common language? This question points to an important problem: how neutral is the common neutral language? Is Engelhardt's language not the specific moral language of a specific moral community? Is this language itself not the expression of a commitment to a certain "hyper good," in particular, the demand for universal and equal respect and self-determining freedom—primal values in the liberal tradition?[41] Such questions assume that the values of mutual respect, self-determination, and rights to privacy are not decontextualized standards but themselves expressions of community-bound agreements, especially in the United States.

Sociologist Renée Fox was one of the first to critically examine the socio-cultural context of bioethics.[42] She points out the negative consequences of the domination of individualistic values. By stressing the autonomy and rights of individuals, other significant considerations (for example, community, solidarity, the common good, duties, and responsibilities) have been neglected, as have critical philosophical questions concerning the value of medical progress and personal and public health in communal life. The implicit value framework of individualism sets the agenda for bioethical activities and debates. The focus is on microethics. As Abraham Edel has remarked, "Applied ethics tends to focus on the individual's dilemmas as individual problems."[43] What is lost with the self-evident assumption of this specific value-context is the relevancy of social, economic, and environmental problems for bioethics. The political and institutional setting in which ethical problems emerge is not considered relevant for bioethical discourse.

Reduced Critical Potential

A third criticism is that bioethics has lost a critical and independent perspective. As bioethics became more acceptable, it lost its critical edge. Bioethicists were accepted by scientists and health professionals as colleagues and insiders. Daniel Callahan explains that this not only required keeping theologians at a distance and having philosophers focused on applying neutral moral language, but that it also demanded a specific ethical approach, mediating between rejecting and blessing scientific progress. This so-called "regulatory ethics" aims at guidelines and rules for the prudent use of new technologies and interventions. Callahan admits that this type of ethics fits perfectly well into the economic and political system that dominates American society.[44] The agenda of bioethics accommodates the social and cultural context in which it has emerged.

In fact, according to Callahan, this approach was necessary for the societal and political acceptance of bioethics as a new discipline. Sources of funding and support should not be alienated. Being a critical outsider will not bring in grants and donations. Tina Stevens shows how the positioning of the Hastings Center as an "independent" institute required two strategies. First, it had to demonstrate that it could provide answers to dilemmas. Problems could be solved and not merely analyzed. Callahan pointed out that this implies "staying away from the dark corners of doubt and uncertainty."[45] Academic analysis is not much appreciated in this pragmatic perspective. Marginalization of theological and philosophical discourses is another outcome. They are labeled as vague and fuzzy. There is no room for examining underlying theoretical and fundamental issues, so-called "larger questions," such as the impact of technology on human life or the kind of society promoted by scientific progress.[46] Sometimes, it is associated with a framing of theological discourse as focused exclusively on abortion, embryos, and euthanasia; everybody will agree that these are contentious issues that should be avoided since they only lead to conflicts. At the same time, this framing can conveniently ignore other issues raised by theologians, such as the inequality of the market economy, the lack of justice, the inaccessibility of health care, and the pervasive vulnerability of many citizens.[47]

The second strategy to emphasize the advantages of a narrower bioethical approach presented mainstream bioethics as a neutral narrative. It is important not to assume a political ideology, and to reject activism and advocacy. In practice, bioethicists should not be too critical of scientific and medical advances. Otherwise, analysis of biomedicine and science will be marginalized and penalized. In the end, the Hastings Center came to function, in the words of Stevens, as "pro-technology agency."[48]

The result of these strategies was impressive: a flourishing field of bioethics and mushrooming centers and institutions. On the other hand, the price paid was "institutional capture."[49] Bioethics cannot raise issues that go against the interests of the medical profession and the scientific authorities. Bioethics itself has become part of the neoliberal establishment. Its practitioners have become cooperative insiders. Like embedded journalists have been accused of biased reporting and war propaganda, embedded bioethicists will be regarded as allies rather than opponents; they are supposed to admire the advances of science and the wonders of medicine without serious critical questioning. In fact, bioethics has rarely said no to new technologies and medical interventions. Bioethics organizations have not taken stands against injustices.[50] As Evans points out, "The ethics of the bioethics profession is the same as the ethics of science and medicine."[51] While bioethics originated from criticism of science and medicine, it has played a useful role in turning protest and critique into a form of ethics management. Through deemphasizing power differences, and transforming public concerns into private queries and political controversies into ethical challenges, it has assisted in the pacification of significant social controversies and potential conflicts. Bioethics has functioned as an effective "immunization" against radical critique.[52]

Contracted Perspectives

The last type of criticism is that the emergence of the dominant paradigm has accelerated the decline of other perspectives. Broader questions related to human values can no longer be asked. When the focus of mainstream bioethics is on practical solutions and individual autonomy, it will be difficult to raise other types of questions.

An example is the debate about genetically modified organisms (GMOs), which are often promoted as a possible solution to the problem of food shortages, especially in developing countries. It is argued that GMOs enhance the choices of consumers by making food cheaper and healthier. From an ethical perspective, it seems that genetically modified food has tremendous benefits. The main question is whether it has risks and possible harms. The debate, therefore, is focused on the issue of safety. Scientific experts argue that there is no convincing evidence that modified food is dangerous or risky. The conclusion often is that there is not a good argument to oppose the introduction of genetically modified food.

Within this framework, however, some questions cannot be asked or are not considered as relevant. One question concerns justice. Food biotechnology is often promoted by a few transnational companies that have a monopoly over seeds and make farmers dependent on buying such seeds.[53] These companies have also patented food products available in biodiversity and cultivated for ages by traditional communities. Another issue is that genetic manipulation of life is morally objectionable in many worldviews and religions. Animals and plants are not machines or commodities that can be manipulated and patented since that does not respect the dignity of life. At the same time, the justification of patents is based on the idea of private rights while in many cultures and indigenous communities the products from biodiversity belong, first of all, to the community; they cannot be privately owned. Finally, for many people, food is not a simple resource or commodity that can be exchanged on the market. Even if there are many choices, the fundamental issue is that we need food; without it we cannot survive. Food is also the expression of cultural diversity.[54]

Narrowing the ethical debate on GMOs into a balancing of benefits and harms, and primarily focusing on matters of safety, reduces the relevancy of such broader questions. Working with the dominant paradigm assumes that one accepts the rules of the game and proceeds on the basis of mainstream principles. Once this is done, one can have intensive discussions about individual autonomy, benefit and harm, and risks within this framework, but it is no longer feasible to question the rules of the game themselves. Such questioning will not be regarded as an ethical debate but as something else: political criticism, economic analysis, or cultural discontent.

Revitalizing Bioethical Discourse

Albert Jonsen complained in 2000 that bioethics has become boring.[55] It has become part of the establishment. It is "too domesticated." Jonsen apparently means that bioethics has lost its critical edge and has become an insider in the health care practice and policy context. John Evans has a somewhat different interpretation. He argues that bioethics is currently in a crisis. It is well established in the areas of research ethics and clinical ethics but increasingly challenged in public policy and cultural debates, as the case of Charlie Gard illustrates. Jonsen suggests two remedies: bioethics becoming more interdisciplinary and more global.

In both cases, bioethics can learn from other disciplines, cultures, and traditions. Evans refers to Potter who coined the term "bioethics" and advocated for a broader view of ethics, including social and environmental concerns. In this way, bioethics can overcome its narrow focus on biomedicine and human autonomy. Evans recognizes that bioethics has evolved into a specific and distinct profession with a "unified system of abstract knowledge."[56] He agrees with Jonsen that bioethical debate has become "an establishment activity" but articulates that its system of abstract knowledge is no longer respected.[57] It is contested by social movements and activists. His remedy is to accept that bioethics as a profession and academic field has a limited focus, specifically on research ethics and health care ethics consultation. In these two areas, it represents the values of the public and is effective in streamlining processes using the system of principlism.

This is not the case in the domain of public policy ethics. Here bioethics must improve its methods if it wants to maintain its legitimacy. It must create principles that are less vague, no longer pretend that principles are universal, and rely on empirical studies. However, in the fourth domain of bioethics, what Evans calls "cultural bioethics" (the public debate on ethical issues in society), bioethics should abandon its work.[58] Bioethics can no longer make a meaningful contribution here. Interestingly, Evans argues that theology has a specific role in this domain. While excluded from the other three domains of bioethics, theology and religious discourse should specifically raise larger questions in the domain of cultural bioethics, although this last domain is no longer the proper field of professional bioethics, according to Evans. Rather than being discarded, theology should be reinserted now that we have entered a new era of bioethical debate.

Jonsen and Evans share the concern that contemporary bioethics needs to be revitalized by broadening its scope. This can be done by bringing in ideas from other disciplines and cultures, according to Jonsen, and specifically from theology and religion, according to Evans. Such revitalization will not be necessary for all domains of bioethics in Evans's view since bioethics is consolidated in the areas of research and clinical consultation. The question is why larger issues will not be relevant in these areas. Research has been increasingly commercialized, giving rise to issues related to conduct and integrity. Globalization of clinical trials has produced such ethical challenges as global justice, double standards, and protection of vulnerable populations. These new issues can properly be addressed by expanding the dominant framework of principles. Jonsen's point of view that bioethics in general needs to be rejuvenated by innovative and critical thinking, therefore, seems to be more plausible than Evans's effort to protect some domains and preserve them as the proper territory of professional bioethics.

Broadening Bioethical Imagination

How can philosophy and theology contribute to a revitalization of bioethics? In the early days of bioethics, philosophers criticized modern society and culture. Herbert Marcuse, for example, argued that technology is a new form of social control. It assists capitalist societies in alienating and exploiting people. And medicine and health care were the specific targets of such critics as Ivan Illich and Michel Foucault. Illich introduced the concept of medicalization. He accused the medical establishment of having become a major threat to health in that professional health care had come to expropriate the power of individuals to heal themselves. It not only created diseases (iatrogenesis) but also transformed all aspects of human life into technical problems.

Illich's thesis of medicalization foreshadowed developments in bioethics. Transforming such personal experiences as suffering, sickness, disability, and death into medical concerns for which treatment should be sought was possible because other frameworks of interpretation and meaning, especially religious ones, were marginalized.[59] Around the same time, Foucault introduced the concept of biopolitics as an instrument to analyze and criticize the power of medicine and the life sciences.[60]

The impact of these critical ideas on the growth of bioethics in the United States was limited. They were more favorably received in Europe, possibly because the social context of science and medicine was attributed a significant role. Influential studies questioned the role of medicine in modern society. In the United Kingdom, for example, Thomas McKeown demonstrated that conditions of housing, nutrition, and hygiene contributed more to the general improvement of health since 1880 than advances in scientific medicine.[61] Another reason why critical voices were more outspoken in Europe than in the United States was that traditions of critical analyses of medicine existed for a long time in several European countries. Since 1870, numerous philosophical studies on medicine were published in Poland, Germany, and France. Initially, epistemological issues were explored about the conception of medicine as a natural science. Later, the emphasis shifted to anthropological concerns, criticizing the dualistic and Cartesian view of human beings as an amalgam of body and mind, rejecting a purely objective approach, and demanding the introduction of the subject into the life sciences and medicine. It can be argued that this tradition of philosophy of medicine, in fact, paved the way for the subsequent interest in the ethical dimensions of health care.[62] From a European perspective, the emergence of bioethics is not a new event but instead is continuous with earlier efforts to critically philosophize about medicine. The difference, of course, is the practical orientation of bioethics. Rather than critical reflection and interpretation, problem solving has become a paramount concern.

Given this history of philosophical involvement in medicine and health care, philosophy can help to broaden the bioethical imagination under two conditions. One is providing a critical voice, particularly scrutinizing the context in which ethical issues arise. Presently, such concepts as medicalization and biopolitics are receiving more attention.[63] A second condition is a broader approach to philosophy itself. It should not position itself merely as moral philosophy but should also articulate other areas of philosophy, such as philosophy of science, epistemology, or philosophical anthropology. This broader approach was the purpose of the establishment of the European Society for Philosophy of Medicine and Health Care (ESPMH; http://www.espmh.org) in 1987. It was primarily motivated by the need for critical reflection on the role of medicine and health care in present-day society and culture. Its founders wanted to remedy the tendency of philosophical analysis and ethical evaluation to neglect to scrutinize the almost exclusive search for technocratic and econometric solutions for problems in health care and welfare systems.

Theologian James Gustafson once commented that even if we do not like the answers, we cannot avoid the questions.[64] Matters of life and death, suffering, disability, disease, and aging have always been addressed by theology and philosophy.[65] It is obvious that fundamental questions concerning such matters cannot be definitely answered. Not all queries are amenable to solutions. Nonetheless, theology and philosophy can bring some clarification and enlightenment to the meaning of suffering or what we may hope for when, for example, our child has an incurable disease. Gustafson argued elsewhere that theology can clarify "basic presuppositions" concerning the human condition and contribute to bioethical imagination through providing a wider moral perspective. Examples include showing that morality is more than a focus on human self-fulfillment and that the well-being of creation is more than just human well-being, setting the last concern in a broader context of the common good and the health of the planet.[66] Such considerations are not restricted to theology. Recognizing human finitude and human limitations, as well as the need for self-criticism, is a challenge for everybody, religious or not. In the present patterns of ethical action and thought, these ideas are identifiable, according to Gustafson, as "functional equivalents of theology."[67]

Lisa Cahill concurs with her colleague. In her opinion, theology can specifically contribute in articulating common moral values in public discourse and at the global level.[68] Its focus is on the social conditions that create bioethical problems. Theology has the power to identify, expose, and challenge such conditions. It criticizes the exclusive emphasis on individual autonomy by elaborating the role of community and relatedness. It also denounces the idea that the discourse of mainstream bioethics is neutral and shows how on the contrary it is governed by the values of individualism, science, and the market. Cahill furthermore points out that religious discourse is not only reflected in academic theological activities. It is perhaps more often involved in grassroots movements and civil society networks. While the impact of theology on the theoretical discourse of bioethics is limited, its engagement in practical activities of public policy, social practices, and social networks is underestimated in bioethics.[71]

Reflection on the human predicament, life, death, and suffering as well as community, relatedness, vulnerability, and solidarity have longtime been objects of philosophical and theological thinking. Voices from theology and philosophy will disclose broader perspectives than those that are currently offered in the dominant paradigm of bioethics. Even if these perspectives cannot directly be translated into policy recommendations, they serve as a lens. Thomas Shannon summarizes these perspectives thusly: "What religious traditions have to offer is a broader vision of both individuals and society, a more inclusive vision of justice, an inclusive vision of the common good, and a view of human dignity that argues that individuals should receive from the community as well as participate in its well-being."[70]

Conclusion

The title of this chapter is ambiguous. The text describes how bioethics has emerged and has been accepted as a new discipline, critically assessing the medical and scientific establishment. It also shows how bioethics became part of the establishment itself. The founders of bioethics were theologians and philosophers. In the process of developing bioethics, however, these two disciplines were rapidly sidelined and extruded. A powerful paradigm has materialized based on a common morality with generally accepted principles that can be applied to ethical problems and queries in all fields of scientific and technological activity. Bioethics professionals became firmly embedded in clinical practices and research efforts.

The message, then, is that the consolidation and concentration of bioethics came with a price. Bioethics lost some of its initial features and, in the words of Jonsen, the critical spirit was "dampened."[71] Presenting itself as a neutral and apolitical narrative, bioethics moved critique of the dominant scientific and economic worldview outside of its domain. That makes it impossible to recognize how bioethical discourse is dominated by the "religion of the market," as Cahill has pointed out.[72] It is also difficult to critically question the primacy of autonomy and emphasize the connectedness and mutual vulnerability of human beings.

The next argument made is that there is a need to broaden bioethical imagination and discourse. It is important to go beyond the engineering model of bioethics. The application of ethical principles is not impartial and value-free. It requires interpretation and deliberation. Moral problems are not given but emerge from prior understandings. In practice, however, scientists and health care professionals often formulate problems without questioning these understandings. Bioethicists should be more critical of moral facts and problems because what counts as a moral problem is not self-evident.[73] These concerns are also important for the issue of agenda-setting in bioethics. Usually, attention is selectively focused on specific topics within the framework of microethics.

As chapter 1 illustrated, in the public and global bioethics debate more is at work than the narrow perspectives of mainstream bioethics. In the case of Charlie Gard, it was not only the consideration of best interest or benefits and harms of a potential new treatment that inspired public concern. The public was motivated by feelings, experiences, and desires, such as hope and togetherness. Obviously, there was a different interpretive perspective at work that was more related to what traditionally have been the concerns of theology and philosophy. What has been lost in contemporary bioethics is perhaps what Charles Taylor has called the "background picture of our spiritual nature and predicament."[74] Can it be that for many people, regardless of religious affiliations, the worldview of bioethics is not satisfactory? That scientific facts are not neutral, that human beings are not simply autonomous decision makers, and that medicine is not merely bodily intervention? If that is the case, bioethics discourse does not, or at least does not fully, express concerns that are relevant for many people, and that used to be articulated in theological and philosophical discourse. Moral perception might differ, and moral cases are ambiguous. This leads to the question of what interpretive perspectives determine our understanding of what is morally significant. What leitmotiv is at work when a specific case is propelled to the epicenter of the public apprehensiveness?

The common assumption of bioethics discourse is that of a rational debate based on scientific evidence and facts. Moral judgments are based on logical reasoning and ethical principles. There is no place for emotions and feelings. The perspectives from which moral values emerge are usually not examined. That was clear in the case of Charlie Gard. The ethical intuitions that motivated grassroots bioethics in that case—such as concerns of hope, togetherness, and relatedness, the idea that seriously handicapped Charlie has dignity and is not a monster, that his parents are pilgrims driven by hope—were not considered relevant for the debate.

If it is critical to recognize what has moral significance, moral perception plays a crucial role in ethical discourse. How do we perceive particular situations as suffering, unjust, racist, or dishonest, and thus as morally significant?[75] Moral perception is related to particular situations; it recognizes what the morally significant features are. This requires moral sensitivity and experience. Otherwise one would not see injustice and indignity when and where it occurs.

Perception, therefore, comes prior to deliberation. Moral perception can lead to moral action but also to responses such as emotions instead of action. In considering moral judgment, we need to go back therefore to our prior understandings. What kinds of interpretive framework are guiding us? Also, if moral actions are more related to moral emotions than to moral reasoning, the same question arises. Moral reasoning is often not the outcome of previously formed judgments, but rather the post-hoc rationalization of judgments. Judgments are frequently the result of moral intuitions, as argued by Greene and Haidt.[76] These intuitions arise from traditional interpretative frameworks. This brings us back to the issue of theology, religion, and philosophy discussed in this chapter. Religious and philosophical traditions express worldviews, with stories and symbols that shape perspectives.[77]

The understanding and interpretation of cases, as I argue in the next few chapters, is frequently driven by metaphors such as ghosts, monsters, pilgrims, prophets, and relics. The initial response to cases can be based on astonishment, wonder, and amazement but also disgust and repulsion. But what is called the "yuck factor" is not sufficient. Interpretation and explication are needed, and this is what the metaphors provide. They refer to broader perspectives. In health policy nowadays, metaphors are based in the language of business and economics.[78] Considering health care as a market transforms moral perception. The emphasis is on consumers, producers, and choice. The role of vulnerability, dependency, and empathy is reduced. Moral agency is instrumental rationality. Individuals are assumed to make rational choices, maximizing benefits and minimizing costs. Actions based on such values as solidarity and mercy do not fit into this perception.[79] In bioethics, the role of metaphors is rarely investigated, but it is undeniable that they are used.[80] Referring to patients as "monsters" or "vegetables" immediately locates a case within a particular framework. Metaphors invoke specific perspectives.

The bottom line is that in bioethical discourse, metaphors and the interpretive frameworks they invoke are often related to religious and philosophical traditions. There is a long history of interactions between religion and medicine. Illness has closely been connected to sin. Healing has been a religious effort of salvation. This history has not been eradicated by the emergence of modern scientific medicine. Metaphors from these traditions have not disappeared even if they are not explicitly used. Bioethics as the new science of applied ethics underestimates the role of Christianity in forming Western values and how it has redefined human identity and association. The significance of the value of individual moral agency could not have emerged without the long history of being shaped within the Christian tradition.[81]

A frequently used metaphor, especially in the early days of bioethics, is "playing God." It is often denounced as irrelevant and obstructionist. However, it raises important points about the limits of human knowledge and intervention, as well as the aims of science, technology, and their uses. It does not state a moral principle, but as Allen Verhey argues, it invokes a perspective: "reminding humanity of its fallibility and finitude."[82] Modern medicine is sometimes characterized by hubris and arrogance, taking knowledge as mere power to master nature, and human beings as makers and designers, as creators of perfect babies and enhancers of human bodies. The metaphor highlights that this effort to imitate God often results in failure and deception. Human agency is limited, and the finitude of human existence is often denied.[83]

Mary Midgley has repeatedly argued that we must attend to the myths, metaphors, and images within science.[84] The imaginative visions that are expressed in such metaphors are central to our understanding of the world. They suggest particular ways of interpreting the world but they also reveal what values are guiding scientific approaches. Jeffrey Stout has emphasized the importance of interpretive frameworks within moral discourse. Ethical reasoning is always situated within specific traditions and is dependent on particular forms of life that have developed within a historical context.[85] Because imagination usually is creative rather than imitative, it refines and enriches the moral perception.

The role of metaphors and the value perspectives invoked in bioethics is the thesis of this book. Chapter 3 elaborates on the metaphor of ghost in present-day bioethical debates. Ghost phenomena—such as fake authors, drugs, practitioners, and diseases—are not only frightening because they are usually invisible, but they also imply deceit and deception. They compromise the fundamental value of trust in health care. Chapter 4 discusses the metaphor of monster that refers to the values of respect and dignity. Chapters 5, 6, and 7 focus on pilgrimage, exemplifying the role of hope in health care; the metaphor of prophecy, which plays an important role in genetic testing and screening that aims to predict the future; and the notion of relics, which refers to the possibilities of healing and intergenerational connections. The book concludes with an examination of the global perspective that evokes the basic notion of togetherness as well as concerns for vulnerable and "undesirable" beings.

CHAPTER THREE

Ghosts

Why do exceptional cases (such as Charlie Gard) receive so much attention in public debates? As stigmata cases they reveal underlying value perspectives that are frequently not addressed. These perspectives are reflected in fundamental metaphors that express imaginary visions about what is worthwhile. Bioethics—with its focus on individual cases, rational explanation, and justification—commonly does not explore such visions. As a new field of academic and public concern, it developed from a dissatisfaction and uneasiness with modern health care. As previously argued, bioethics marginalized its founding disciplines of theology and philosophy during its evolution. It became applied ethics, employing moral principles and neutral moral language and focusing on the pragmatic resolution of difficult cases. In this process, broader value perspectives were sidestepped.

Nonetheless, grassroots bioethics shows that different interpretative frameworks are at work in real-world ethical dilemmas. They involve more than objective facts and rational analysis and instead include such influences as imagination, feelings, and emotions. Many cases, and especially cases that attract attention and affect the agenda of bioethical debate, show that medicine is not merely a scientific, objective, and technical discipline. Values determine its work from the beginning. It is therefore important to explore these value perspectives. One way to do that is to analyze the fundamental metaphors that frame worldviews and the visions underlying bioethical challenges.

Ghost phenomena, as one interpretative framework, pervade contemporary science and medicine. We now have many concerns about ghostwriters. In serious medical publications, the named author sometimes is not the real author. They instead have been hired by a pharmaceutical company to be featured as an author because that suggests academic reliability and scientific authority. There are a growing number of journals that appear to be scientific and objective but, in reality, are commercial enterprises that only publish for money. Diseases can also be manufactured in order to create new markets for drugs and interventions so that the boundaries between real, false, or imagined diseases are blurred. The same phenomenon exists with medication. It is estimated that in poor countries, 10 percent of all medicines are fake.

Although publications, journals, diseases, medications, and interventions appear miscellaneous, they are all connected as manifestations of ghost phenomena. The metaphor of ghost is appropriate because it points out that contemporary health care is not simply a matter of scientific reliability, objective knowledge, and patient trust. It also includes deceit, as well as creepy and invisible phenomena. Ghosts are invisible forces that are often deceptive and harmful. They are scary even when we know they are not real. And while they affect individuals, ghost phenomena are not individual problems. They emerge in specific environments and circumstances. Addressing the ghosts of science and medicine therefore requires a critical focus on the contexts in which they appear: commercialization, publication pressure, promotion of drugs, and accessibility of medicines. Ghost phenomena undermine public trust and scientific credibility. They incite us to critically question the value context of science and medicine: how objective is science, how reliable are scientific facts and evidence, and how independent are scientists?

Ghostwriting and Ghost Publications

In August 2004, Adriana Fugh-Berman, a professor of pharmacology and physiology at Georgetown University in Washington, DC, received a draft of a scientific article about anticoagulation therapy with warfarin for prevention of stroke. The study was sponsored by the pharmaceutical company AstraZenica, which had developed a competitive drug for warfarin. Fugh-Berman was already named as an author in the draft. She refused to participate, but by coincidence, she was asked later to review the same article, with another recruited "author," that had been submitted to the *Journal of General Internal Medicine*. This is how the practice of ghostwriting was detected.[1]

Publications with authors who are not the real authors are not uncommon. The *Art of the Deal* (1987), the proud accomplishment of Donald Trump, was, in actuality, written by a ghostwriter. However, in scientific publications, ghostwriting is relatively new. The discovery that a substantial number of publications have not been written by the author(s) named was embarrassing for the scientific profession. Scientists are supposed to be honest and objective. They present and communicate the results of research and analysis in publications for which they take responsibility. They do this in order to share reliable knowledge with colleagues and to improve treatment of patients. Readers feel deceived and manipulated when it turns out that publications have been produced by invisible writers.

Humanity has always been interested in ghosts, in general. Recent polls indicate that 45 percent of Americans believe in ghosts, while in Britain, 34 percent of the population believe that ghosts exist.[2] Often associated with dead people, they can appear in different forms and travel back in time to haunt us. Ghosts can bring a moral lesson or seek revenge. They are, first of all, a reminder that there is another dimension to life. According to Susan Owens, ghosts are "mirrors of the times." They reflect our preoccupations.[3] The point is that ghosts are difficult to grasp; they are phantoms without shadows, invisible forces that create uncanny feelings and darkness. Even if they are visible, they are translucent and often associated with trickery and deception. They can do terrible harm.[4] Ghost phenomena in health care, then, call forth similar associations. They refer to opaque operations, invisible maneuvers, deceit, and trickery. They are haunting us, disturbing our assumption that scientific and ethical arrangements are clear and properly applied.

Some years ago, plagiarism was detected in an essay of one of the students in my graduate bioethics program. Learning that this was a reason to dismiss him from the program, the student was not as shocked by the plagiarism as he was that he had ordered the essay online and had paid a substantial sum of money for nothing. It is well known that students can purchase essays, research papers, theses, and even dissertations online.[5] That similar writing services are used by researchers and scientists themselves was a surprise, however. Court proceedings in the United States over recent years have disclosed documents that reveal pharmaceutical companies extensively using ghostwriting to promote their drugs. For example, Merck hired ghosts to write articles favorable to Vioxx, its painkilling drug.[6] Academic scholars were cited as authors of these studies though they had not been involved in the research or in the drafting of the manuscripts. Study results have been misrepresented with significant harm to patients. Deaths of some trial participants have been omitted from the publications. The drug had to be withdrawn from the market in 2004.

It is not clear how common ghostwriting is because it is a hidden underworld of medical publishing and not commonly reported. Incidentally, it may be detected, as were the cases of Fugh-Berman and Vioxx, and it is estimated that approximately 10 percent of publications in biomedical journals are ghostwritten.[7] However, detailed studies of promotional campaigns for drugs (as revealed in legal proceedings) gave much higher figures (over 50 percent in some cases).[8] With this backdrop, Sismondo and Doucet argue that 40 percent of journal reports of clinical trials of new drugs (and the number is even higher for conference presentations) are ghost managed.[9]

In most cases the ghosts are identifiable. They are professional writers working for medical communication agencies. But their names as real authors are concealed. The publication features an author who usually has not contributed to the text. It falsely suggests that a prestigious academic authored the paper to give the appearance of providing an independent and reliable opinion. These publications do not honestly report scientific findings but are part of a marketing strategy. The ghost (or his employer) is usually paid by a pharmaceutical company that has sponsored the research. It is not visible who is behind the report or what various interests are at stake. The whole system of publication planning and commercial writing companies is ghostly.[10] In the case of Fugh-Berman, the company sponsoring the manuscript and hiring the writing company had developed a drug to compete with warfarin. In the publication, it wanted to show that there were various drawbacks to the use of warfarin, preparing the ground for an alternative drug.[11] In this subtle way, even without direct promotion of a specific drug, readers were manipulated and deceived.

The obvious problem with ghostwriting is that it can be harmful to public health. Since these publications are misleading, they do not give an honest presentation of actual risks and benefits. They undermine scientific objectivity since conflicts of interest are not disclosed. With a well-known scholar as a fake author, a publication falsely suggests disinterestedness and objectivity. The implication is that bioethical debate cannot take scientific facts at face value. It should be critical of the general assumption that scientific publications, especially in prestigious journals, are sources of authority that are reliable and objective. Ghost publications cover up the potential biases of medical knowledge since they masquerade the interested parties.[12] These strategies of deception and concealment are threatening the trust that people have in scientific knowledge. For many scholars, this is the main reason why ghostwriting is wrong: it erodes the basic idea, not only among scientists themselves but among the public, that science is reliable and trustworthy.[13]

Several proposals have been made to exorcise the ghosts. First, acknowledgement of the real author will eliminate ghosts. This is unlikely to happen since the purpose of the publication is to lend authority to a specific opinion. This is why a fake author from academia is enlisted. Clarifying that he is not the real author will eliminate the marketing effect. Another option is disclosure. The author should mention possible conflicts of interests, for example, that he or she has received an honorarium from a specific company to author the paper. This is also not likely to happen nor does it resolve the basic problem of conflicting interests. Furthermore, even though medical journals have introduced criteria for authorship and declarations of conflict of interest, not all of them have misconduct policies.[14] Finally, sanctions for academic misconduct is yet another option, but many universities do not have policies prohibiting ghostwriting that would support that initiative.[15]

The metaphor of ghosts puts the blame on the industry that is secretly hiring writers who are not mentioned. It suggests that scientists are victims of manipulative practices. This is not a true reflection of reality. Scientists themselves are available as fake authors and eager to be involved in publications.[16] Rather than focusing on pharma as the threatening and horrifying ghost, one should question the context in which the ghosts are emerging. Historically, ghosts are believed to appear in dark environments: ruins, decaying buildings, and graveyards. They now appear because the idealistic edifice of science and medicine is corrupted. Ghosts are produced by the commercialization of medicine. The distinction between commerce and academic research is blurred.

There are pervasive financial relations between industry, scientific research, academic institutions, and medical establishment.[17] Normally we assume that doctors and scientists have a primary interest that drives their decisions and actions (e.g., for doctors, the welfare of the patient; for scientists, the discovery of reliable knowledge). But in practice, secondary interests often prevail—financial gain but also ambitions for prestige, awards, and recognition. These conflicts of interests produce scientific misconduct. This is not incidental but has become a chronic pathology in science.[18] Scientific data are falsified and fabricated, as a long list of scandalous cases illustrates.[19] On average, 2 percent of scientists admit to falsifying research at least once, and up to 34 percent admit other questionable research practices, such as changing the results of a study following pressure from sponsors.[20] Nowadays, honesty is not fashionable.

Scientific misconduct is usually explained by the seductive influences of money. Medical research is now increasingly funded by the industry. Universities are dependent on grants and sponsors. Practicing physicians receive gifts from pharmaceutical companies. In one study, 28 percent reported receiving payments for consulting or serving as a conference speaker.[21] In 2014, US physicians and teaching hospitals received $6.49 billion from pharmaceutical and medical device manufacturers. More than $400 million was for "entertainment" (i.e., food and beverages), travel, and lodging.[22] Authors of clinical practice guidelines that are important for the practice of many physicians often have financial interactions with pharmaceutical industries whose drugs are considered in the guidelines.[23] Medical education is sponsored by the industry; on average, medical students in American medical schools are exposed to one gift or sponsored activity per week.[24] Education has transformed into marketing and advertising.

Medical journals, as media of publication, are in a difficult position. In some cases, they are no longer independent. They receive substantial income from the pharmaceutical sector through advertising and reprints. It is argued that they should no longer publish industry-sponsored clinical trials.[25] At the same time, the academic climate has changed considerably. Evaluation and promotion of faculty depends on the number of published papers and their impact factor, as well as grants and patents. Researchers are under pressure to publish, and that is what they do. In this context, commercial management of the scientific literature is helpful and the nature of scientific publications has fundamentally changed.[26] They are no longer aimed at dissemination of knowledge but are vehicles for commercial marketing and scientific self-promotion.

Pharmaceutical companies, as well as scientists, know that this is something to hide. Conflicts of interest and reiteration of biases need to be disguised. Ghost publications can be regarded as bizarre since they are contrary to the values embedded in scientific research. They are stigmata cases that reveal that scientific activity has radically transformed under the influence of market ideology. But the symptoms at the surface are often not related to the rotting environment from which ghosts materialize. Because bioethics usually is concerned with individual rather than contextual determinants, it focuses on responsible conduct and integrity of individual researchers. This approach is worthwhile but insufficient. Time and again, it is apparent that self-regulation and appeals to responsible conduct do not work—ghosts cannot be asked to exorcise themselves.

Ghost Journals

Predatory journals are another current plague for scientists and ethicists. These are journals that look like scientific journals but do not care about science or scientific quality but solely about profits. They are an academic nightmare.[27] They seduce many scholars to submit manuscripts because they promise quick publication after payment of a fee. Generally, there is no serious review of submitted materials, indexation in legitimate databases is falsely claimed, and fake impact factors are used. Scientific journals usually rely on independent peer review to assure that knowledge is trustworthy and verifiable. For predatory journals, this is not a primary concern. Submission of a fake and flawed paper to predatory journals showed that 60 percent of the journals accepted this paper and claimed peer review had been done.[28]

This ghost phenomenon is the result of the explosive growth of open access journals. This business model in journal publishing does not rely on the subscription of readers but instead on the publishing fees of authors. It is not difficult to set up an online journal; the challenge is to find authors who are willing to pay for being published. In this context, it is imperative for the journal to accept as many papers as possible. Emphasis on the quality of content, important to attract subscribers, becomes an obstacle to generating profits.

Another guarantee for the quality of journal publications used to be the editors and members of editorial boards. Distinguished and experienced scientists commonly run prestigious journals. Of course, this system is not perfect and many flaws have been detected recently, but ghost journals do not care at all. They have no infrastructure to guarantee scientific quality. The peer review system is absent or deficient; within a few days one receives a standardized message that one's publication was accepted. True editorial boards are missing. Members are either unknown or nonexisting scholars, or persons are mentioned without their knowledge. Editors of these journals are not selected based on scientific accomplishment but on the willingness to pay. One study showed that a fake application for an editor position resulted in immediate acceptance and even appointment in forty predatory journals.[29]

It is estimated that currently there are eight thousand predatory journals.[30] Bioethics is not an exception. In this field, there are approximately twenty-five predatory journals, with fancy names such as *Journal of Clinical Research & Bioethics*.[31] And from predatory journals stem ghost conferences as well.[32] Once a student happily informed me that her abstract was accepted for presentation at the 20th International Conference on Medical Ethics and Health in Paris in March 2018. The conference website looked impressive, and the conference was organized by the World Academy of Science, Engineering, and Technology (WASET). Unfortunately, this is a known fake publisher organizing predatory conferences. It schedules thousands of conferences every year, often in the same location on the same day. Anyone paying the registration fee can present a paper. However, the "academy" is registered in Azerbaijan and managed by a family in Turkey.[33]

Criteria have been proposed to recognize fake journals,[34] but distinguishing between predatory and legitimate journals is still not easy. And even more deceptive are "hijacked" journals.[35] They look exactly like respected journals that have already existed for some time. They can even have the same name. An example is the longstanding *International Journal of Philosophy and Theology* (https://www.tandfonline.com/toc/rjpt20/current), founded in 1938 in Louvain, Belgium. Since 2012, another journal with the same name and produced by the so-called American Research Institute for Policy Development can be found at http://ijptnet.com. The bank account of this journal is in Bangladesh.[36] Jeffrey Beall, librarian at the University of Colorado-Denver, once published an online repository of open access journals with predatory behavior, the so-called Beall's list, in order to help identify the culprits. In January 2017, the list was temporarily removed from the internet, probably because of legal threats from predatory publishing companies, but reappeared again a few years after that in other formats.[37]

Predatory journals demonstrate several features of ghosts: they are deceptive, present illusions, and are also harmful. They undermine the quality of medical research. We can no longer be sure that published studies are reliable; it is unclear whether they have been sufficiently tested, vetted, and analyzed by the scientific community.[38] Pseudo-journals also create a false academy. Appointments and promotions of university scholars are nowadays based on publication output. The number of publications is mostly determinative. But when output is driven by payments, what confidence can we have in the quality of scholars? Finally, loss of trust and confidence in any scientific publication will affect public trust in science in general.[39]

How can we exorcise these ghosts? It is too easy to blame the publishing companies. The ghosts appear within a context. Scholars and researchers have increasingly been pressured to publish. Decisions about funding and career advancement are based on publication activity. In most countries, the quantity rather than the quality of scholarly work is considered. Open access journals are an ideal way to broadly disseminate scientific knowledge, but they also create the opportunity to boost output if a commercial motive takes priority and scientists want to pay for their publications.[40] In the current academic climate, scientists may have all kinds of reasons to publish in predatory journals.[41] Quick and easy publishing is not discouraged in this climate.

Remedies, therefore, will have to scrutinize academic practices and endorse more stringently the values underlying scientific work. Scholars should give more attention to the selection of journals. Publications need to be read and assessed for content and quality, rather than simply counted.[42]

Ghost Diseases

In some traditional cultures, diseases were directly attributed to ghosts. The spirits of deceased people were assumed to haunt specific persons and make them ill. An example is the mysterious disease called *kuru* that affected the Fore tribe in New Guinea. It impaired movements, making people weak and shaky, until they became paralyzed and died. Nobody with kuru survived. The disease had attracted medical attention since 1955 but was difficult to explain or identify causative factors. According to the natives, it was the result of sorcery, and eventually it was linked to cannibalism, a practice that existed in New Guinea until the 1960s. The neurological degeneration was the result of an infection by a "slow virus" with a long incubation period. Fifty years later, the disease died out.[43]

Ghost diseases still emerge from time to time when people fall ill for mysterious reasons. Genital retraction syndrome (or *koro*) has been reported in China, Nigeria, and the Democratic Republic of Congo. In these cases, the penises of some men have suddenly disappeared; people believe the penises to have been stolen and penis thieves have been arrested.[44] In 2017, North Korean defectors referred to a ghost disease that affected many people, but it turned out that it was the result of leaked radiation from nuclear test sites. Another example is a mysterious disease that affected people working at the US embassy in Havana, Cuba. They had headaches, dizziness, and hearing loss. The US State Department attributed this disease to "sonic attacks" by the Cuban government. Extensive examinations and investigations revealed that people were seriously affected though without any clarity about causes and mechanisms. Recently scientists argued that the unusual sounds were produced by crickets.[45]

Most ghost diseases nowadays are different. They are not mysterious events but instead are fabricated. This is the phenomenon of "disease mongering," which is "a widening of the diagnostic boundaries of illness for reasons of economic benefit."[46] It is regarded as "the selling of sickness."[47] The concept of disease is stretched so that a larger market for treatments is created. In everyday life there often is no clear boundary between normal and abnormal.

People have complaints such as anxiety, depression, heartburn, fatigue, or baldness that can make them feel ill. In some severe cases, such complaints might be associated with diseases or abnormal conditions, but this is usually not the case. Complaints are temporary and expressions of the variability of the human condition.

Disease mongering means that the unclear boundary is exploited. Ordinary life ailments (such as hair loss or shyness) are turned into medical problems. Mild symptoms are sold as manifestations of a serious disease, and risk factors such as hypertension and osteoporosis are conceptualized as diseases in and of themselves.[48] It is argued that normal varieties of human existence are expressions of diseases that are not recognized. The prevalence of such diseases is underestimated and better diagnoses should be encouraged. Doctors should be better educated in determining these hidden diseases. The suggestion is that all cases should be treated, primarily with medication that is available. This is exactly the point of disease mongering: it is creating a need for treatment and thus expanding the market for drugs. It is not interested in fatal or short-duration diseases. More profitable is the fabrication of diseases that are chronic and lifelong. Rather than promoting curative drugs, the focus is on lifestyle drugs (for instance for allergies and acid reflux) that you need to take every day for the rest of your life.[49] Ghostwriting, in fact, is often employed to shape publications that promote the selling of these products.[50]

One infamous example of disease mongering was the launch of "chronic halitosis" as a new disease characterized by bad breath. Listerine, named after Joseph Lister, the 19th century British pioneer of antiseptic surgery, was promoted by its manufacturer as an antiseptic mouthwash in an aggressive marketing campaign in the 1920s, successfully arguing that it was a cure for this new disease.[51] Presently, discussions are unfolding about whether "fragile skin" should be regarded as a disease.[52] Other well-known examples are erectile dysfunction, female sexual dysfunction, restless legs syndrome, and premature ejaculation.[53]

A long-time ghost disease controversy concerns attention-deficit hyperactivity disorder (ADHD). This diagnosis now applies to 4.6 million American children between 6 and 17 years of age. Almost 60 percent are taking medication. Prescription stimulant sales have more than quintupled since 2002. It is argued that ADHD is not a simple and recognizable condition. Its diagnosis depends on an observer's interpretation. There is always a "zone of ambiguity," and parents, teachers, and physicians may disagree on whether the symptoms are mild, moderate, or severe in impairing the functioning of the child. Prevalence rates range from 1 percent to 20 percent, and nowadays a new market is targeted: it is argued that about 10 percent of adults have ADHD.[54]

"Mongering" refers to undesirable activities. A warmonger is a military hawk, instigating or attempting to stir up war. A disease monger is somebody who advocates or tries to propagate diseases, and the diseases advocated are like ghosts. It is unclear whether they really exist or not; they are frequently illusions that produce fear in people. In an earlier stage of disease mongering, doctors were blamed for making people ill. Critics argued that diseases were produced by the power of the medical profession, and a major threat to health was the medical establishment undermining the power of individuals to heal themselves.[55] However, the drivers of medicalization today have changed. Medicalization is now promoted by commercial and market interests.[56] The real ghosts nourishing this phenomenon are therefore easy to identify: the pharmaceutical industry that is using ghostwriters and ghost journals to generate ghost diseases. The pharmaceutical industry, rather than the medical profession, is the engine that defines human challenges as medical problems demanding treatment with medical drugs. And the purpose of disease mongering becomes obvious: its aim is to promote the sales of drugs.

Medicalization is closely associated with pharmaceuticalization: medical problems are basically pharmaceutical problems.[57] The argument, of course, is that this is all for the benefit of humanity and so nothing is wrong with disease mongering. The industry claims to be concerned with the welfare of patients. The growing increase of drug sales reflects the progress of medical science, recognizing the needs of the patient, more sensitive diagnostics, and more innovative medication. Appeals are made to the autonomy of patients. The industry just provides information to educate people so that they can make more informed choices. The message is always the same: some conditions are underdiagnosed, and effective drugs are available.

These arguments show how disease mongering is a ghost phenomenon. Promoting diseases whose reality is questionable appeals to such values as individual well-being and responsibility of health. It leaves out the commercial context in which human conditions are transformed into diseases. Bioethics should not take for granted the facts and evidence provided by medical expertise but should question the underlying value system. It should broaden its scope beyond individual experiences and their medical interpretation.

The harm of disease mongering is not only deception; the ghosts produced are also exploiting people's anxieties and fears. They reinforce the sense of vulnerability, especially when people are suffering and aging. At the same time, they articulate faith in scientific progress, promoting the idea that there is a pill for everything. Drug therapy is advertised as the most adequate treatment option while other approaches are downplayed. Viagra is the solution for erection problems although they are strongly associated with psychological factors.[58] For ADHD, various options exist: changing sleeping and eating patterns, classroom interventions, and behavioral therapy.

Many studies show that behavioral treatments are more effective than drugs alone, but it is common sense that stimulants are the easiest approach. And it is argued that these options reflect different values. Drugs focus on efficiency. Behavioral interventions emphasize the value of engagement; they do not locate the problem in the individual but within the context of interaction with the family and social environments.[59] In cultures that highly appreciate individual performance, social tolerance for heterogeneous behavior is diminished. But in this cultural environment, it is not the individual that is the main problem. Fabricating diseases, therefore, is reinforcing a reductionist approach in health care. It is reiterating fundamental values of contemporary medicine; problems are individual and located in the body.[60] Complaints can only be explained by somatic models. Sexual interaction, for example, in the case of premature ejaculation is reduced to coital penetration.[61]

Similar to other ghost phenomena, modern health care is not haunted only by the pharmaceutical industry—creating and promoting pseudo-diseases is the result of alliances between industry, physicians, patient groups, and news media. Various parties are shaping public opinion and changing public perceptions of ordinary and mild human problems. It is doctors who diagnose disorders and prescribe drugs. They have entangled themselves with the industry and are sensitive to commercial temptations. They are also much less in control now that medication can be purchased on the internet and also promoted on television. Direct-to-consumer advertising is legal in the United States (though not in most other countries), and promotion and advertising are effective ways to increase prescription frequency. For example, fabricating the syndrome of "late-onset hypogonadism" and promoting testosterone as the remedy has led to a tenfold increase in testosterone prescriptions. Advertising encourages "patients" to put pressure on physicians to prescribe while evidence is weak and possible risks are undervalued.[62]

The news media plays a significant role in disease mongering as well. Media coverage of health issues is often sensationalized, disseminating extravagant claims and fueling false hopes. The majority of messages exaggerate potential benefits, minimize potential harms, do not mention costs, and do not disclose conflicts of interest. The role of the health care industry is mostly invisible while its "miracle drugs" are hailed as breakthroughs.[63]

Finally, the role of patient advocacy groups is dubious in the corporate construction of diseases. Research recently concluded that 83 percent of patient-advocacy organizations are receiving financial support from industry (with 39 percent reporting that they are receiving at least $1 million per year). In 36 percent of these organizations, representatives of drug and biotechnology device companies are on the governing board. The majority do not have policies for conflicts of interest.[64] How can such organizations be considered independent? It is known that companies have created their own "patient groups."[65] Patient-advocacy groups may therefore be an important vehicle for the industry to monger diseases.[66] On the other hand, there are also groups that resist medicalization or have been harmed by certain drugs and are litigating against manufacturers.[67]

Disease mongering has important implications for the doctor-patient relationship. Rather than patient compliance, the emphasis nowadays is on physician compliance.[68] Patients, extensively informed via the internet and supported by online illness communities, expect physicians to acknowledge their expertise and to recognize their definition of complaints as physical illness. They basically want the medicalization of their condition.[69] Receiving a diagnosis not only removes the mystery of their condition but also has practical consequences for receipt of benefits, reimbursement, and special assistance. The tension between lay and medical perspectives has created a new type of ghost diseases, so-called contested diseases. People experience real and identifiable symptoms but their complaints are not recognized by the medical profession as legitimate diseases or are dismissed as mental illness. People start negotiating over the diagnosis and treatment, and physicians argue that they cannot detect any biomedical mechanism that explains the symptoms.

A famous example is Morgellons. Patients have open skin lesions and infections with the sensation of bugs crawling under their skin. Physicians interpret this condition as delusional parasitosis, a mental illness. This diagnosis is strongly rejected by patients, who argue that they have a real disease, although medically unexplained. Patients organized themselves, created an online community, started media campaigns, promoted research, and labeled the condition after a medieval French skin disease. They succeeded in increasing visibility and legitimacy for this condition as a disease.[70]

While disease mongering is different from writing scientific reports and publishing in journals, it is similar as a ghost phenomenon. It highlights that medical reality is fuzzy and unclear. Facts and evidence are contested. Subjective and objective interpretations of individual experiences are often mixed. Ghost diseases impact and compromise the rational decision-making of doctors as well as patients. The implication for bioethical discourse is that medical and scientific expertise should be scrutinized for its underlying values. Ghost phenomena should be exposed. The scope of issues to be analyzed from an ethical point of view should be expanded.

Ghost Medication

Traditionally, the term "ghost pills" referred to the empty shells of tablets fabricated for the slow and steady release of medication over a period of time. The outer structure of the pill would not disintegrate, and this is what was seen as a ghost pill, sometimes noticeable in stool after it passed through a person's system.[71] Currently, health care, especially at the global level, is haunted by a different type of ghost pill: fake medication. The World Health Organization estimates that in developing countries, 1 in 10 medical products is falsified or substandard. This means that people take medicines that do not work or are harmful. Most commonly affected are antibiotics and antimalarials. These are significant drugs that are often vital for recovery and even survival. It is estimated that 72,000 to 169,000 children die each year from pneumonia due to fake antibiotics. An additional 116,000 people die from malaria for the same reason.[72]

Where there is more market surveillance, regulation, and control of drug distribution channels, there is lower risk that a product is falsified. And since it is illegal to make and disseminate fake drugs, the real extent of the problem has been hard to track. Global reporting on falsified and substandard medicines has only been done since 2013. The problem is evident in all countries, although much more frequent in the developing world,[73] and it is generally acknowledged that the problem is on the rise.

One factor is the internet. Sales through internet pharmacies are rapidly growing, and an overwhelming majority of online pharmacies operate outside legal regulations. It is reported that 77 percent of Viagra samples purchased online are falsified products, for example.[74] Global trade is also a factor. Many countries do not produce medication, but medical products are composed of ingredients from across the globe.[75] Another driver of fake medication is organized crime. In 2012, this illegal market was estimated to generate $432 billion a year. It is a profitable business without much risk because many

countries do not have drug legislation. The risk of detection is low. Law enforcement is soft, and penalties are ridiculously weak.[76] Frequently, health
workers in resource-poor counties have such low salaries that they are susceptible to corruption. Criminal networks that used to dominate the illicit
drug market are now, according to the World Health Organization (WHO),
increasingly targeting the medicine trade.[77]

Ghost medication is not a new problem. Faking and manipulating medicines is as old as medicine in general. When I was a medical student, a professor of tropical medicine told us that when he was working in West Africa
in the 1950s, he noticed that people were extensively using small bottles of
Dutch *jenever* (gin) as medication. It was labeled as medication for a whole
range of conditions. Fake penicillin started selling immediately after the end
of the Second World War.[78] Some decades ago, the most falsified drug was
Viagra, but these days all classes of medications can be falsified, from lifestyle
drugs to lifesaving drugs. A notorious example is the blockbuster anticancer
drug Avastin. In 2012, nineteen medical practices in the United States had
purchased the drug from an overseas supplier. The fake version contained
salt, starch, and various cleaning solvents.[79]

It is obvious that fake drugs are harmful. Harm is caused by therapeutic
failure, and the death toll presumably is around 1 million people annually.[80]
Since diseases such as malaria are not properly treated, people die while they
could have been saved with genuine drugs. People die because of adverse drug
reactions. Due to improper or toxic ingredients, fake drugs may directly kill
patients. They lead to increasing resistance to treatment, particularly anti-
microbial resistance, which is now a growing global problem. Paying for fake
medication is also a waste of money. And finally, they lead to a decline of
confidence in health systems.[81]

The ethical challenge is to determine what the real problem is. Fake medication is a general label that refers to poor-quality drugs. But the label covers
different categories.[82] Falsified drugs are intentionally and fraudulently produced (they usually have no active ingredients, or they contain an incorrect
dose of ingredients or incorrect ingredients). Substandard drugs are the result of unintentional or negligent errors of manufacturing (usually containing
insufficient active ingredients). Counterfeit drugs falsely represent the identity and source to make it appear as a genuine product (branded or generic).
WHO lumps all categories together under the acronym SSFFC (substandard,
spurious, falsely labeled, falsified, and counterfeit medical products).

From the perspective of bioethics, the real threat is the category of falsified drugs. They intentionally deceive people and harm them. Counterfeit drugs do not necessarily endanger human life; they may cause harm if they do not have the same pharmaceutical ingredients as the real drugs, but if they are a good counterfeit, they have the same substances and effects without being distinguishable from the genuine brand or generic drugs. Of course, they present the ethical problem of infringement of patent rights and the practice is regarded as piracy. They may cause loss of revenue for the industry. Many companies have their own monitoring and security systems, collecting information worldwide, but the industry does not share the data. They are afraid of reputational damage and loss of confidence in their products.[83]

The problem of counterfeit drugs is of a different ethical magnitude than the risks of falsified drugs. It is argued that counterfeit and falsified drugs should not be conflated.[84] By lumping all categories together, WHO does not take that position. In fact, it does not want to alienate the pharmaceutical industry that often supports its activities. For exorcising the ghost of fake medication, it will be important to emphasize the public health perspective rather than that of commercial trade. Appeals to criminalize the illegal trade and to draft a global treaty to ban fake drugs are primarily in the interest of the pharmaceutical industry, especially if they put counterfeit, substandard, and falsified drugs in the same category. Emphasis on counterfeiting can prevent major attention being given to the real threat to public health, which is assuring the quality of medicines. It may also obstruct the availability of legitimate drugs.

In 2009, Dutch customs seized a shipment of the antiretroviral drug abacavir from India to Nigeria. They were alerted by the pharmaceutical company claiming a patent on the drug in the Netherlands. The drugs had been purchased by the UNITAID program to supply generic medication to Africa. The argument was that the shipment involved fake medication, which justified legal action.[85] The case illustrates the point that reference to counterfeit medication can be used to obstruct the provision of generic medication. And thus the reason why there is a market for counterfeit medication in the first place. Medicines are too expensive for many people. For example, the costs of Avastin is $2,400 per injection. The fake drug was offered at much lower prices. Review of trials with Avastin in 2011 showed that there was no real benefit, but some patients and physicians still wanted to use it. That was the reason why some medical practices in the United States purchased the medicine online. WHO has pointed out that 2 billion people do not have access to medicines that are crucial for their health.[86] Therefore, the fundamental problem is the price of medical products, and the vacuum is filled with fake drugs. Lowering the price of pharmaceuticals will reduce the demand for fake drugs.[87]

Ghosts disguise various types of poor-quality medicines. They also are a symptom of two characteristics of present-day bioethical problems. First, although individuals are harmed, the problem is at a more general level. Bangladesh, for example, faces a growing problem with fake drugs. Television station Al Jazeera interviewed a mother who lost her son due to lethal falsified medication.[88] The same day in that hospital, three other children died from the same drug. Focusing on the case of an individual will not explain the ethical problem: it will be necessary to connect the individual case with the underlying problem at the global level. That analysis will also make clear that Bangladesh, with an inadequate infrastructure, a dysfunctional government, unreliable regulations, lack of enforcement, widespread poverty, and corruption will never be able to solve the problem on its own. It will need the cooperation of other countries that are less impacted by fake medication. The implication is that normative analysis requires a perspective beyond individual cases.

Second, the roots of bioethical problems, because they are global, are often associated with international policies. Over the last few decades, policies imposed by the World Bank and the International Monetary Fund have negatively affected existing infrastructures for public health. Featuring deregulation and privatization, regulatory and control systems for the provision of medication in many countries have been weakened and counterfeit-medicine markets have been created. The example of Peru illustrates how neoliberal policies have produced a chaotic growth in the number of pharmaceutical establishments and a serious problem of fake medicines.[89] Normative analysis should, therefore, focus on the underlying structures and mechanisms that facilitate the appearance of ghosts.

Ghost Interventions

When Tom Intili was told that he could participate in a clinical trial of a surgical procedure for Parkinson's disease, he was very optimistic. At 50 years old, he'd had the disease for ten years. He knew about promising experiments with innovative new drugs and thought this might be a chance to improve his condition. A year after the brain surgery, Tom felt much better. He no longer needed his wheelchair. Then his neurologist told him that he had undergone sham surgery. The preliminary results of the trial showed no difference between treated and untreated patients and so the drug company stopped the trial. Tom's condition immediately reversed to what it was before the trial.[90]

Sham surgery copies all aspects of a surgical procedure except the thera-peutic intervention. The patient undergoes anaesthesia or sedation, his skin and tissues are incised, he is kept in the operation theater as long as the ac-tual surgery should take, the team and machines make the usual noises, and even the smells that usually arise are produced. In sham brain surgery, burr holes are drilled in the skull. All aspects of real surgery are mimicked. It is a phony operation because the patient thinks that he has undergone surgery to receive a treatment but he has not gotten any treatment at all, just the surgi-cal works. It appears to be a therapeutic intervention without real therapy. For Parkinson's disease, for example, experimental cells are not implanted in the brain. The overwhelming majority of Parkinson researchers today are in favor of such fake operations. It is the only way, they argue, to prove that a surgical intervention is effective. They even want the sham to go further: not only make holes in the skull but even penetrate the brain tissue.[91]

Many times miracle cures for Parkinson patients have been reported, with-out long-term results. Patients travel around the globe to have cells trans-planted into their brains. To be sure whether an intervention is effective, it is necessary to eliminate bias as well as the placebo effect, according to the re-searchers. In clinical studies this is done by comparing two groups: the treat-ment group that receives the actual intervention and the control group that receives a placebo. For present-day science, double-blind placebo-controlled studies are now the gold standard. It takes the placebo and its effects very seriously. The placebo effect is particularly strong in Parkinson's disease. New therapies give desperate patients hope and optimism. If surgery wants to prog-ress, it needs to take the same approach. Furthermore, if it can be concluded that certain surgical interventions are not effective compared to placebo interventions, it can save many efforts and resources. For example, studies showed that knee surgery for osteoarthritis was not more efficient than the placebo procedure. The researchers suggested that the billions of dollars spent on these surgeries was wasted.[92]

However, sham surgery is contested. It is an invasive procedure. Making a hole in the skull is not the same as prescribing a placebo pill. It is potentially harmful since anaesthesia can be risky, let alone penetration of the brain. Tom was under anaesthesia for six hours. It is also unnecessarily risky when there are no therapeutic benefits to the patient (except the placebo effect), and alternative, less harmful, study designs are available. Furthermore, in-formed consent of patients is problematic. In serious and disabling diseases such as Parkinson's disease, patients often assume that there is a potential direct benefit. Studies offer high hopes and expectations; patients are eager to participate. This so-called therapeutic misconception compromises informed consent. The major ethical objection is that sham surgery is active deception. It suggests that a real intervention is done while patients (at least some of them) are fooled. On the basis of these arguments, it is concluded that sham surgery is morally unacceptable.[93]

This ethical conclusion conforms to the position of the American Medical Association (AMA) regarding placebo medication. In 2006, the AMA decided that the use of placebos in clinical practice should be prohibited. It is unethical to deceive patients by prescribing placebos. It is only allowed when the deception is removed: the doctor should explain what placebos are and why he is prescribing them, and the patient should agree to receive a placebo. This strict position was softened in the revised AMA Code of Medical Ethics in 2016. Placebos for diagnosis and treatment are now allowed under three conditions: the physician should enlist the patient's cooperation, the patient should give consent to administer a placebo, and placebos should not be given to mollify a difficult patient.[94] The basic suspicions are still the same: placebos may undermine trust, compromise the patient-physician relationship, and result in medical harm to the patient.

Revealing the use of placebos will indeed make clear that there is a ghost intervention. At the same time, some argue, it will eliminate its efficacy. If the effect of placebos is based on the expectations and beliefs of patients, telling them that they will receive inactive substances will remove the potential placebo effect. The question is whether placebo effects can be obtained without deception. The AMA argues that placebo effects exist even if patients are openly informed about the use of placebos.[95] Today it is difficult to prescribe sugar or bread pills (patients will find out quickly), but instead active treatments are prescribed with the intention to promote a placebo response. An example is the use of antibiotics for the common cold. In such cases, there are placebo effects without placebos. Any drug or action can have a placebo effect.

The history of medicine has been the history of placebos. For most of the time, doctors did not have effective medication. All kind of substances, not only inert ones, were used to promote the healing process. Until the 1950s, the general view was that placebos were harmless and could be comforting to patients. Since then, however, the placebo came to be considered more powerful. An influential but heavily criticized study of Henry Beecher in 1955 concluded that 35 percent of patients had a positive response to placebo treatment.[96] This appreciation for placebos was the result of the increasing acceptance of the randomized clinical trial as the standard for evidence in medicine. In such trials, placebos are used as methodological tools, assuming that they are inert substances. Ironically, it shows the effects of placebos. The more powerful the placebo, the more need for double-blinded trials.[97]

The emergence of bioethics produced a different evaluation of placebos. Since they involve deception, prescribing placebos does not respect the moral principle of personal autonomy. As intrinsically misleading interventions, they have also harmful effects. Placebos can have adverse effects and even lead to addiction. Patients may find out that they have been duped and lose confidence in physicians. In the long run, placebos also reinforce the idea that there are pills for everything.[98] In this more critical climate, it is doubted whether placebo interventions really have meaningful therapeutic effects. It is argued that the placebo effect is a myth. Since placebos are inert, they cannot cause any effects.[99] They are ghosts haunting biomedicine.

Interestingly, contemporary studies of the placebo phenomenon acknowledge that the placebo effect is not adequately explained. It is related to patient expectations, beliefs, and imaginations. Also, the impact of the physician is emphasized. Michael Balint, the author of one of the classics of family medicine, wrote that the doctor is the most important drug.[100] But it is primarily the context of the relationship between health professional and patient that facilitates therapeutic responses. More than which drug or intervention is offered, it is the trust of the patient in the health provider as well as the entire healing context that promotes healing processes. For healing, the symbolic meaning of treatment as well as the setting within which it takes place are more important than what is provided. The ethical boundary is clear: when an effective standard therapy exists, it is unethical to give placebos. At the same time, how doctor and patient communicate will affect the healing response.[101] This has nothing to do with deception. Long ago, a study in family practice showed that patients who received a positive consultation got much better than those who had a negative consultation, whether or not treatment was provided.[102] An optimistic, reassuring, and empathic health professional contributes to healing.

Conclusion

In the medical world, there are authors who have not written "their" publications, journals that are not really scientific periodicals, diseases that are fabricated by companies to sell drugs, medications that do not contain any medical substance, and surgical interventions that are sham operations. Many ghosts are haunting contemporary medicine. In numerous cases, it is difficult to recognize that we deal with these ghosts. They are all real phenomena in the sense that we see them happening, they have a concrete impact on patients and health professionals, and they influence not only how we think about health and disease but also how we feel. At the same time, these phenomena are not real. They suggest that diseases exist while, in fact, ordinary symptoms are only getting a different label. We think that we had surgery but, in fact, it was a fake operation. A graduate student submits a manuscript to a journal, hoping to have a scientific publication, while in fact, the journal is only interested in her money. Ghost phenomena make us aware that there is another, often dark, dimension to human life. Even in the respected field of health care and medical science, there is a whole realm of hidden manipulations, deceit, and invisible maneuvering. These ghosts undermine trust, put into question scientific objectivity, and make it doubtful what are facts and what is evidence.

Many ghosts emerge from specific contexts. Nowadays, commercial interests principally determine these contexts. Perhaps most of the bioethical problems of today are the result of the seduction of money rather than the power of science and technology or the medical profession. The philosopher John Stuart Mill already warned that the love of money is "one of the strongest moving forces of human life," but his admonition is not taken seriously in bioethics.[103] Commercial interests easily compromise other interests, such as the patient's good, the public's good, or society's good. As long as bioethics is focused on individual decision-making, it will never get a grip on the underlying processes that drive ghostwriting or disease mongering.

Bioethical analysis should ask why and how ghosts appear. Exorcism rituals that highlight transparency, reporting, and disclosure will never dispel the ghosts. They assume voluntary cooperation based on the idea of individual virtues and the influence of education and persuasion. Some ghosts will perhaps cooperate, but in general they have more vices than virtues. Scientists and physicians themselves are heavily involved in promoting drugs on behalf of companies. Three-fifths of the chairs of medical school departments in the United States receive personal income from drug companies.[104] Many scientists suspect and even observe scientific misconduct, but the reporting is low (whistle-blowers are frequently sanctioned).[105] Clinical guideline panels are dominated by experts with financial relations to interested companies. Bioethics itself is contaminated. Members of Institutional Review Boards who are supposed to independently assess research proposals, commonly have relationships with the industry, but they do not disclose this. There also is a growing number of commercial research review boards.[106] The database created in August 2013 by the Physician Payments Sunshine Act (a section of the Affordable Care Act), requiring commercial companies to report payments and gifts to physicians in the United States, shows that more than 800,000 physicians have received industry payments.[107] Medical education continues to be heavily subsidized by the industry despite efforts to reverse this trend.[108] Studies show that voluntary reporting does not work, and voluntary disclosures are not reliable.[109] Even full disclosure does not eliminate the basic problem; it will not remove the ghosts due to financial conflicts.

The conclusion from these ghost phenomena and the unsuccessful efforts to exorcise them is that the scientific system of objective assessment is broken. Bias is permeating the entire system. Research data are manipulated and misrepresented. The medical literature is no longer reliable. The current emphasis on transparency and disclosure is only a fake movement that allows the ghosts to reappear all the time. Restoring integrity and trust will require more drastic measures. The remedy is to clean up the context that generates the ghosts: the commercial entanglement of medicine and science. To restore the credibility of knowledge claims, independent sources of information need to be available with information material that is not sponsored and biased by corporations.[110] Since conflicts of interests are mostly invisible, they should be identified and reported in a publicly available database of conflicts of interests. But the most important consideration is that such conflicts are not necessary.[111] Managing them as best as one can wrongly assumes that they are inevitable, that they cannot be eliminated. Financial conflicts, however, are optional. They are the result of a choice and not inherent in the activity of a physician or scientist. They compromise independence and the possibility to produce disinterested and impartial research. Preserving independence as a scientist and doctor has a price, but it makes clear to the public that there is no hidden agenda. It clarifies that interventions and decisions are based on the patient's health or the public's health and not on often invisible commercial interests. From an ethical perspective, perhaps the best experts are those that are not conflicted. They will usually not be haunted by ghosts.

The metaphor of ghosts helps to regard such disparate entities as publications, journals, diseases, medication, and intervention as bizarre cases that exemplify a particular type of stigmata. They point to the experience that the underlying value system of science is compromised. Objectivity and independency as fundamental values are in these cases no longer upheld. Consequently, the various ghost phenomena are attracting much attention inside and outside the scientific and medical community. They are a travesty of what is considered as the ethical core of scientific and medical expertise: integrity, honesty, and reliability. Scientific professionals and organizations are therefore rightly worried since the phenomena are not simply superficial incidents. Particularly as ghosts they are real and frightening because they affect the basic values of the scientific enterprise.

For this reason, ghost phenomena are also fascinating for bioethics. But they present a fundamental challenge. First, they show that science is not abstracted from its context. As a human activity, it is vulnerable and open to failures, distortions, and fabrications. However, the changed contexts of contemporary research and health care are seldomly addressed. That scientists are clearly aware of the significance of these contexts is disclosed in their concern that ghost phenomena undermine public trust in scientific activities and findings. Second, the phenomena show that some values are often prioritized over others. Science and medicine work with a specific world-picture, a particular way of interpreting the world.[112] They are focused on individual concerns and accomplishments, rather than the relatedness among individuals and their embeddedness in a community and context. Now that ghost phenomena demonstrate that this world-picture is compromised because its basic values are not implemented in various practices, opportunities are provided to expand the moral horizon. Bioethics may question the underlying imaginative visions of science and medicine that shape how we see ourselves and the world and provide alternative visions. This requires a change in the way bioethics defines itself. If it considers itself primarily as a rational and objective discourse, similar to (other) scientific approaches, it is vulnerable to ghost phenomena itself. A more critical and self-reflective discourse of bioethics will be needed.

CHAPTER FOUR

Monsters

"We had this monster living under our roof, and we didn't know it." This was the comment of the family who had taken care of 19-year-old Nikolas Cruz after his adopted mother had died. On February 14, 2018, Cruz killed seventeen students at a high school in Parkland, Florida, with a semiautomatic weapon. It was known that Cruz was obsessed with guns and violence. He had purchased at least ten firearms. He also was on psychiatric medication. A year prior, he was expelled from the school for disciplinary reasons. Five months earlier, the FBI had received information that on YouTube Cruz announced, "I'm going to be a professional school shooter." Cruz passed the background check when he bought the AR-15 rifle.

In the contemporary world, the use of the metaphor of monster is very much alive. Monsters have dominated the human imagination since the beginning of recorded history. They present the bizarre, the odd, the frightening, and the unexpected. They also refer to what is horrible in all cultures; monsters present universal nightmares. Some monsters are fictitious; others are real. But the use of the metaphor has common characteristics. Calling a being a monster produces dread, horror, and disgust but at the same time fascination and curiosity. It has moral connotations. A monster is a warning: it can bring evil and disaster, and it shows that something is wrong with society. Furthermore, using the metaphor has consequences. It is often related to stigma, discrimination, rejection, and exploitation.

In medicine and health care, it is no longer politically correct to use the monster metaphor, but implicitly it continues to determine the context in which abnormality and anomaly are defined. Even if the language of monsters is no longer used, every entity or being outside of what is regarded as the standard range in a specific culture faces stigmatization, discrimination, exclusion, and marginalization. Examples include conjoined twins, albinos in some African regions, and children with Down syndrome in Western countries. There is also constant fear that, presently, monsters are created by modern technology. Chimeras, transgenic animals, and Frankenstein food are all labeled as monstrous. Robotics and artificial intelligence raise anxieties about future monsters that will dominate human beings.

As I have argued so far, metaphors help us to make sense of the surrounding world; they suggest an interpretative framework to articulate particular values. Bizarre cases reveal fundamental value disputes. Here I analyze the moral context and implications of designating something or somebody as a monster. The metaphor of monster is specifically used to reconsider the boundaries between the ordinary and anomalous, the normal and pathological. It indicates that some limits have been transgressed, and it is this feeling of transgression that clarifies why certain issues in science, technology, and medicine attract more public concern than others. While this metaphor is usually neglected in bioethical debates because it is regarded as emotional and irrelevant, my argument is that including it is fruitful since it stimulates bioethics to focus on the limits of medical interventions, thus offering the opportunity to broaden the scope of ethical reflection, to move beyond "applied ethics" and develop a critical stance toward science and technology based on a wider set of values.

Monsters have a long history in the public and scientific imagination. There are various types of monsters with various characteristics. The primary function of this metaphor in ethical debates is to delineate the domain of humanity. It creates a separate category of beings, emphasizes boundaries with the unnatural and inhuman, and warns against the hubris of reason. Because these boundaries are always flexible and shifting due to scientific and technological innovations, the metaphor of monster demonstrates that bioethics cannot simply be application of principles; reasoning is intimately connected with emotional responses such as outrage, disgust, and indignation (as expressions of the monster metaphor). But the metaphor also poses a broader challenge to bioethics: reflecting on what it means to be human.

The Continuing History of Monsters

Monster stories that scare people have always existed in human history. Monsters are threatening because they have an unusual size and form. They can have a bizarre anatomy with deformations and malformations, merging human and animal features. The idea of giants, for instance, was stimulated perhaps by fossils of mammoths and dinosaurs.[1] Birth defects, and especially conjoined twins, were not uncommon and might have animated human imagination as well.[2]

In antiquity, many writers reported on monsters. Herodotus described Troglodytes who lived in caves in Ethiopia. Homer wrote about the Cyclops, a one-eyed giant who ate the men of Odysseus. Other historians distinguished various types of monsters, such as Blemmyae (without a head) and Cynocephali (with the head of a dog). In ancient mythology, monsters were hybrid beings, such as the Chimera, mentioned by Homer, being a lion in the front, a goat in the middle, and a snake behind. They usually lived in remote and dangerous areas, monster zones far away from the civilized world. Monsters were not only terrifying but also outcasts and pariahs.

Monsters presented a problem for Christianity. They were mentioned in the Bible: the Behemoth and Leviathan in the book of Job, sea monsters in the book of Daniel, and giants (Nephilim) in the book of Genesis. Jonah spent three days in the belly of the sea monster that devoured him (Matthew 12:40). St. Augustine argued that monsters exist and are human; they are descendants of Adam. But why are they tolerated by God if they are evil creatures? Monsters were considered to teach moral lessons. They reveal God's creation and the diversity of nature, but they also remind human beings of their sins and instruct them to be more virtuous. Monsters, therefore, have a moral and spiritual meaning.[3]

During the Renaissance, attention became more focused on birth defects, as existing and observable morphological anomalies that could be systematically studied with direct observation. These anomalies were regarded as errors of nature, giving rise to a new discipline of teratology, the study of monstrous births. But there still was a mix of mythological monsters and congenital abnormalities. One of the most famous books on this subject was *On Monsters and Marvels*, published in 1573 by Ambroise Paré, royal physician and admired as one of the fathers of modern surgery. His medical approach was aimed at a classification of monstrous births—the book included many illustrations—and the causes identified were very diverse. According to Paré, monsters can be caused by the glory of God, the wrath of God, or by demons and devils. They can be the result of too much or too little seed, smallness of the womb, hereditary illness, injuries and accidents of pregnant women, and maternal imagination. For Paré, monsters illustrated the generosity of God's creation.[4]

In the 17th and 18th centuries, there was an incredible fascination with monsters, and the search for scientific explanations, initiated by Paré, only intensified. Many efforts were undertaken to classify them. Real monsters were more strictly separated from imaginary monsters. In his classification of natural things, Carl Linnaeus included the species *Homo monstrosus*, related to *Homo sapiens* but different.[5] The emphasis was on careful observation and description, trying to identify patterns that could explain the diversity and complexity of monstrosity. The most observable monsters, then, were monstrous births. They showed cases of deformity clearly outside the normal course of nature (most often as "double monsters," or conjoined twins). These cases were always individual, unexpected, and rare, but they were actual events that could be reported and described.[6] Many scholars, but also kings and nobles, started to collect monsters for so-called "cabinets of curiosities" to show the diversity of the natural world. Several cabinets became the origin of various museums of natural history. Today, people still visit one such famous collection of deformed and macabre specimens at the Hunterian Museum in London. Also, many compilations of interesting cases were published (e.g., Ulysse Aldrovandi's *Monstrorum Historia* in 1642) and there were journals specializing in monstrosities (e.g., *Miscellania curiosa*). The focus of activities was on causation more than on deciphering meaning. The basic idea was that we can learn from monsters. They are exceptions that can tell us something about normal developments. Monstrosities are experiments of nature. They are first of all study objects: variations and expressions of human genetic diversity. They reveal the laws of nature.[7]

The 19th century saw a shift from supernatural to natural causation. The new language of anomalies replaced the vocabulary of monsters. Scientists studied birth defects and morphological deformations from genetic and evolutionary perspectives.[8] They examined previous cases to determine what was real or fictitious. For example, "stone children" could be explained as calcified remains of an extrauterine pregnancy (lithopedion). Humans with different varieties of horns, as well as two-headed boys, were found to have really existed. Blemmyae were likely imagined from the experience with anencephalic or acephalous births.[9] Many facial deformities were identified that were interpreted as animalization (suggesting the existence of cat-women, fish-boys, or pig-faced ladies). At the same time, egg-laying women or women giving birth to serpents, dogs, or rabbits were clearly fictitious. The strangest miracle of all time—a Dutch countess giving birth to 365 children at the same time in 1276—could be explained as a hydatidiform mole.[10]

What continues was the moral context. One of the most famous human curiosities in the 19th century was Julia Pastrana. She was suffering from a rare genetic disease and had excessive hairiness over most of her body and overdevelopment of the jaws. She was a healthy and intelligent woman. Nonetheless, she was continuously and mercilessly exploited by her husband, who exhibited her for money as the ape-woman and advertised her as a hybrid between baboon and human. Even after her death, her mummified body was toured around for decades. She was recently buried in 2013,[11] but her case illustrates how the odd, the bizarre, and the unexpected were exhibited and exploited in freak or monster shows well into the 20th century.[12] Another notorious example is Sarah Baartman, a South African woman who toured many countries as the "Hottentot Venus."[13] She was an exotic spectacle and the object of general curiosity and public entertainment, but she often performed under threat by her employers. She died in poverty in Paris where the famous anatomist Georges Cuvier dissected her body, comparing the Negro race with the monkey tribe. Her remains were exhibited in a Parisian museum until 2002, when she was officially buried in South Africa.

These stories illustrate how human bodies were once exhibited in museums as interesting specimens. Instead of having artifacts, museums preferred to exhibit real specimens of indigenous people. Only recently have these remains been returned to indigenous communities for proper burial. At the end of the 19th century, public display of human monsters and prodigies became problematic. It was no longer acceptable in a growing number of countries. Many regarded displaying human beings during their lives, but also after their deaths, as an insult to human dignity. The skeleton of the Irish giant Charles Byrne, who had successfully displayed himself during his life, is now on display in the Hunterian Museum, for example. Byrne had arranged to be buried at sea since he did not want to be dissected and exhibited after death. But John Hunter snatched his body and prepared the remains for his museum.[14] The skeleton is now the focus of controversies about exhibition ethics. Persons who have been objects of spectacle during their lives continue to be museum specimens even against their wishes.[15]

Finally, monsters have been significant subjects of art.[16] They have been painted, for example, in the macabre work of Hieronymus Bosch in the 15th century. More recently, they figure in the cinema as vampires, zombies, and werewolves. Literature has famously produced monsters—from Faust to Frankenstein—and science fiction thrives on imaginary monsters. All in all they reveal the deep anxieties of society. Monsters also affirm the human need to distinguish the normal from the abnormal.

Categories of Monsters

The historical overview suggests that four types of monsters have been distinguished.[17]

Fictitious Monsters

Fictitious monsters are the supernatural entities created in mythology. Examples include the Minotaur, the Sphinx, and Medusa in ancient Greek myths, and very often they are hybrids such as those were (e.g., part animal, part human). In the Middle Ages, various monsters, such as werewolves and dragons, existed at the margins of God's creation, often in specific monster zones where civilization was absent. Often they represent evil and are manipulated by demons and Satan. They are the entities fighting with St. Anthony in the Egyptian desert.

Fictional monsters are frightening, but they also defy human beings to resist and challenge them. They are associated with stories of heroic deeds and monster killers, such as Beowulf in the old Nordic myths, Theseus who killed the Minotaur, and St. George slaying the dragon. Monsters must be confronted and defeated if humanity wants to survive.

Nowadays, monsters figure as imaginary and scary entities in literature and movies. Zombies are especially popular monsters.[18] They are in between the dead and the living. They look like human beings but lack conscience. Even if they resemble a person we know, that person is gone—similar to humans in a persistent vegetative state. They incarnate an empty, senseless, and mad death, announcing the end of humanity. As creatures of the night, zombies show how humans can be enslaved and lose everything that is typical of being human. They symbolize the loss of our capacity to think and to feel. Under the varnish of civilization hides a bloodthirsty beast. Zombies make us aware that monsters are buried deep within us. Destructive violence and animalization are never far away. We may think that the progress of civilization has disguised the materiality of our existence and the basic animality of our body. But this civilized condition is very fragile. In an era where humans can easily destroy the planet and inequality is growing across the world, zombies present a sentiment of decline and degradation of morals. Since humans often lose the battle against zombies, they also illustrate that humanity will perish if it is not united in the fight against evil.[19]

Natural Monsters

Natural monsters include existing zoological aberrations, biological abnormalities, but also exotic races. They were classified in so-called "bestiaries" in medieval times and were the primary objects of collection of British surgeon John Hunter and others like him. In most cases, they were deformed creatures due to congenital abnormalities. They became the basis of later scientific studies, especially in teratology and evolutionary biology.

The most frequent and ancient example of a congenital abnormality considered monstrous was conjoined twins. In the 17th century, the Colloredo brothers traveled around in Europe. Lazarus was a normal and handsome man, but his brother Baptista grew out of his body as a parasitic twin. This strange deformity impressed many people, and they were even received at the royal court in London. In the 19th century, the Tocci brothers became a sensation. They had two heads, two necks, four arms, but only one lower body and one pair of legs. Already when they were very young, they were exploited by their father and exhibited for money in many countries. Later, they retired and married two women.[20]

Another historical example of natural "monsters" was the Gonzales family. Petrus Gonzales was born in the late 1530s in the Canary Islands. He had hypertrichosis universalis, a rare condition also known as Ambras syndrome. Most of the surface of his body was covered with hair, especially his face. He was considered as a humanoid animal and was acquired (or adopted) by Henry II, the King of France. He became well educated and acted as a real nobleman. He married and had several daughters, also extremely hairy. Many portraits of him and his family exist. He has been included in almost all catalogs of monsters and wonders since the 16th century.[21]

Human Monsters

Human beings themselves can develop into monsters. This category is what Stephen Asma called "inner monsters."[22] In antiquity, Medea is a frightening example. She helped Jason and the Argonauts to obtain the Golden Fleece. Jason, however, betrayed her for another woman, and in revenge, Medea became a monster and killed her own children. This myth points out that there are uncontrollable internal forces within human beings. In modern times, the unconscious has become the home of monsters. Human beings can become amoral persons without any empathy or conscience. They no longer belong to what is commonly regarded as the domain of humanity.

Examples are continuously emerging of humans that are no longer human beings. Particularly concerning was the case of Josef Fritzl in 2008 in Austria. For twenty-four years he held his daughter captive in a secluded part of his house. He abused and raped her, resulting in the birth of seven children. His wife was unaware of what happened over those years.[23] Fritzl is a stark reminder that monsters are not only fictitious or natural. They can be human beings whose humanity is questioned. Their monstrosities challenge the notion of what is typical for human beings: civilization.

Stéphane Audeguy warns against three types of modern monsters in particular.[24] First are psychopaths. They present spectacles of horror and demonstrate how fragile the boundary is between humane and inhumane. Current attention to pedophilia shows contemporary society's concerns with human predators. The infamous Belgian case of Marc Dutroux who kidnapped, tortured, and sexually abused six young girls in the 1990s was an eye-opener worldwide, not only for this type of criminal monster but also for the incapacity of the legal system and police to deal with these monstrosities. Dutroux was convicted earlier for the same crime in the 1980s but released after three years with a government pension.

The second type of modern monsters are the mass murderers. Past decades are marked by such mass killings as concentration camps, the gulag, Cambodia's killing fields, and genocide in Rwanda. Mass murderers themselves can have a terrifying normality though. They are perfectly normal citizens, good fathers and husbands, but complete moral aberrations. This type of behavior is what Hannah Arendt has called "the banality of evil." During his trial in Israel, Nazi officer Adolf Eichmann was depicted as a monster, but in reality, he was a normal, law-abiding citizen, a mediocre bureaucrat obeying orders, but without any conscience.[25]

The third type of modern human monsters described by Audeguy are mutants. Like many natural monsters, they are the result of genetic mutations: new creatures produced by deliberate hybridization and mutation. This third type brings us to the last category of monsters.

Fabricated Monsters

In ancient mythology, monsters were made by gods. In medieval Christianity, they were the work of God who created them for a specific purpose. Later, they were interpreted as errors of nature, showing the diversity and variability of the natural world. They could be studied and described, but also dissected and analyzed. The new science of teratology furthermore allowed experimentation. The founder of this science of monsters, Etienne Geoffroy Saint-Hilaire, started with experiments on embryos in the 1830s to explore how and why monstrosities originated and developed. He assumed that monsters were not the result of disorders but creatures of nature that exemplified a different order. By modifying embryos, one could study how deformities were generated.

This was exactly the mission of Camille Dareste, a French zoologist who founded the science of teratogenesis. In his view as well, the monstrous was the extraordinary result of natural causes. His objective was the controlled and systematic production of monsters in the laboratory. Most monstrosities have indeed been produced artificially since 1855. Dareste bragged that he produced several thousand monsters.[26] For the general public, Mary Shelley's *Frankenstein* (1818) highlighted the creation of monsters in the laboratory. It showed that research can grow out of control. The humanoid monster fabricated by scientist Victor Frankenstein was an aberration but not monstrous from the beginning. Interactions with humans made it evil.[27]

However, the metaphor of monster is not only related to fear of the horrific things that science may create. There is also the fear that human beings themselves produce monsters because their inventions can simply be misused. An example is the discussion about the guillotine. It was invented by a medical doctor to make decapitation faster. Created in the name of humanity, the device, as a simple mechanism of execution, was intended to produce less suffering. The machine was soon regarded as monster, facilitating the death of thousands of people. A similar discussion today concerns drones. It is argued that they prevent the deaths of soldiers and have less collateral damage. At the same time, they introduce a different type of warfare that is quick, dirty, secret, and outside of any public or legal framework.[28] With the sophisticated technologies of today, humans will be able to generate severe threats to themselves and to the planet.

Lastly, humans do more than produce monster machines. They repeatedly try to transform people and groups into monsters. Fashioning a new human being was the purpose of the social experiments of Nazism and Communism, resulting in the monstrosity that is genocide. Human beings created monstrous institutional systems that turned individuals into abusers and torturers. They promoted dehumanizing social and cultural frameworks in which others—such as different faiths, people of color, and women—were demonized. It has often been human behavior that has fabricated monsters.

Characteristics of Monsters

What do we mean if we use the term "monster"? The etymology of the word starts with the Latin *monere*, which has two sets of meaning: reminding, showing, revealing, and also alerting, warning, admonishing, foreboding, presaging. The monster, therefore, shows something but also has a prophetic significance. It is ambiguous. It refers to a revelation and at the same time a warning.

First, monsters are the object of wonder, amazement, and astonishment.[29] They are marvels that show themselves. They illustrate the omnipotence of God, the mysteries of nature, or the variability of biological development. People are fascinated with extraordinary, strangely formed bodies. More than objects of study, they are first of all curious phenomena, oddities giving delight and pleasure to people. This fascination explains why they attracted so much attention and were collected, displayed, and exploited as spectacles in freak shows, circuses, and museums. London's popular Madame Tussaud's wax museum, in fact, began with the Chamber of Horrors, a gallery of guillotined heads from the French Revolution.[30]

Second, monsters simultaneously provoke disgust and horror. At first, they look familiar but immediately give an impression of difference in physical appearance and behavior. They are often ugly and deformed, eccentric and exotic, outside the usual order of nature. They also present images of imperfection, degeneration, and decay. Because they depart from what is normal and expected, they give rise to aversion and distaste. Not knowing what to expect and anxious about possible dangers, people are fearful of and often dread monsters. Frequently, they are assumed to be hostile to the human race. Often distanced from civilization, monsters represent the unknown and the dark.

Third, monsters are associated with meaning. They do not occur incidentally. They are a sign of the times and a portent of the future. Since they are something unnatural, out of the ordinary, and deviating from the normal, they are not merely strange but also possible omens, a distinct sign that something is not right with the world and that evil is near. They not only inspire horror but also bring disaster. Monsters are menacing because they are malevolent and destructive. Some have argued that the term "monster" is also related to the Latin word *monumentum*. Monsters are reminders, memorials, that cause us to reflect and to transform our behavior to avoid disorder and catastrophe.[31] As emblems that need interpretation, monsters bring a message that witnesses need to decipher. The intense reporting of monster cases in the past had a specific purpose: it urged moral self-examination.[32]

Fourth, monsters demonstrate transgressions. Monsters are hybrid beings; they can take any shape or form. What characterizes monsters is not only anomalous morphology but also the breaching of boundaries: between human and animal beings, between the living and the dead (e.g., zombies), between male and female (e.g., hermaphrodites), between species (e.g., chimeras), and between humans and machines (e.g., robots). (Such boundary crossings are likewise visible in conjoined twins: are they one or two individuals?) Monsters are therefore liminal beings; they are in between existing categories. They are a threat to human and social life because biological and moral boundaries are transgressed. This is also the reason why they are a lesson: monsters signal the vulnerability all human beings have of lost humanity.[33]

Uses of the Monster Metaphor in Ethics

Lawyers arguing for the right to die in the case of Karen Ann Quinlan (1975) referred to her as "the anencephalic monster."[34] In other words, medicine sometimes creates monsters lying in a vegetative state in nursing homes and implies they do not have a life worth living and deserve to die. Generally, the use of the term "monster" in bioethical discourse is rare these days. It has pejorative connotations that ethicists want to avoid. Labeling somebody or something as a monster is considered an emotional and irrational response. As an expression of the "yuck factor," it is an immediate negative reaction out of revulsion, disgust, or discomfort; no arguments are needed to reject an idea, situation, or condition as harmful or evil. With its emphasis on rationality and argumentation, ethical discourse wants to go beyond this level of emotion and intuition.

As argued in previous chapters, emotions and feelings cannot be so easily dismissed in ethics. Philosopher David Hume argued that the source of ethics is not the brain but the heart. Making moral judgments depends on passions and sentiments, not on human intellectual faculties. Moral approval and disapproval are not the conclusions of moral reasoning. Reason alone cannot produce action, according to Hume. Moral judgments, however, do influence action. Therefore, they cannot be derived from reason. In his *Treatise of Human Nature*, Hume argued, "Morals arouse passions and produce or prevent actions. Unaided reason is powerless to do such things. So the rules of morality are not conclusions of our reason."[35]

Moral judgments find their origin in human sentiments and passions (i.e., feelings). Reason can discover facts and analyze concepts and their relations, but it cannot explain what actions are desirable. Mary Midgley has applied these ideas to the use of the metaphor of monstrosity.[36] It is not merely a matter of gut feelings or exclamation of yuck that can be dismissed as irrational. People who regard something as a monstrosity—for example, developments in biotechnology that result in cloning or the production of chimeras—do not merely express repugnance and outrage but also articulate intrinsic objections. And there may be good reasons for such objections.

What is the ethical significance of the monster metaphor? In other words, what moral message is conveyed? The first message is reductionism. Labeling people as monsters objectifies them and reduces them to less than human. They are regarded as human artifacts that can be owned, bought, exchanged, and exhibited instead of real people. Sarah Baartman, Charles Byrne, and the Tocci brothers were all treated as spectacles, objects of curiosity to be displayed and viewed. The gaze of others objectifies them. The consequence is that personhood disappears or is dissolved. The monster is "the Other," not recognized as a moral person.

What is missing in most monster stories is the first-person perspective of the very being labeled the monster. In many cases, such as monstrous births, it is impossible to have any first-person perspective, but in other cases it has simply not been recorded. And where monsters are concerned, what will be the purpose of treating them as persons? Objectification, instead, easily leads to exclusion, stigmatization, and discrimination. If anencephalic fetuses or patients in a persistent vegetative state are monsters, they are disgusting objects that should be discarded. While in earlier times, the Middle Ages in particular, monsters were an acceptable part of the world since all was God's creation, later, with the rise of scientific approaches, they needed to be collected for study or be removed from the world as anomalies that are no longer acceptable.

The second message concerns ethical limits. Invoking the metaphor of monstrosity implies that a disrespect for limits has occurred, that existing concepts of nature and species have been disregarded, and that ethical boundaries were transgressed. Indignation or outrage about certain developments, conditions, or situations indicates that the notion of human nature or that of human rights is violated. What is unnatural or inhuman is monstrous. The Louvain philosopher Herman De Dijn argues that new technology produces monsters that expose and subvert the order of nature and society.[37] This is a problem for ethics since moral experiences are connected with often traditional and deep-seated ways of life and living together, frequently associated with religious and philosophical frameworks. They entail fundamental distinctions and restraints that need to be respected. They express the view that what is valuable is intrinsically vulnerable. Specific forms of disrespect are therefore unacceptable: abuse of children, ignoring persons in need, racism, exploiting defenseless people, not burying the dead. Human life is inevitably contingent and vulnerable. That explains, according to De Dijn, that what are gradual distinctions from the perspective of science are radical differences from the ethical perspective.[38] Another way of saying the same thing is that in the real moral world (in contrast to the world of moral fiction with its emphasis on rational principles and arguments) certain things are sacred.

The argument that something (e.g., human life) is sacred has two implications: respectability and inviolability.[39] Respect is based on the value that entities or objects have in themselves. Because human life has intrinsic value it commands respect, not because it is subjectively valuable for someone or instrumentally valuable for some purpose or use. Something can become sacred through its history, through how it has originated. A painting is valuable since it is the result of a creative process. Endangered species are protected against extinction since they are the product of a long natural evolution. Intrinsic value, therefore, is related to the process of creation or origination rather than to its products or results. It depends on how something came into existence. How it was produced matters for its value.

When the idea of intrinsic value as the basis of respect is applied to human life, it entails equality. The life of every human being has the same inherent value. There cannot be differences in respect even if a person is deformed, abnormal, or monstrous. Another consequence is that intrinsic value is not an individual accomplishment. Thus, human life is not a human product; it is the outcome of natural or divine creation. The second implication of the idea of sacredness is inviolability. If something is sacred it is untouchable. The idea marks a boundary that cannot be transgressed. When human life is inviolable, deliberately ending it is intrinsically bad. Calling something sacred is at the same time a call for protection. We are horrified when a species is lost because a creative achievement has vanished forever, whether it was created by God, nature, or evolution. References to sacredness do not often feature in bioethical discourse. They are, however, commonly used in public debates concerning new technologies. The idea of the sacred is not a mysterious idea that is coming from theology or religion. Although it has religious connotations, it is used within a secular context.[40]

The third message of the monster metaphor warns against hubris. It is a protest against human arrogance. Monsters not only challenge the moral and cosmological order of the universe but they seriously threaten this order. That is why they are dangerous and harmful. They represent forces that are no longer controllable and that have the potential to overpower human beings themselves.[41] The concept of monster is a moral one, referring to horrific human conduct or efforts to create beings that have lost their humanity. Using the term is a symbolic protest against the ideal to liberate humanity from limitations and restrictions. Human arrogance and misjudgment will lead to loss of moderation and unknown, unmanageable menaces. This is a lesson expressed in a painting of Francisco Goya in 1799: the dream of reason produces monsters. Reason alone should not govern us. At the same time, we should not abandon reason and only trust imagination.[42]

Another metaphor used in this context is Dr. Frankenstein. What is created out of wonder and curiosity with intentions to improve humanity can result in inhuman monsters.[43] There is a fundamental revulsion for tinkering with life. Humanizing animals, mixing human and animal traits, thus creating chimeric embryos, can result in creatures that terrify people. Gods previously produced fictional monsters but now scientists are taking over. This not only occurs in the laboratory. In such movies as *Outbreak* and *The Andromeda Strain*, infectious agents can transform into monsters. Human constructions can become monstrous. This idea has also been applied in the realm of ethics itself. The concept of individual autonomy has been called an "anthropological monster."[44] Regarding human beings as self-interested, self-determining subjects ignores the basic importance of the cooperation between humans, the interconnectedness of humans, and the interrelations between human beings and the environment. Ultimately, bioethics itself is considered monstrous. It generates horror stories due to its radical dismantling of traditional morality and medical ethics.[45]

Contemporary Monsters

The monster metaphor has two different contexts. One concerns natural and human monsters that do exist, and the other concerns fictitious and fabricated monsters that are brought into existence or can be created but do not yet exist. The suggestion in the first context is that such monsters better not exist; in the second context, it is better not to bring them into existence.

Luckily, many existing monsters are not born alive or die soon after birth, as is the case with anencephalic fetuses. Still, there is a need to detect monsters during pregnancy, and detection mechanisms may fail so that once in a while monsters are born. In India in 2014, for example, a healthy baby girl with two heads was born. This came as a surprise because the mother was unable to afford an ultrasound.[46] In many countries, the number of babies born with conditions such as Down syndrome has significantly decreased with the advanced technology of prenatal screening. The majority of women with a positive test terminate their pregnancies (in the United States, 67 percent; in Denmark, 98 percent). The syndrome is now almost eradicated in Iceland.[47] Medical technology is now so advanced that many monstrous deformations are removable. A baby boy born in Uganda with four legs and four arms was successfully operated on to remove the extraneous body parts.[48]

Although persons with deformities and abnormalities are no longer (or very rarely) explicitly labeled as monsters, they continue to provoke horror and repugnance and are often the object of stigma and exclusion. Even when not used explicitly, the monster metaphor still works when persons are considered as anomalous, out of the ordinary, and even deviant. The assumption is that it is better for them not to exist. A contemporary example is albinos in Tanzania. Relatively frequent in that country (affecting 1 in 1,400 people), albinism is regarded as monstrous; albinos are objects of hatred and ostracized from society. Since 2006, sixty-seven albinos have been killed. They are also objects of crimes since witch doctors claim that their body parts bring health and success. Many are attacked, sometimes by their own relatives, who cut off their arms, hands, fingers, and jaws to sell them to traditional healers.[49]

Another example of deviancy is Down syndrome, first described by Langdon Down in 1866 as mongoloid idiocy and interpreted as an atavism, a phenomenon of racial degeneration.[50] Until 1970, it was described in the *Encyclopedia Britannica* under the heading of "monster." For a long time, somebody with Down was not regarded as a person. Joseph Fletcher, one of the founders of bioethics, argued that idiots are not human. He proposed a "profile of man," a list of criteria for humanness. The first was minimum intelligence. Retarded children may feel well or happy, but that is not sufficient to be a human being. In Fletcher's view, if children fall below a minimum quality standard of human health and potential, they should not be brought into the world.[51]

Most reports on persons with Down syndrome used to present a bleak picture of profound mental retardation with a pessimistic prognosis.[52] Although much more is now known about the variability of the syndrome, and people with Down live longer and healthier lives than before, the condition is mostly framed as a universal category and as a risk, problem, threat, or abnormality. This negative outlook is often presented in information materials. Observation of screening practices in the United Kingdom shows that Down syndrome is rarely addressed in explicit detail. It is made absent in the antenatal clinic and implicitly condemned as abnormal.

Having a disability as a specific way of being in the world is stigmatized. From the very beginning, implicit assumptions about what kind of lives are worth living configure the care provided.[53] These observations also illustrate the significance of objectification, discussed earlier. In everyday practices, persons can be made absent and invisible through focusing on their bodies. Concentrating on abnormalities of bodily shape and form or on diagnosing and categorizing conditions and behaviors make personhood absent. Anomalies or genes receive more attention than the individual and unique person. In this process, the person is excluded and transformed into the Other that is strange and odd.[54] The same mechanism is used to deal with patients in forensic and correctional practices. Patients are depersonalized as monsters.[55] It also explains why it is argued that anencephalic babies should be used as organ donors. They are monstrous objects, comparable to patients in a persistent vegetative state aptly labeled by Fletcher as "vegetables." Not utilizing such bodies would be a waste.

This way of thinking is not exceptional. In the early days of the euthanasia debate in the Netherlands, psychiatrist Jan Hendrik van den Berg published a small but very influential booklet on medical power and medical ethics. The impact was primarily due to cases and photos: a child with hydrocephalus, a child with phocomelia, a patient after hemicorporectomy, an adult in a vegetative state, and demented persons in bed. This horror gallery was used to argue that medicine can be cruel, and that doctors in these circumstances have a duty to kill, that is, to eliminate monstrosities.[56]

The common thread in these cases is horror. Certain beings are labeled as monsters for various reasons. They have a horrific appearance or they have abhorrent behavior placing them outside of humanity. They can be so deformed or defected that there is disgust for what their life is or will be. Persons who look different, or who are disabled or mentally ill, may provoke dismay and uneasiness. They are regarded as symbols of otherness, and are therefore marginalized; their humanity is denied.[57] Crucial for horror is the conflict between humanity and the inhuman, between the normal and the abnormal.[58] The monster metaphor is here used to articulate differences, particularly to establish boundaries between the normal and abnormal. This is the essence of stigmatization: it is stereotyping individuals or groups, linking them to undesirable characteristics and making it easier to treat them with disrespect, to exclude and discriminate them.[59]

Science and technology present a different context. The point in this context is not the labeling of existing beings as monsters. The primary concern of the metaphor here is that monsters are deliberately or unintentionally brought into existence or their creation is planned. The metaphor is used to show that such new creatures should not be devised and constructed. In this context, references are frequently made to *Frankenstein*. Mary Shelley's 1818 book gave birth to a literary and social tradition of criticism toward scientific progress. It produced a new creation myth. It promoted the frightful image of the mad scientist. But first of all, it was a warning against scientific hubris. Reliance on scientific rationality brings ruin and darkness.[60] The impact of Frankenstein is particularly strong in bioethics. With all good intentions, scientists may jeopardize the future of humanity if they do not respect the difference between what science can do and what it should do.[61] Similar and more modern warnings invoke the metaphors of the *Titanic* and thalidomide. The first deconstructs the unjustified belief in new technology ("the unsinkable ship"), and the second demonstrates that modern medicine can really create modern monsters (in this case eight thousand children born with phocomelia).[62]

Although the emphasis is on not creating monsters, medical and technological interventions have already fabricated many monsters, mostly unintentionally. Efforts to enhance human beings can have counterproductive results. For example, South Korea has the highest rate of plastic surgery per capita in the world. Almost one-third of all women in Seoul have undergone surgery of eyes, noses, and jaws while aspiring to be more beautiful. In a growing number of cases, and after numerous cosmetic interventions, the result is unnatural. Surgery produces a "plastic-surgery monster." A new word, *sung-gui*, was introduced in the language to refer to such monsters.[63] Another use of the monster metaphor, in this case in Poland, is to designate children born with in vitro fertilization as monsters.[64]

In the media, monstrous constructions draw much attention. In 1997, photos of the "ear mouse" circulated on the internet—a laboratory mouse with an ear grown on its back. The ear looked human but was actually from cartilage cells from a cow. Twenty years later, Japanese researchers succeeded in growing a human ear on the back of a rat using human stem cells. The justification of such experiments is that bioengineered tissues can benefit people who have facial abnormalities. However, the question is whether such genetic manipulation is not itself creating monsters. It is fabricating new forms of life, transferring genetic materials from one species to another. Especially, transgenic animals have become the target of criticism.[65] These animals have been produced as possible sources for organs and pharmaceuticals. In these endeavors, animals are regarded as bioreactors, as means for the efficient production of commercial substances. Ethical considerations focus primarily on risks, for the animals and for humans. That animals might have intrinsic value and deserve respect for their integrity and welfare, and that boundaries between species should be respected, is often not taken into consideration.[66]

Genetically modified organisms (GMOs) are a hot topic right now as well. Critics denounce food grown from GMO seed as "Frankenfood." Currently, about 11 percent of the cultivated area in the world is covered by GMO crops, particularly in the United States, Canada, Argentina, Brazil, and India. On the other hand, in the European Union, 57 percent of the population is not willing to support GMO food because they feel it is unsafe, lacks clear benefits, and promotes inequality in farming. Nineteen European countries currently ban genetically modified crops.[67] Biotechnologists and agribusinesses, however, argue that modified food is the best option to guarantee global food security, although even that is contested.

There is evidence of increased yields in some crops as GMOs but this higher productivity is often associated with reduced profits for farmers because their costs have increased. GMO seeds cost twice as much as non-GMO seeds. Significant concerns also exist regarding human health and biodiversity, even though GMO proponents argue the food is safe and without risks for health and environment. Research findings on these topics differ and are often ambiguous. Data is mostly provided by the industry as well and so it is not clear how objective and reliable it is. Assessment is primarily based on safety and balancing benefits and costs.[68]

This emphasis on consequences misses the point of the Frankenstein metaphor. Regardless of risks and safety, there is a deeper concern. Genetic modification implies transfer of genes from one species to another. It also involves the creation of new or mixed species by human beings. In many worldviews and religions, this is morally objectionable. Life is not a simple commodity that can be managed as a resource to be changed and enhanced by biotechnology. Animals and plants are not machines that can be designed and patented. Considering the gene as the core of living beings does not do justice to the dignity of life.[69]

The metaphor of monstrosity is more powerfully used regarding new fu-

ture possibilities to create monsters. These monsters will be of our own making, and we now have the opportunity to prevent their design and fabrication.
One example is human cloning. After the cloning of Dolly the sheep in 1996,
people suggested that cloning of human beings was within reach. In fact,
within a few months of Dolly, the Raelian Movement, a religious sect, announced that they had started a company to research human cloning. In 2002,
they claimed that the first human clone was born. These developments pro-
voked a strong international response. Many countries proceeded to ban human
cloning. They frequently cited the metaphor of Frankenstein. Cloning signifies that human beings want to have full control over their future. For Jean
Baudrillard, it is the drive to liberate humans from sex and death. It is primarily the quest for immortality. But this enterprise is fatal. Humankind is
playing with its future, and the risk is that limitless experimentation will
permanently eradicate the human race.[70]

Many people reacted to the idea of human cloning with repugnance and
disgust. Leon Kass argues that this response of repugnancy is the emotional
expression of profound wisdom.[71] Repulsion is not based on horror or strangeness but on the sense that violation of boundaries occurred. Cloning is not a
neutral technology that can be good or bad depending on its uses. Motives
and intentions are not decisive. Nor are the results and outcomes of technological interventions. Cloning is also more than an option for individuals to
reproduce. It has serious social and cultural implications that go beyond the
level of individual decision-making. It relates to the meaning of parent-child
relationships, of personal identity, sexuality, love, and intimacy. Repugnance
as a response questions the assumption that cloning is a justifiable way to
avoid the risks of genetic diseases, to enhance human health, and to extend
human life. It points out and affirms that by not taking into account broader
perspectives of meaning, cloning is dehumanizing. Cloning regards somebody not as a unique person but instead as just copied genetic material, even
if we know that this is factually untrue. Therefore, repugnance is an outcry
against moral myopia, the propensity to concentrate the ethical discussion
only on issues of personal freedom, safety, and benefit-harm ratios.

Other areas of concern are robotics and artificial intelligence. Robots are
increasingly applied in manufacturing, transportation (driverless cars), entertainment, the military (drones), and health care (surgical robots, nanobots
for diagnosis and treatment, assistive robots for care). Robots are fascinating
since they are automata that can very much look like human beings and/or
take over many human tasks. What worries people is that as robots become

more humanoid, they are replacing human beings. With the incorporation of robotic technologies into the human body as well, the distinction between what is natural and artificial becomes unclear. Are such cyborgs and androids human beings or autonomous machines?

Another, and more profound, fear is that when robots become more intelligent, human beings will no longer be able to control them and, worse, they might threaten the dominance of humans. The rapid growth of artificial intelligence will lead to a new generation of robots that are learning, autonomous agents that no longer need any human command or control. These technological advances, as many scholars point out, will present an existential threat to humanity, potentially ending human civilization in the near future.[72] An example is the development in the military area of lethal autonomous weapon systems that select and engage targets. While military robots already have changed the ethics of war, this new generation of weapons will eliminate human responsibility and leave decisions over life and death to machines.[73] This is precisely the scenario of the Frankenstein metaphor. The designer who built robots for mostly beneficial purposes has created machine-monsters that might get out of control and become a threat to humanity itself.

A final example of monstrous possibilities concerns resurrection technology. The idea is to resurrect extinguished species through genetic engineering.[74] In 2013, Australian scientists started the Lazarus project to revive a fascinating frog species that went extinct in 2002. The created embryo died within a few days, but the idea to reverse the extinction of species gained much support. It is a way to undo one of the most distressing effects of the global environmental crisis. However, efforts to resurrect the past also raised serious questions. Cloning may never bring the original species back. It is questionable whether the resurrected species will find an adequate habitat or adapt to current ecosystems. Survival is not merely a matter of genes. There is also the question why we want to bring species back to life. Why resurrect dinosaurs? It seems that a considerable element of sensationalism plays a role. But can we be sure that we do not recreate monsters and revive, for example, dangerous viruses? And if science can resurrect extinct animals, why not human beings? Approximately 350 people have their corpses frozen by cryogenics immediately after death in hope of future resurrection.[75] More severe cases are attempts to fabricate dangerous viruses in the laboratory. The Spanish flu of 1918–1919 is known as a "horror pandemic," killing tens of millions of people. In 2005, researchers reconstructed the H1N1 virus. It was considered a recipe for disaster.[76]

The common thread in the second context of the monster metaphor is disquiet and indignation rather than horror. The fabrication of new creatures by science and technology is unsettling, distressing, and worrisome. It produces anxiety about what might happen. The story of Frankenstein illustrates that there are always benevolent intentions with which to begin. But actions aimed at the good of humanity can easily go out of hand. Victor Frankenstein was a brilliant scientist who wanted to help humanity. But in fabricating a new species of humanity, he created an unnamed monster he could no longer control. Ironically, in the current ethical discourse the monster is now itself called Frankenstein. The scientist and the monster are fused and confused. Both share the same monstrosity.[77]

Some have pointed out that there is another aspect of the story that needs attention, too. Frankenstein's monster only became evil because it was abandoned. Frankenstein ran away from his creation and left it alone. He did not assume any responsibility for what he fabricated. The story, therefore, not only warns of hubris but also for irresponsible behavior.[78] Monsters are threatening because they are "interstitial."[79] As categorical violations, they cross the boundaries of commonly used classifications, for example between human and animal, and between living and dead. Although in most cases they do not exist and are still fictitious, they excite real emotions.[80] At the same time, monstrosities demonstrate technological creation; they are fascinating and marvelous since they show the power of science and technology.

So why are we attracted to something that is repulsive? Why are we fascinated by beings that are horrifying?[81] Perhaps it is related to the discovery of the unknown, the impossible, which is simultaneously a violation of the moral order. Perhaps it is the return of what is usually repressed, giving us a glimpse of what is otherwise hidden and secret. Contrary to Freud's interpretation of the uncanny, it is not the return of the primitive and the archaic but the return of repressed possibilities and potential wonders, blurring the boundary between fantasy and reality.[82]

The employment of the monster metaphor in the two contexts distinguished above is different, but the fundamental ethical thesis is the same. In the first context, the metaphor is used to introduce boundaries. Suggesting that some beings are monstrous, and thus belong to a special category of otherness, violates the notion of human dignity by transforming individuals into objects of disgust and disgrace. In the second context, the metaphor is used to protect boundaries and to uphold the notion of human dignity, arguing that humanity should be respected. In both cases, distinctions are made between what is human and inhuman. On the one hand, it is argued that beings who are considered abnormal should be identified and detected, thus preventing their existence. On the other hand, beings who are anomalous should not be designed and brought into existence. In both cases, the ethical implication of the use of monster metaphor is that monstrous beings should not exist.

Conclusion

This chapter demonstrates that the metaphor of monster has been used since antiquity. In ancient mythology, monsters were fictitious beings haunting humanity. Since the Middle Ages they became real beings, such as monstrous births that were amazing and frightening at the same time. More recently, the image of Frankenstein is employed to warn against the fabrication of monsters by science and technology.

Today, the imagery of monsters is operating in two ways. Stigmatizing specific categories of human beings—such as persons with albinism or disabilities—produces a boundary between the normal and abnormal. Although people are no longer explicitly designated as monsters, the underlying assumption usually is that they should not be considered as full human beings because they are deficient, deviant, or anomalous in at least some crucial dimensions. In specific cases, people are explicitly labeled as monsters to create distance in care and treatment. The monster metaphor operates here as a means of objectification; it transforms a being into an object and removes it from the category of human beings. This is a way of taming the monster: by making it into an object and spectacle, it is marginalized and relegated to the margins of humanity or even beyond.

A different way of applying the metaphor is in the context of efforts to enhance beings with cosmetic interventions and genetic modification, for example, creating transgenic animals and GMO food. The metaphor focuses explicitly on future scientific and technological schemes involving human cloning and artificial intelligence. The reference to monstrosity in this context is frequently combined or replaced by the metaphor of Frankenstein. Both metaphors have the same purpose. They provide fundamental objections to particular scientific and technological practices and innovations. The wonders of medical progress are connected to monstrosities. In assessing potential scientific marvels, consequentialist arguments referring to potential risks and dangers will not be sufficient. What is at stake, according to the metaphors, is that boundaries and barriers that are considered sacred, are surpassed and eclipsed, so that our very humanity is at stake. This is why Michael Jones asserted, "All monster stories, beginning with Frankenstein . . . are in effect protests against the Enlightenment's desacralization of man."[83]

Talking about monsters and Frankenstein is usually dismissed as irrational and emotional rhetoric that does not provide good arguments. Revulsion is not the conclusion of rational deliberation. It is therefore not admissible in bioethical discourse. Such disdainful response does not recognize that a different vision of ethics is at work. Ethics is presented not as a primarily rational exercise but first of all as a way or vision of life. It is not merely a theory or a method to apply principles to troubling situations. Ethics is a quest to understand what it means to be human. It involves moral experience and sensibility. Ethical discourse cannot divorce feelings and reasons.[84]

It is true that monsters provoke horror; they elicit emotional responses: fear, disgust, and revulsion. This does not imply that using the metaphor of monsters can be disregarded as a simple ploy to appeal to emotions. Using the metaphors of monstrosity and Frankenstein is an effort to recall at least three lessons for ethics. The first lesson is that ethics is more than arithmetic of consequences. For Fletcher, ethics is making decisions about what to do and why. We cannot escape decision-making. "We have to decide, even if it is a decision not to decide."[85] There is no room for virtues, care, contemplation, or reverence. Humans are responsible for everything; they are the masters of the universe. In Fletcher's view, ethics is only concerned with outcomes and effects. It needs to consider the balance between means and ends, acts and consequences. The only things, according to Fletcher, that ethically count are the results of our decisions.[86] This is precisely the view of ethics that is criticized by Kass in his discussion of human cloning. Ethics is reduced to a matter of motives and intentions, benefits and harms, rights and freedoms while it is, first of all, a matter of meaning.[87]

The second lesson is that science is not a value-free enterprise. The metaphor of monsters is not the expression of antiscience attitudes. Instead, it points to an ideology of reductionism and materialism that is prevalent among scientists. Science is driven by a worldview determined by reductionism, dualism, atomism, and mechanism. It is not free of values. It assumes, for example, that human behavior is determined by genes, that it can be explained in terms of brain activity and nerve cells, that the inner self is an illusion, that natural selection teaches us that selfishness is the basis of motivation, that only physical objects are real, and that life is a merely physical phenomenon.

Using the metaphor is a way to denounce this restricted ideology. It makes an appeal to restore the unity of our view of human beings and their being embedded in the natural world.

Doing so refers to a central theme in the philosophy of Mary Midgley.[88] The current image of the human being is deeply disconcerting because of the dominance of scientific materialism. The richness and diversity of human life are no longer recognizable. The imaginative vision of science has promoted the idea that science is omnicompetent, able to answer all questions and provide solutions to all problems. It assumes that there is no limit to the reach of science. Midgley argues that this world-picture of scientism is wrong.[89] There are different ways of seeing the world. Values structure our world, the world in which human beings are living; it is not the world of physics. Furthermore, the pervasive images of science are damaging for moral thinking. They reinforce rationality, objectivity, facts, analysis, and individuality and ignore the importance of emotions, imagination, synthesis, and cooperation. It is necessary to restore the unity of human life and the human being.

Ethics need to respect that unity. It should articulate the role of imagination in knowledge; it should encourage synthesis and provide overall visions. Ethics should also clarify that while the focus of science is on explanation, its specific aim is understanding, to grasp the meaning of what is happening. Hence the importance of experiences, feelings, emotions, and the first-person perspective. At the same time, ethics should criticize the continuous tendencies of science to fragmentize and atomize the world and its inhabitants. The scientific worldview is a myth, but it has produced a world of objects without subjects. Midgley argues that contrary to scientism, human behavior can only be understood "by reference to people's own thoughts, dreams, hopes, fears and other feelings."[90] Human beings are subjects; they are "creatures with needs, tendencies and directions of their own."[91] They are necessarily embedded in nature, and therefore vulnerable and dependent.

One would expect that bioethics would embrace this call to focus on the unity of the human being. In this sense, one would also assume that it would be more receptive to what the metaphor of monstrosity wants to convey. The history of bioethics can be reconstructed as an effort to reintroduce the subject into the world of health, disease, and care. At the same time, bioethics has exorcised religion and philosophy, following the ideology of modern science, although, as Midgley pointed out, many scientists (e.g., Newton, Faraday, Maxwell) have been deeply religious.[92] Bioethical discussion often follows the idea that science can bring salvation: it can achieve immortality; create posthuman beings free of diseases, disabilities, and aging; conquer the universe; and colonize outer space.

The third lesson that the metaphors of monstrosity and Frankenstein give to ethics is the importance of limits. The metaphors introduce the moral language of abomination every time there is a potential transgression or a threat to the established cosmological, social, or moral order.[93] They voice the fear that our status and dignity as human beings is jeopardized. As Jeffrey Stout points out, "What makes the action abominable is its degrading character— its capacity, within some social settings, to make the agent himself seem (or be) less human."[94] Stout mentions as examples cannibalism, bestiality, and necrophilia, but one can add child abuse or torture. He denies that revulsion and repugnance are merely immediate emotional experiences that cannot be further explained. They are derived from an antecedent commitment to specific categories of our cosmological, social or moral order, defining the line between, for example, human and inhuman. These commitments and assumptions can be discussed; they focus on human dignity and human finitude. However, they are often neglected because they bring bioethical discourse within the discursive realms of philosophy and theology.

Finally, using the metaphors of monstrosity and Frankenstein implies a warning about human hubris and the dangers of transgressing certain boundaries. This warning has two dimensions. First, it is a protest against the technological imperative. What we can do is not the same as what we should do. There is always a necessary tension between ethics and technology. The basic rule of ethics is that "can" should not imply "ought." Contemporary technoscience, however, with its myth of "salvation by technical fix," naturally assumes that there is inevitable progress and that human vulnerability and finitude not only can but should be overcome.[95] The metaphor of monstrosity signals that we should not act on possibilities but should instead acknowledge the values at stake and act on the basis of what are the most important values.

Ethical discourse will always narrow the possible courses of action since it is concerned with values and not merely with capability. Moreover, the metaphor urges that there is more at stake than potential risks. Technology frames the world, and nature in particular, as a resource that can be exploited but also modified and improved. It imposes a divide between nature as object and human beings as subjects (or persons). However, it promotes a similar attitude toward human nature itself: the human subject becomes an object. For example, in a medical-technological perspective, the body is regarded as a biological object (*Körper*) rather than as a lived body (*Leib*) as a way of being in the world. Due to its reductionism, the technological imperative might scale down the range of values at stake. It risks the annihilation of the unity of the human being.[96] That such risk is not hypothetical was argued decades ago by Jacques Ellul. The technological order has become so powerful that human beings are adapted to the technical milieu. Choices are limited since the individual being is already incorporated into the technical process. According to Ellul, it is uncertain whether human beings will be able to master technology. This mastery requires that human beings again become subjects rather than objects. It also implies a different type of skill than technical perfection. Rather than reproducing and perfecting the technical universe of objects, machines, and material things, mastery should promote the values of human flourishing.[97]

The second dimension of the warning is a reminder about the nature of ethics. If ethics is regarded as a way or vision of life, it is related to the notion of the morally possible and impossible. Moral injunctions indicate what is a recognizable and coherent way of life. They provide, in the words of Stuart Hampshire, "the skeleton of a man's morality . . . a skeleton of an attainable, respectworthy and preferred way of life."[98] This implies that some courses of conduct are ruled out because they are inhumane, cruel, humiliating, or unjust. Moral prohibitions are barriers to action, especially in such areas as the taking of human life, sexual relations, family duties and obligations, the administration of justice, burial of the death, duties in times of war, conditions for telling the truth, rights of property, care for the vulnerable, and duties of friendship. The notion of morality, according to Hampshire, requires that there are some strong barriers to what is allowed. Certain behaviors are prohibited since they are "disgusting, disgraceful, or shameful, or brutal, or inhuman."[99] This is precisely the point of the monster metaphor. Some actions are morally impossible; they provoke horror and outrage because a moral barrier has been transgressed or will be transgressed. Such barriers are not based on rational considerations but moral sensibility, embedded in a vision of ethics as a way of life.

One does not have to endorse the monster metaphor to notice its possible effects on ethical debate. First, it clarifies why certain issues are agenda-setting. Fascination and horror in regard to scientific and technological innovations indicate that ethical assessment is not a straightforward deductive application of principles but has to recognize that a plurality of values is at stake. Second, taking the metaphor of monsters seriously implies that the scope of ethics will be wider than usual in mainstream bioethics. Metaphors, as argued in chapter 3, are understood within a context—in this case, the ideology of scientism, which assumes that science always leads to progress so that what is achievable should be done. Bioethics should critically examine this context and analyze how far it ought to be extended or restricted.

Accepting ethical boundaries is not necessarily related to religious belief. Moral prohibitions constitute secular ethics. The notion of morally impossible action does not express a theoretical concern; it is also not related to a specific morality but is an aspect of respect for morality itself. Some acts are monstrous since they undermine and corrupt ethics as a way of life, as a particular ideal of humanity. Consequently, the metaphor of monstrosity brings back some of the perspectives that have been lost in the development of bioethics, as described in chapter 2. It articulates moral experience, the need for interpretation, and the significance of emotions as well as a critical and independent point of view. Doing this, it contributes to the broadening of bioethical imagination.

CHAPTER FIVE

Pilgrims

More and more patients travel around the globe in order to receive medical treatment outside of their native countries. It is estimated that 12 to 14 million people make medical trips each year. More than fifty countries today offer programs for health travelers. Top destinations are Costa Rica, Mexico, Israel, India, Singapore, and China. Patients frequently travel for cosmetic surgery, dentistry, cardiovascular surgery, orthopedics, cancer treatment, and reproductive interventions.[1] This phenomenon is known as medical tourism. Patients travel because interventions are less expensive or can be done faster than at home. They also want access to experimental treatments that are not yet available in their countries. Most often, patients are crossing borders because they hope for cures. Like the parents of Charlie Gard who wanted to use an experimental treatment in the United States, they do not wish to forego any opportunity that promises improvement and cure.

This phenomenon of medical tourism more closely resembles the ancient tradition of pilgrimage. Patients sometimes define themselves as pilgrims in search of salvation. In contrast to the metaphor of ghost (science can be deceptive) and the metaphor of monster (medicine and scientific research can be frightening), which both express the need for ethical debate that is broad and takes into account the larger questions that are often neglected, the metaphor of pilgrim highlights the benefits and blessings of medicine and medical science. Scientific developments, and especially contemporary health care, are regularly regarded as producing miracles.

The first part of this chapter examines the phenomenon of medical travel and will compare it with the ancient practice of pilgrimage. It then discusses medical miracles that are driving the quest for salvation and healing as well as the element of hope, which is not often explicitly discussed in health care or in bioethics. Two contemporary developments that frame the discussion are the adoption of right-to-try legislation in the United States and the quest for health and immortality that inspires the movement of transhumanism.

Medical Tourism

People have always traveled searching for health and cures for diseases. In ancient Greece, Epidaurus was one of the most famous healing centers. The sanctuary of Asclepius in Epidaurus was especially known for healing blind and lame people, and pilgrims would leave votives, often in the form of replicas of healed body parts.[2] In the 19th century, wellness resorts became attractive, focusing on lifestyle and well-being—healing spas for fitness, relaxation, and rest in places like Vichy in France, Baden-Baden in Germany, and Bath in England.[3] With the advance of scientific medicine in the 20th century, patients moved to more developed countries with sophisticated health care systems, for example, to have open heart surgery in the United States or separation surgery for conjoined twins in the United Kingdom. What is new now is that the flow of patient mobility has shifted from more developed to less developed countries.[4]

While the search for health has always existed, health is now considered a commodity that can be bought and consumed on the global market. This is celebrated as an opportunity that enhances individual choices and empowers patients. Studies are usually positive, and specific guidebooks offer practical information and "road maps."[5] At the same time, medical tourism is a new phenomenon; only since 2010 has there been a growing number of related research publications.[6] There is a lack of systematic data and hard evidence, for example, about clinical outcomes. While many studies focus on the commercial prospects, not a lot is known about the implications for patients and health care.[7]

It is clear that the phenomenon is facilitated by processes of globalization. If mobility is taken as one of the most typical features of globalization, medical tourism is a signature manifestation of globalization. It has emerged because international travel is more accessible and affordable, and because communication through the internet is easy. Globalization has homogenized many areas of life, but health care systems are still different across the globe. Quality of care, medical services, and prices can differ substantially. Such discontinuities can be exploited when health becomes a commodity. Hungary, for example, has more dentists per capita than any other country and can therefore profile itself as a destination for dental work.[8]

A significant role is played by the ideological context of globalization. Medical tourism is not a spontaneous affair but the product of neoliberal policies.[9] Countries like India and Thailand have been pressed to attract foreign revenues, to reduce public spending and subsidies, and to promote private sector health care delivery. The 1997 financial crisis in Asia forced countries to open their health care to global markets. In Thailand, for example, private health care collapsed due to the crisis. Private hospitals then started targeting international patients.[10] Governments have developed specific policies to attract foreign patients. Singapore inaugurated Biopolis in 2002, and Dubai Healthcare City was launched in the same year.[11]

For an ethical assessment, it is relevant to distinguish the concerns that motivate patients to travel. One concern is economic. Health care abroad can bring substantial savings. For example, a knee replacement costs $34,000 in the United States, $11,500 in Thailand, and $7,500 in India.[12] Another consideration is that treatment is faster than in the home country. Waiting lists can be bypassed, for instance, for transplantation. But people can also circumvent bioethical legislation and receive services that are not available or even prohibited at home. Examples are surrogate motherhood provided in India and assisted suicide in Switzerland. A final concern is that patients with often chronic and desperate conditions are seeking experimental procedures and innovative treatment, such as stem cell therapy, for which there is insufficient evidence and that are not approved at home.

Ethical assessment will furthermore look into the benefits and problems of the phenomenon. Within the paradigm of neoliberalism, medical tourism has many benefits. It helps countries generate foreign exchange. It promotes economic growth. Foreign patients provide income that can be invested into national health care. Technological innovation in the medical tourist sector will trickle down so that the quality of health care in general may improve. Medical tourism may also counter the outflow of health workers, often referred to as "brain drain."

Unfortunately, there is not much evidence that the purported benefits are materializing.[13] The effect on the brain drain to other countries is very limited while at the same time there is an exacerbation of the internal brain drain, attracting health professionals from the public to the private sector within countries. Also, spillover to the public health care system is limited. Some philanthropic programs have been initiated, but they do not systematically increase access for local residents.[14] Many medical tourist facilities are owned by foreign companies, and overseas stockholders will not invest profits in destination countries. Theoretically, governments can use revenues to invest in the public system, but on the contrary, they subsidize the medical tourism industry and reduce the budgets for public health.[15] It is argued that different approaches are possible, but that requires going beyond the framework of neoliberalism. Cuba was one of the earliest countries to promote medical tourism. The revenues, however, were invested in public health services so that the benefits went to the population as a whole.[16]

These criticisms point to the problematic implications of medical tourism. Frequently, three problems are highlighted. First are the implications for care. In an effort to underscore the advantages of medical tourism, Josef Woodman argues that "many of the finest medical practitioners are technicians."[17] They don't need to have personal skills. This exactly reveals that the emphasis in medical tourism is on technology and medical interventions rather than care and compassion. It is selling products that will be delivered at one point in time. There are limited opportunities for communication and building a relationship. Continuity of care is not possible. Most medical travelers do not seek pretravel health advice.[18] Usually there is lack of postoperative care and long-term follow-up.[19] Possible complications are the responsibility of the health care system of the traveler's home country. Moreover, the credibility of information is questionable. Medical tourist websites promote and exaggerate benefits and downplay the risks. Sites often do not provide reliable and objective information but are adverts for commercial purposes.[20] This raises the question how informed the decision-making really is.

A second problem is related to implications for health.[21] Medical tourism can grow because health is deliberately promoted as a commodity that can be traded. It is regarded as an object of profit rather than a human right or a public good. Health care is approached as a service industry. The basic criterion for this type of care is the ability to pay rather than medical need. This neoliberal perspective significantly changes the priorities in health care, focusing on technological interventions and cures instead of primary care and prevention that are most relevant for local populations in most countries. It furthermore creates new global risks, such as dissemination of multidrug-resistant agents and superbugs.[22]

The third problem is the implications for health systems. It is generally assumed that medical tourism has a negative impact on the provision of public health. It helps to create a two-tiered health system. Sophisticated and advanced health care is available for the wealthy (mostly tourists) while health care for the local population is poor and underfunded.[23] Private elite hospitals attract health care workers from the public health system so that resources are diverted from the public to private system. In Thailand, salaries in private hospitals are six to eleven times higher than in public institutions,[24] and many medical tourist destinations already have a shortage of physicians. For example, India has seven physicians for every 10,000 people, Thailand has four, while the United States has twenty-five.[25] The internal brain drain further reduces access to physicians for the local population. It is estimated that the resources needed to provide services to one foreign patient may be equivalent to those used to provide services to many more local patients; this ratio is one to four or five in Thailand, for example.[26] Finally, it is pointed out that there is a risk that the training of health professionals will focus on curative procedures that may be appealing to tourists and neglect primary care for local populations.[27]

The phenomenon of medical tourism challenges the common approach of mainstream bioethics. I do not want to argue that an ethical analysis with rational balancing of moral principles is not possible. In fact, there are many examples of applied ethics analyses of medical tourism.[28] My point is that such an approach is narrow. The focus is often limited to the four principles of mainstream bioethics. It is also limited because it abstracts from the neoliberal context of the phenomenon and the underlying value perspectives. From a Western point of view, with its emphasis on individual autonomy, medical tourism is appreciated since it enhances patient choice and empowerment. It liberates people from the restricted domestic framework of health services. People have more choice, cheaper access, and better value. This is precisely what the ethical principle of respect for personal autonomy demands. Of course, it is a narrow perspective since informed global citizens are primarily regarded as individual consumers. People have more options to satisfy their health needs and desires.

Medical tourism illustrates that the global market has no boundaries. At the same time, it is evident that choices are expanded only for a limited group of people. Traveling patients need to have sufficient money to engage in the practice. That means for most people this option does not exist. Furthermore, is medical travel really a voluntary option? Many patients have little choice since they have no health insurance, for example. Economic pressures bring patients to explore these options in the first place. Or, patients with life-threatening and severely disabling conditions are desperate to find relief. Krystyn Adams and her colleagues found that people engaged in medical tourism acknowledge that desperation results in quick decision-making.[29] In these circumstances, the notion of informed consent seems flawed.

The conclusion is that medical tourism is not simply a matter of individual choice.[30] It cannot be disconnected from broader questions regarding public health and health equity. Millions of Americans do not have health insurance.[31] For uninsured and underinsured people, medical tourism is an attractive option. In European countries, waiting lists for major interventions are long. In the United Kingdom, for example, in 2016 almost 200,000 people had to wait longer than eighteen weeks before having surgery (including conditions such as broken limbs, traumatic injuries, cancer, and eye problems). And that number is increasing.[32]

Although medical tourism offers a way out of the waiting game, the question is whether it is ethical that wealthy nations, such as the United States and the United Kingdom, are using the scarce resources of developing countries, such as India.[33] People are motivated to become medical tourists because the health care system in their home country is deficient. This could be the result of deliberate policies to restrict access to health insurance or to limit the number of physicians. Underinvestment in medical education in the UK, for example, is resulting in waiting lists and thus instigates patients to travel abroad. India, however, has widespread poverty, hunger, malnourishment, and limited access to clean water for many people. Compared to the UK, life expectancy is low. At the same time, spending for public health in India has steadily declined since 1985.[34] The country now has the most privatized health care system in the world. How is medical tourism morally justified in such conditions when it will further diminish access to health care for vulnerable and marginalized populations? These questions show the importance of the principle of justice, but at the same time they also engage a broader set of principles: solidarity, vulnerability, nondiscrimination, and social responsibility.

Ethical analyses of medical tourism, even when applied ethics is enriched with a broader set of principles, will still be unsatisfactory, I contend, as long as the value context of the phenomenon is not addressed. Medical tourism exists because of inequalities within and between countries. It is associated with deficiencies and failures of health care systems, and it will further exacerbate the unequal distribution of health resources in destination countries. Foreign patients will compete with the local population for scarce health care resources. An increase of medical tourists will reduce the level of health services available for domestic users. A dreadful example is Singapore. The government now encourages citizens to seek more affordable care in Malaysia since health care prices have increased notably.[35] Medical tourism will furthermore distort priorities with its preference for technological solutions. There are now more CT scanners and mammography equipment in private institutions in Bangkok than in all of England.[36]

Such inequalities are the result of decades of neoliberal global policies that requested countries to reduce public services, promote privatization, and facilitate health care as a commercial product that can generate financial profits. Analyzing medical tourism with the help of ethical principles usually does not go to the roots of the moral concerns. Patients are traveling around the world as global customizers because their needs and desires are not sufficiently met within their local settings. They are primarily regarded as consumers who should have free choices, rather than vulnerable individuals in need of care and protection. The neoliberal perspective does not make a difference between the consumption of health care and other commodities, such as clothing and smartphones, which can be produced and exchanged anywhere in the world. Medicine is first of all a package of services to be bought by rational and autonomous individuals. Paradoxically, while medical tourism seems to be driven by the values of the neoliberal market ideology, the phenomenon illustrates that health care is not a commodity, and that other values are at work. Patients are driven by the hope that their condition can be ameliorated faster and cheaper than at home. They are looking for cures and treatments, even when they know that the chances of success may be low. In the belief that miracles may occur, they, like the parents of Charlie Gard, do not want to forego such chances.

Tourist or Pilgrim?

People are involved in medical travel for a variety of reasons. Interest in cosmetic surgery, for example, has grown substantially (e.g., rhinoplasty, liposuction, or removal of tattoos). There is also a growing reproductive tourism industry. Though the term "tourism" is not appropriate here since it suggests leisure, pleasure, and relaxation. In most cases, medical travelers are patients and they suffer from medical conditions for which they seek help and relief. It is argued that the association with tourism trivializes the fact that people travel for serious reasons. Other terminology that has been proposed includes a more neutral term such as "transnational medical travel"[37] or "medical refugees." Middle-income Americans, especially, are escaping from unaffordable health care and evading impoverishment.[38] Citizens of countries where some reproductive interventions are prohibited may use the term "reproductive exile" to indicate that they feel forced to travel to seek assisted reproductive technologies in other countries. Traveling for reproductive care is not a neutral experience. Patients wish that they could have stayed at home.[39] Likewise, people traveling to Costa Rica for cosmetic surgery see themselves as refugees from a dehumanizing and greedy system. They want to escape the injustices of the US health care system and enjoy the kindness, competence, and personal approach of Costa Rican health providers.[40]

From the beginning, however, health travel has been combined with recreation and tourism. Cross-border travel for medical treatment has also been deliberately marketed in connection to tourism. The pleasures of destinations are consistently underlined. Gorgeous Getaways advertises cosmetic and plastic surgery holidays in Thailand and Malaysia. Surgeon and Safari offers medical treatment in South Africa and their reproductive tourism is advertised as a "romantic holiday."[41] The term "medical tourism" appropriately emphasizes the commodification and commercialization of contemporary health travel. It is not simply health care but at the same time a business transaction, a matter of advertisement, internet marketing, industry involvement, and inducement of consumer demand.[42]

Certain forms of medical tourism, recognizing that they involve patients, are better characterized with the metaphor of pilgrimage. Priscilla Song has proposed to label persons in search of stem cell treatment as "biotech pilgrims."[43] This metaphor can also apply to reproductive tourism and transplant tourism. In these cases, people travel because they are desperate; they want to find a cure for a chronic and disabling disease, to find a transplant to save their life, or to fulfill their desire for children. Cosmetic surgery tourists seek personal transformation, not only of their bodies but also in a wider sense of "rebirth" and spiritual renewal.[44] Wellness tourism also has characteristics of a pilgrimage. Hoping to improve health and well-being, travelers are seeking transcendence: finding their inner selves, increasing self-understanding, gaining a sense of renewal, and thinking about what life means.[45]

The metaphor of pilgrimage can explain some dimensions of the phenomenon of medical tourism. Reproductive tourists want to evade restrictive legislation. They are usually aware of ethical issues (e.g., surrogate mothers in developing countries risk being exploited) and that services may be unsafe or dubious. But nonetheless, they consider their condition as painful and distressing. They do not regard themselves as tourist but as patients. They find the tourist label insensitive. It reiterates their experience that infertility is often not considered as a serious condition. It also does not acknowledge the stress and burdens of undergoing treatments. It trivializes the efforts they undertake to satisfy their need.[46] At the same time, they are motivated by hope. They undertake a burdensome and expensive journey. The internet is the main source of information. Patients can therefore be victims of misinformation. They are attracted by the miracle babies promised by fertility technology.[47] The centers and physicians providing fertility care want to sell their services. What patients need is independent and reliable information that is often not available.[48]

Similar considerations apply to what is called "transplant tourism." Desperate patients are looking for any possibility to prolong their lives. They are a vulnerable population and not always aware of the risks of transplantation practices in different parts of the world. Transplant commercialism is often related to organ trafficking, too. Vulnerable populations such as prisoners and impoverished people are exploited to "donate" organs. Thousands of Kenyans each year make a medical pilgrimage, mainly to India, for transplantations. Doctors are referring these patients in exchange for kickbacks.[49] Transplant tourists from other countries make a "kidney pilgrimage" to Pakistan since poverty-stricken villagers are willing to sell their organs.[50] Personal experiences of organ recipients are not well known since they are engaged in often illegal and morally reprehensible practices.

The pilgrimage metaphor is especially apt for stem cell travel. Driven by desperation, patients with debilitating chronic and life-threatening diseases travel to countries such as China and Russia where hundreds of hospitals and clinics offer stem cell "therapies." Tourism is not an appropriate term to describe this form of traveling. These medical journeys involve hardship and expenses. In their quest for cures, patients have faith in the possibilities of new technologies. Their travail is much better expressed by the metaphor of pilgrimage.[51] Patients travel because stem cell interventions are not approved at home. Stem cell clinics offer hope of improvement and even cure in exchange for significant payments (more than $47,000 on average).[52] They usually downplay risks and promise success for a wide range of disorders, sometimes referring to "previously incurable diseases," such as multiple sclerosis.[53] Personal experiences of patients are well known. There are many online communities exchanging information, personal blogs narrating travel abroad, and public efforts for crowd funding. Most reports are optimistic and positive about the effect of treatment. Websites of stem cell clinics feature patient testimonials claiming success and miracle cures. Scientists, however, argue that there is a lack of independent scientific evidence of efficacy, that there are only anecdotal reports of therapeutic benefit, that there is growing evidence of harm, and that there is no theoretical rationale for interventions and their supposed effect.[54] Against this background, offering unproven treatment is regarded as exploitation of generally very vulnerable patients facing death and severe disability. They are victims of fraud, quackery, and charlatanism.[55] Employing the metaphor of pilgrimage, however, produces another story. Patients with terminal and devastating conditions have little time to wait until clinical trials provide sufficient evidence and new drugs are approved for prescription. They are eager to use all possibilities and are less interested to contribute to the progress of medicine as research subjects. Journeying to places that promise healing is furthermore a manifestation of patient autonomy in combination with the role of hope that is often ignored in the practice of health care.

Characteristics of Pilgrimage

The use of the metaphor of pilgrimage in relation to medical travel is contested because of its religious connotations and uncritical faith in possible benefits.[56] However, pilgrimage is a global phenomenon found in all cultures and not necessarily related to religious traditions. Today there are many secular types of pilgrimages, for example, to Graceland to commemorate Elvis Presley, to war cemeteries, and to places in Africa associated with the slave trade (so-called roots pilgrimages).[57] Religious pilgrimages are on the rise, though, too. Tens of millions of people travel each year to Lourdes, to Santiago de Compostela, to the Virgin of Guadalupe, to Mecca and Medina, and to the islands of Shikoku or Varanasi. In general, pilgrimage can be regarded as a "common human phenomenon."[58] It always involves travel and movement, a physical journey to a place, often remote, that has a special significance.

Pilgrimage is an expression of faith, no doubt, and pilgrimage sites can be manifestations of divine power. It is not unusual to hear reports of apparitions or miraculous cures of illness that make a destination the symbol of promise and hope. Healing is, therefore, a key ritual. The spring water in Lourdes, for example, is believed to have restorative power. The journey also requires preparation. Practical guidebooks for pilgrims have circulated for ages, and although every pilgrimage is an individual and personal experience, it is often done together with others. Pilgrimages create solidarity and fellowship; they are associated with a sense of community and common belonging to the same fate. There is a mix of motives making a clear separation between tourism and pilgrimage impossible. And complaints about the interplay of pilgrimage and commercialization are as old as the phenomenon itself.[59]

Another characteristic of pilgrimages is that pilgrims frequently leave votive offerings: miniature models of body parts that have been healed, cast in wax or wood.[60] Numerous ex-votos have been found in Epidaurus as tokens of gratitude. Furthermore, in many instances, pilgrimages are problematic for orthodox religious authorities. They often involve competing discourses that contest official interpretations.[61] In folk religion, individual people decide what they find worthwhile and what they pursue, whether or not endorsed by authorities. These pilgrims are concerned with daily problems; they want relief of ailments and are convinced that miracles happen, while the church discourse is primarily concerned with the otherworld. In Catholic pilgrimages, for example, pilgrims can be in direct contact with saints that provide healing without assistance of the clergy. Miracle discourse is in continuous tension with sacrificial discourse.[62] Immersion in the waters at Lourdes is for pilgrims a possibility for cure while the religious authorities emphasize that it is a renewal of the sacrament of baptism.[63]

Finally, typical for pilgrimages is "liminality," the transition from one social space into another, for example the passage from illness to health, from ordinary, everyday life to a transformed, renewed, and healed existence. Journeying to a distant holy place, the pilgrim is "separated from the rule-governed structures of mundane social life, becoming both geographically and socially marginal."[64]

Medical tourism demonstrates phenomena that are quite similar to traditional forms of pilgrimage. Of course, the religious language of holiness, apparitions, and saints is not used but other similarities are striking. In his guidebook, Woodman states that medical travel is a life-changing experience.[65] Patients travel to a specific destination that is often remote and peripheral (at least from American and European perspectives) but has a specific meaning since it offers something extraordinary.[66] Fetal cell transplantation, for example, is offered in China because scientific innovation there is not impeded by ethical and legal restrictions on the use of cells from late abortions. Patients also undertake the travel abroad hoping that their condition will improve. They are motivated by faith in the advances of medical science that will produce healing and transform their situation. This search for transformation and salvation is not expressed in terms of miracles, although references to miracles are not uncommon, but as producing "small victories."[67] The miracle discourse is minimized and the scientific approach emphasized.

Medical tourism is furthermore characterized by community building. Patients create kinships and networks, often online, to exchange experiences, hopes, and frustrations but also to explore promising treatments and investigate their safety and efficacy. Patients are not individual consumers searching for treatment options, they are "participants in a shared quest of greater significance."[68] Making the journey is a shared process, a sign of social relatedness in a transnational quest for regeneration. Like in traditional pilgrimages, patients leave testimonials and photographs online after treatment, as votive offerings in modern cyberspace.

What is also striking are the competing discourses. Against the scientific orthodoxy that treatments are unproven, it is argued that patients cannot wait; they want to escape chronic disability, inexorable decline, or certain death. They have nothing to lose. They want cure and treatment, not research or the promotion of scientific knowledge. It is argued that there are alternatives to the hegemonic discourse of randomized clinical trials as the only way to establish evidence.[69] Contrary to what is often assumed, patients may be well informed. It is their deliberate choice to travel and take the risk. They want to be in control. It is paternalistic to assume that they are misinformed. It is not simply the case that desperate patients are the victims of fraud and pseudomedicine. As in traditional pilgrimages, orthodox and expert interpretations are contested. It is more helpful to consider medical tourism as a clash between scientific and medical authority and patient autonomy.

Finally, the characteristic of liminality applies in this context as well. Medical travelers break away from ordinary life, and for a variety of reasons. They move between two social worlds; they leave the familiarity and reliability of their usual environment to face the uncertainties of unknown health care systems and cultures, often far away. In the case of stem cell transplantation in China, they also break with an ethical context in which late abortions and the use of fetal cells are highly problematic.

Miracles

Miracle discourse is flourishing in modern medicine. In September 1948, rheumatologist Phillip Hench gave a series of injections of a new compound to a 29-year-old woman hospitalized at the Mayo Clinic. The woman had suffered from rheumatoid arthritis for four and a half years. She had severe pain and was wheelchair bound. The injections had an astounding effect. Within a few days, her pain and stiffness disappeared, and the patient was able to walk around and even go shopping.[70] The new drug, cortisone, was touted as a miracle cure. Hench was awarded the Nobel Prize in Physiology and Medicine in 1950. Around the same time, many new drugs were discovered and applied in clinical practice: penicillin (since 1944), streptomycin (1943), chloroquine (1943), and lithium (1949). Because of their impressive effects on previously untreatable and sometimes lethal conditions, they were applauded as wonder drugs.

This has not changed in contemporary medicine. New drugs are frequently marketed as miracle cures. An example is Herceptin. Introduced in 1998 as a new medicine for breast cancer, it was welcomed as a wonder drug by cancer patients. It was first approved for use in cases of advanced breast cancer. Subsequently, it was promoted for women with early stage cancer after surgery and chemotherapy completion, even before it was approved, with the argument that it halves the chances of getting cancer again. Clinical trials, however, did not show that overall survival with Herceptin was significantly different from nontreatment.[71] Nonetheless, patient organizations, often sponsored by the manufacturer of the drug, pressured policymakers and insurance companies to provide and reimburse the marvelous medicine as fast as possible.[72] An editorial in the *New England Journal of Medicine* even called the results of clinical studies "simply stunning." They suggested, "maybe even a cure," the studies will "completely alter our approach to the treatment of breast cancer."[73] At the same time, an editorial in the *Lancet* called for caution. Referring to the same studies, it concluded that "the available evidence is insufficient to make reliable judgments. It is profoundly misleading to suggest, even rhetorically, that the published data may be indicative of a cure for breast cancer."[74]

More than a decade later, research showed that one year of adjuvant Herceptin treatment after chemotherapy improves long-term disease-free survival.[75] But this is only true for a specific group of patients (HER2-positive patients with a gene mutation that promotes the growth of cancer cells, which is the case in only 1 of every 5 breast cancers). Also, the ten-year disease-free survival is 69 percent for one year of treatment, compared to 63 percent for observation without treatment. This is an important benefit, but it is not a miracle cure for breast cancer. The example of Herceptin shows that for patients with serious and life-threatening diseases, advances in medicine are greeted as miracles because they open up unexpected possibilities for recovery and cure, even if the real benefits are relatively small and not miraculous from a scientific point of view.

New drugs and interventions are often heralded as "groundbreaking," "astonishing," and "miraculous" in the news media. Journalists most frequently use these superlatives but so do physicians, industry experts, and patients.[76] Especially innovative interventions are promoted with miracle discourse as well. A Polish man paralyzed from the chest down since 2010 was able to walk again (using a frame) after nerve cells from the patient's olfactory bulb were transplanted into his spinal cord in 2012. A British researcher involved called the achievement "more impressive than man walking on the moon." Four years later, it was reported that the patient could also ride a tricycle.[77] The story was all over the news media and internet as a breakthrough for paralyzed patients.

Even more fascinating are stem cells. They have a "magical aura."[78] Since they can renew and differentiate, these cells have the potential to regenerate damaged or dysfunctional tissues. Stem cell applications are advertised by clinics across the world as therapies; diseases should no longer be regarded as incurable since stem cells produce a miraculous transformation, even if there is no scientific proof at all. Fetal cells, harvested from aborted fetuses, have the same aura. The Chinese neurosurgeon who started injecting these cells into patients with spinal cord injuries and amyotrophic lateral sclerosis (ALS) is almost celebrated as a saint. His amazing procedures are producing miracles. Patients expect the "beginning of a new life."[79]

The point is that in many cases, such miracle discourse is stimulated by scientists themselves.[80] In 2012, the US Food and Drug Administration (FDA) initiated a new program, created by Congress, to expedite approval of new promising drugs: the "breakthrough therapy" designation. This label suggests exceptional benefit. No wonder that these drugs are described in the press and welcomed by patients as miracle drugs.[81]

In the early days of the so-called Genetic Revolution, with the start of the Human Genome Project in 1990, dreams were promulgated that genetics will solve the world's food problems, provide better pills and vaccines, eradicate genetic diseases, cure cancer and Alzheimer's disease, identify the genes responsible for various aspects of human behavior, and facilitate the enhancement of human beings.[82] In June 2000, the White House celebrated the completion of the first draft of the human genome. President Bill Clinton announced that this was the beginning of a new era. Humans were now learning the language in which God created life. With this new knowledge, "humankind is on the verge of gaining immense, new power to heal . . . In coming years, doctors increasingly will be able to cure diseases like Alzheimer's, Parkinson's, diabetes and cancer by attacking their genetic roots."[83] Similar promises and expectations were expected from research with human embryonic stem cells. "Medical miracles do not happen simply by accident," said President Barack Obama when he signed an executive order to allow federal money for this type of research, reversing the policy of his predecessor. Stem cells are especially commended since they can restore tissue functions and regenerate body parts. This will solve future problems of organ failure and neurodegenerative diseases due to aging.[84]

Public opinion polls indicate that 79 percent of adult Americans believe in miracles. The percentage is higher for people with religious affiliation, but even most people without such affiliation believe in miracles. In one survey (in 2007), 23 percent of respondents replied that they had witnessed a miraculous physical healing. It is also argued that belief in miracles is on the rise. American doctors also believe in miracles: 74 percent believe miracles occurred in the past, while 73 percent believe that they occur today. In a more secularized country, such as the United Kingdom, 59 percent of people believe that miracles take place. One in six say that they, or someone they know, have experienced a miracle; half of those were nonreligious.[85] These findings are not remarkable since they confirm the predominance of miracle discourse in contemporary science and medicine. They are significant, however, because traditionally, miracles have always been regarded as superstitious, delusionary, and irrational from the point of view of science and philosophy.

The argument against miracles was made especially strong by philosopher David Hume. He argued that there never had been or could be sufficient evidence to establish that a miracle occurs. No testimonial evidence is sufficient to justify belief in a miracle. Another argument is that miracles arouse passions that impede clear thinking. One more argument is that miracles are mainly reported from "ignorant and barbarous nations."[86] Hume's position is often reiterated in subsequent debates. It depends, however, on what is regarded as "miracle." For Hume, it was a violation of the laws of nature. In his view, the uniformity and regularity of nature does not allow any exceptions. This uniformity makes science possible. It is objected that such a deterministic view is a metaphysical presupposition and thus a belief that is not rationally justified. Charles Sanders Peirce, for example, argued that miracles are unique events that are outside the scientific worldview. Natural laws are statistical and probabilistic.[87] This has been a common position prior to Hume: miracles are not contrary to nature but beyond the powers of nature.[88]

The discussion so far has focused on a specific type of miracles: religious miracles. These are wondrous events brought about by divine intervention. This notion of miracle is universal; it has been used in all cultures and all times.[89] What is happening today is the emergence of another type of miracle: scientific, or more specifically medical, miracles. These are very unlikely events, such as a spontaneous remission of cancer that is extremely improbable but nevertheless occurs. Both religious and scientific miracles have common characteristics that are important from the perspective of patients. First is that what happens is wonderful, amazing, and marvelous. (The word "miracle," in fact, is derived from the Latin verb *mirari*—to be surprised, to look in wonder, admiration). Second is that the event is unusual and extraordinary. It does not need to be a violation of the laws of nature but can be an extraordinary coincidence. Miracles are "dramatic exceptions to the usual course of nature."[90] A third characteristic is that the wondrous event does not have a natural explanation. The causes of the miraculous healing are not known. It also implies that the healing is unexpected and cannot be predicted. Fourth, miracles are instantaneous; the healing should be rapid, complete, and durable. Finally, miracles are not simply atomic events that are wondrous and strange. The peculiar circumstances in which they happen are determinative. Miracles are declared and recognized by a community of persons. A miracle is not merely a personal experience but a public one. That is why it requires testimony and reporting.[91]

These characteristics are clear for religious miracles. In the Catholic Church, miracles are required for the process of canonization of saints. The Congregation for the Causes of Saints in Rome relies on medical expertise. Panels of physicians examine alleged miracles; they determine whether a cure is inexplicable. If it is, religious authorities then determine whether it is a miracle.[92] Whether a healing is miraculous depends, therefore, on the context; first the state of medical knowledge at the time, but also the religious and political context.[93] While religious miracles, at least in the Catholic perspective, rely strongly on medical evidence, the same is not true for scientific and medical miracles. These miracles are often not proclaimed by experts or based on scientific evidence. As argued earlier, communities of patients as well as private clinics and companies are involved, and miracles are a powerful marketing tool. However, they share the characteristics of extraordinary and unexplainable events that happen suddenly and are publicly recognized by testimonies and witnesses.

The main difference between religious and nonreligious miracles is the attribution of causes. The first type of miracles is the product of religious imagination. God is the agent that brings about the event, demonstrating his power. The second type is the product of scientific imagination. They demonstrate the power of humans, science, and technology. Human beings are capable of inventing new drugs, practices, and treatments. They are a reason to wonder and they provoke admiration and awe, even if the effects cannot be explained or are not yet explained by natural causes. They demonstrate the possibility of the impossible but also our lack of understanding of nature.[94] Miracles do happen, but we do not know why.[95] This scientific imagination, of course, differs from the point of view, widely shared among scientists, that miracles cannot happen since the universe is deterministic. Only if the universe is not deterministic do unlikely events occur since they are statistically probable. That does not necessarily imply that humans produce miracles; they can be a mere coincidence, the result of auto-suggestion or placebo effect.

The attitude of most scientists and physicians is ambivalent. They reject religious miracles but often communicate the potency of new drugs and interventions as "miraculous." In many cases, scientific imagination assumes agency and causality. Medicine and medical technology are these days so powerful that humans using them can produce miracles. Miracles are now medicalized, as Lawrence Schneiderman observed. They are demanded of present-day physicians. People no longer accept the limits of medicine and the inevitability of suffering and death.[96] In this backdrop, miracle discourse is especially strong among seriously ill patients. They have faith in science and technology. The miraculous potential of current medicine gives them hope.

Miracles and Bioethics

The existence and expansion of miracle discourse provides important lessons for bioethics. First of all, it demonstrates the connections between medical science and religion. More than seven thousand cases of healing and cure have been recorded over the past 150 years in Lourdes but only seventy of them were officially recognized as miracles by the Church. Extensive efforts are undertaken to make sure that a medical explanation is ruled out. The Lourdes Medical Bureau as well as the Lourdes International Medical Committee want to make sure that the diagnosis is correct, that the illness is permanent or terminal, that the cure is instant and complete, and that any treatment prescribed is not part of the cause of the cure. Claims of inexplicable healings can only be endorsed on the basis of scientific expertise.[97] The last miracle was recognized in February 2018 while the cure took place in 2008.[98] Jacalyn Duffin, who studied 1,400 miracles (occurring from 1588 to 1999) in the Vatican archives, observes that religion, at least in the process of saint-making, takes medical science very seriously, while medicine usually tends to ignore religion.[99]

The second lesson of miracle discourse refers to the connection between scientific perspectives and popular approaches. Concerning miracles, there is often an opposition between experts and laypersons. Controversies in stem cell tourism arise from different assessments of evidence. For scientific experts, the treatments are unproven and potentially harmful. For patients, they are potential therapies that provide hope in desperate situations. For them, expert arguments are not convincing, and their expectations are corroborated in online communities. The same popular movement supported the parents of Charlie Gard in their struggle with the expert establishment of pediatricians, lawyers, and bioethicists.[100] The miracle discourse illustrates that scientific objectivity is not only different from personal experience but frequently also in contrast with lay perspectives.

The third lesson is that miracle discourse refers to a broader worldview than is usually prevailing in medicine. Miracles provide a different perspective than the scientific one. In a general sense, a miracle is an event that produces wonder, astonishment, and admiration. Since Plato it is assumed that philosophy starts in wonder. We are puzzled by our ignorance and we desire to overcome this state of lack of knowledge. Labeling an event as a wonder is not just a description of the world but it solicits a response since it questions the usual order of things. For a long time, wonders were regarded as special phenomena among more familiar experiences. They indicate that there are things we cannot explain through science or otherwise. A mechanical and materialistic worldview is not satisfactory in addressing issues of healing and meaning in the face of suffering and death.

This was the interpretation of St. Augustine: miracles are not contrary to nature but outside of what human beings know of nature; they are not *contra naturam* but *praeter naturam*.[101] However, since Hume, a distinction was made between marvels and miracles. Both are wonders, but marvels (*mirabilia*) are unusual phenomena, contrary to our ordinary experience but not opposed to the laws of nature. Miracles (*miracula*) are contrary to the laws of nature. Marvels are in principle explainable, although we lack sufficient knowledge momentarily, while miracles are unexplainable and thus irrational. Science and religion can therefore be separated. Wonders have different agency and causality. Manipulating natural processes can produce exceptions to the regularity of nature. Science can be as powerful as religion. Humans themselves can produce marvels, and these are more convincing since in principle they can be explained. A real miracle depends on divine intervention.

Remarkably, this distinction is disappearing today. A majority of people believe that certain things in life cannot be explained by science.[102] The deterministic presupposition of natural laws without exceptions is no longer restricting the scientific imagination. Wonders do not refer to a different world but evoke another perspective on the same world that is interpreted from the perspective of science.[103] For example, miracles are often attributed to surgeons. Surgical intervention can save lives and have spectacular and rapid results. Miracle discourse expresses hope and optimism. The results of surgery can be explained (they depend on technical proficiency, knowledge, and clinical judgment) but nevertheless they cannot be predicted.[104]

Miracles as Obstacles

While more attention is paid today to miracle discourse, it is primarily regarded as a management problem. The main question is how to respond to beliefs in miracles.[105] Patients may invoke the possibility of miracles to justify a particular course of care. For example, they may request to continue therapy that is futile or inappropriate according to physicians. Or patients want to pursue all available treatment options because they believe miracles can happen. In the case of Charlie Gard, his parents demanded experimental treatment that was only harmful to the baby in the perspective of care providers. In such cases, belief in miracles is regarded as obstruction of appropriate medical care. Health care providers generally are not receptive for miracle discourse of patients, especially when it concerns religious miracles.[106] They tend to override patients and parents who want to wait for a miracle and not have therapies withdrawn that are futile from a medical point of view. People waiting for such miracles apparently do not understand the clinical condition; they have irrational beliefs. These can be harmful because proper diagnosis and treatment is delayed or suffering prolonged. They are also vulnerable to exploitation while financial resources could be better used to improve care.[107]

Recent studies argue that ethicists should take miracle language seriously.[108] It is often the expression of hope. Instead of overriding persons waiting for miracles, it is better to engage them and discuss their expectations and hopes. For many patients, religious faith is an important dimension of their life. However, the role of religion in health care is not often explored. For example, for patients with advanced lung cancer, as well as their caregivers, faith in God is an important factor in decision-making. For physicians, on the other hand, it is the least important factor in deciding therapy.[109] The majority of hospitalized patients, however, believe that physicians should consider their spiritual needs. But in most cases physicians do not do that.[110]

There can be several reasons why religious patients wish to continue treatment. One is the hope for a miracle. But there can also be openness to wonder and surprise, believing in the healing power of God. It does not imply that for such patients, life need to be preserved at all costs. For them, hope is not so much based on survival but on resurrection.[111] At the same time, it is curious that physicians and medical scientists believe in the miracles of modern medicine. In practice, medical interventions produce at least 10 percent iatrogenic harm, half of health care is not delivered in compliance with guidelines, much care is not evidence-based, and new technologies have so far only provided marginal benefits. It is furthermore estimated that in general practice, medically unexplained symptoms are more common than medically explained symptoms. So why do medical professionals believe that medicine can deliver miracles?[112] Is it because they carry as much hope in them as their patients do?

Hope

Miracle discourse can indicate that health providers often do not (yet) understand what is happening or why particularly innovative treatments are effective. The use of the metaphor of miracle expresses hope and optimism. It recognizes that medical science is imperfect and incomplete. At the same time, faith in miracles does not acknowledge that medicine can become futile and has limits. There is always room for the unexpected beyond scientific imagination. Miracles illustrate how people can overcome adversity and can recover from disaster and debilitating conditions. They are presented in narratives that are shared with others and that provide inspiration and hope.[113] As argued earlier, miracles cannot happen without faith. This is true for religious miracles but equally true for scientific and medical miracles.

Hope has been a powerful force in health care, but it has rarely been studied. The therapeutic value of hope is acknowledged, however, and hope may promote healing and coping and enhance quality of life.[114] Jerome Groopman concludes, "For all my patients, hope, true hope, has proved as important as any medication I might prescribe or any procedure I might perform."[115] Hope provides patients confronted with chronic, terminal, and debilitating diseases with the expectation and promise that healing will be possible.

In the past, hope played a more prominent role in health care. Until recently, medicine was powerless and not able to provide curative treatments. For a long time, one of the moral rules for communication was not to tell patients that their condition was uncurable and terminal. Telling the truth would take away any hope; it conveys the message that death is irreversible.[116] Articulating hope rather than truth is justified with the moral principle of not harming the patient. The interest of the desperately ill person should be protected. Although the future is unknown and beyond control, patients could at least have some agency, attributing meaning to life and suffering, and receive support in relational and interpersonal processes. Recognizing and providing hope when medicine has limited therapeutic power protects patients against desperation. In this context, hope is particularly connected to religious miracles. Pilgrims (for example, to Lourdes) often do not expect healing but they hope for a transformation. Hope is not related to a specific outcome but to issues of meaning.

With the increasing power of medicine as well as the emergence of bioethics, a new moral discourse developed since the 1970s that reversed the relation between truth and hope. As effective therapies became available, the discourse of hope changed. Disclosure of diagnosis, prognosis, and treatment became a precondition for hope. Patients should know the truth (facts and evidence) in order to hope for improvement. The moral justification no longer is nonmaleficence but respect for personal autonomy. Furthermore, hope should be preserved not to protect vulnerable and desperate patients from harm but to assist them to opt for potential cures and to secure their compliance with often experimental and burdensome interventions and drugs. The assumption is that if people know the facts, they would take care for themselves as rational beings, although expectations of future benefits and promising futures are encouraged. Hope is now focused on the possibility of scientific and medical miracles. An example is cancer. It is presented as potentially curable. Individuals can overcome the disease, which has become, in the United States at least, the focus of a special war since 1971. Against this background, the duty of physicians is no longer to preserve hope but actively encourage and instill hope in patients.[117]

Biotechnology initiated another conceptualization of hope. The discourse of hope and miracles is now explicitly used to advertise and promote innovative treatments and drugs. Hope is commodified and employed in a commercial strategy to seduce people. New technologies and innovative practices are popularized and advertised with references to promissory futures. Hope is not primarily focused on the best interests of individual patients but advocated as an asset in commercial health care to earn back the heavy and risky investments made in biotechnology. Regimes of hope are the basis for new markets. This again implies a reversal of truth and hope. Future possibilities are regarded as more important than present truths. Cord blood banking is one example; this is sold to parents as an investment that may be lifesaving in the future.[118] Another example is xenotransplantation. To justify expensive and controversial animal studies, future benefits are overstated.[119] Additionally, reproductive tourism is blamed for selling false hope. Foreign IVF clinics are claiming success rates up to 98 percent, while 2016 data from clinics in the United Kingdom shows that the best chance to get pregnant after each round of embryo implantation is 32 percent on average.[120]

These examples illustrate how the concept of hope is changing. First, the current emphasis on hope is often economically motivated. The dynamics of innovative biotechnology now follow a specific pattern. They require, first of all, a vision of the future. These promises are necessarily exaggerated because they need to attract investment and political allies.[121] Related to this is that innovations are advocated for by practices of "hope management."[122] These often lead to growing cooperation between pharmaceutical companies and patient advocacy organizations.[123] Promoting new technologies is based on "hyperbolic expectations" and "imaginative speculation."[124] It is advocating symbolic language to offer hope rather than emphasize facts that are uncertain and always challengeable. The fact that cell transplantation is not proven is less important than the hope that it may be effective. Articulating hope implies that "present-day evidences, proofs, facts, or truths are giving way to future-oriented abstractions premised on desire, imagination, and the will to the yet 'not present.'"[125] Decisions about using new technologies have become more dependent on emotional processes than rational considerations. They appeal to imagination, affectivity, and future expectations; they reflect authenticity rather than authority. Finally, hope management targets vulnerable populations. It offers hope to patients who usually do not have any therapeutic options left, and do not have much time left. These desperate patients are easily seduced by promises of breakthroughs and statements that diseases are no longer incurable.

The characteristics of current hope management are evident in stem cell tourism.[126] Patients with debilitating disorders and neurodegenerative diseases consider their journey to clinics in China as the last resort in their struggle against their diseases. They object to the term "tourism" since their search for health is not frivolous or for leisure. They have nothing to lose and are driven by hope and a vision of salvation and healing. They refuse to accept incurability and death. They reject the paternalism of experts pointing out that treatments are futile and harmful. In such desperate circumstances, patients are vulnerable to exploitation. Commercial health care agents and agencies sustain a "hope industry," but not necessarily in the interest of patients.[127]

Since its emergence, bioethics has paid attention to the role of hope in health care. It has strongly criticized traditional paternalism as the argument that negative information should not be disclosed to patients with incurable or terminal diseases because it will deprive them from hope. The contemporary connection between advocacy of innovative technologies and management of hope is consistent with the moral principle of respect for autonomy.[128] The assumption is that individual patients themselves are responsible for their hope. They should have the intention to fight disease and disability. Because they belong to a vulnerable population, they should prepare themselves to avoid exploitations. Hope is more effective when patients are more actively participating in their own care.[129] Patients, therefore, must become experts themselves. Medical tourism is advertised as enhancing patient empowerment and individual choice, promising new opportunities.

The approach of mainstream bioethics on the phenomena of medical tourism, miracles, and hope has important consequences. Following the ethical principle of individual autonomy, many patients are now acting as active agents responsible for their own health. Patients with chronic and debilitating diseases explore potential medical interventions; identify experimental treatments in very early stages; share experiences with other patients, family, and friends; and raise funds. Patients are transforming into experts. A new type of patienthood and patient activism is emerging, as can be observed in the Charlie Gard case. Patients use digital media to create communities of hope, sharing experiences, offering mutual support, and operating as witnesses to the benefits of treatments. They are not merely consumers but also producers of information and experiences, most often disseminating optimistic narratives.

When patients need to become experts themselves, however, it is questionable who determines the reliability of future expectations. It is necessary to negotiate between claims of truth or evidence and regimes of hope.[130] For Charlie's parents, for example, hope was more important than evidence. The conflict also illustrates that different concepts of hope can be at work. For health care professionals, hope is medicalized. It is equated with hope for a cure. Problems then arise with patients with terminal disease. These patients are approaching death, and so when they (or their parents) still hope for a cure, hope is regarded as false or unrealistic. But this is a restricted concept of hope: it is regarded as aimed at cure or on extending life as long as possible instead of related to caring relationships or reconciliation with life and death.[131]

Another consequence of the mainstream bioethics approach is that the crucial role of hope and miracle discourse in contemporary medicine is disregarded. It is difficult to blame patients for irrational and irresponsible behavior as long as the language of miracles and hope is encouraged by science and medicine itself. Of course, patients will welcome the message that medical science is continuously progressing, always producing better treatments that are imminent,[132] but bioethics discourse generally ignores the context in which expectations arise and are promulgated. The role of the scientific profession is not frequently addressed. For example, the promise of stem cell therapies is stimulated by scientists themselves. Journalists and media coverage cannot be blamed for the hype of stem cells and regenerative medicine. Medical scientists have helped to create it.[133] The perception of science as salvation, as a producer of miracles, is criticized but often not by scientists. It is no wonder that the general population has sympathy for patients who use experimental treatment despite lack of evidence and potential risks.[134]

Bioethical analysis of medical tourism usually does not address the social context in which the phenomenon has emerged, either. Rather, it is often aligned with the neoliberal ideology that emphasizes individual choice and empowerment. It is not critical of a commercial type of paternalism that provides miracle discourse, hope management, and the promise of benefits without verifiable evidence of safety and efficacy.[135] This is surprising since hype is not a morally neutral activity.[136] Miracle discourse is used not for the best interest of ill people but primarily for selling products in a very early stage, prior to assuring that they are indeed effective and beneficial. Obscuring the commercial dimension is the result of the primary focus on individual decision-making. A broader bioethical approach is necessary for analyzing the context of medical tourism, explaining that choice in neoliberal global health care is an illusion.[137]

Ignoring the social context is also apparent in the localization of hype. It is usually assumed that medical tourism is promoted by faraway countries with inadequate or absent regulation and oversight or with professionals primarily interested in making money and lacking moral consciousness. The idea that Western countries are transparent and have good oversight is taken for granted. That this is a bias is clear from studies showing that hundreds of businesses have emerged in the United States, Australia, Germany, and Ireland selling unlicensed stem cell interventions, advertising treatments as safe and efficacious without any evidence.[138] It is not fair to blame developing countries. It has been argued, in fact, that medical travel is a responsibility of Western countries that do not restrain the travel of their citizens for unsafe treatments.[139]

Right to Try

In May 2018, the United States signed into law the Right to Try Act. This legislation gives people with life-threatening diseases the right to use unapproved experimental drugs. The underlying idea is that terminally ill patients do not have time to wait until new drugs are sufficiently tested and approved. Patients no longer need to travel to China for cell transplantation. The only requirement is that the medication has been cleared in a phase 1 clinical trial that has preliminary safety testing. If it has been cleared, patients can find a doctor willing to try the therapy or ask a drug company for permission to test the treatment.

The driving force of the legislation of this new right is miracle discourse and hope. The main sponsor of the bill in the Senate declared that this is victory for the right to hope.[140] President Trump stated that if the approval process of new drugs is simplified, "we will be blessed with far more miracles . . . In fact, our children will grow up in a nation of miracles."[141] The appeal to miracles and hope presents a powerful vision but is not an argument. It is difficult to be opposed to hope, especially if people are terminally ill and in a desperate situation. This explains why the legislation was supported by both major political parties, and why the pharmaceutical lobby was quiet. Nonetheless, the political and social debate was strongly ideological. It also shows how much the scientific and bioethical discourse is overwhelmed by popular and imaginative images and discourse.

First, the right-to-try debate is moved by a conservative and libertarian ideology. It articulates the ethical principle of individual autonomy. People have fundamental freedom and the right to save their own lives. They should not be told by governments what they can and cannot do, especially if health and life are at stake. Federal regulations and the FDA should not determine access to experimental drugs. The rhetoric of hope and miracles is therefore used to eliminate the role of federal oversight, and the new legislation becomes deceptive. Patients only have a right to ask since pharmaceutical companies are not obliged to provide the requested medication. The only thing that the law will accomplish is that the FDA is no longer involved. This ideological purpose was clear from the start. The Goldwater Institute, a think tank in Arizona sponsored by wealthy right-wing organizations such as the Bradley Foundation and the Searle Freedom Trust, focuses on advancing limited government, economic freedom, and individual liberty. It had advocated the right-to-try law at state levels since 2014, promoting a model bill that forty states had already adopted as legislation.[142]

Second, right-to-try legislation is unnecessary. The FDA has had for a long time pathways to expedite access to experimental treatments. Expedited programs were introduced in 1987 (expanded access), 1988 (fast-track designation), 1992 (priority review, accelerated approval), 2012 (designation as a breakthrough therapy), and 2016 (a program introduced especially for stem cell therapies called regenerative medicine advanced therapy). These changes in policy frameworks aim to assure that new drugs and biological products are quickly available, but on the condition that they are safe and effective. Over the last ten years, the FDA received over one thousand expanded access applications per year. The application process takes forty-five minutes, and requests were reviewed within hours or days. Over 99 percent were approved. The argument that the federal agency is restricting treatment options is therefore false.[143] Pharmaceutical companies control access to their drugs. The real purpose of the legislation is to restrict oversight by the FDA by framing the governmental agency as not protecting the interests of patients. The right-to-try law has been promoted to advance a neoliberal and antiregulatory agenda by using the language of hope and miracles.

Third, right-to-try vocabulary is not only misleading, since it does not give direct access to patients, but it is also dangerous and risky. The legislation assumes wrongly that experimental drugs in clinical trials are safe enough to potentially produce miracles. Patients will more likely be harmed than helped. Roughly 70 percent of early-phase clinical trials of drugs fail; over 50 percent at phase III.[144] There is no guarantee that a product used in expanded-access programs will be effective and safe. In fact, most of the drugs will never be approved for marketing.[145] The legislation is dangerous since it undermines the time it takes science to investigate and understand the fundamental mechanisms of medical innovations before they can be applied reliably. Expanding access may delay data gathering to make evidence-based decisions about approval. It may also deter enrollment in clinical trials.

Finally, right-to-try laws undermine public regulation to protect patients. Oversight (and agencies like the FDA) has been created and expanded for specific reasons. Many scandals involving drug manufacturers, scientists, and medical professionals have demonstrated the need for regulation. Without any governmental oversight, past experiences have demonstrated that abuse and fraud in medicine will continue to occur, particularly now that health care is regarded as an industry in and of itself. While regulators today approve drugs based on less robust clinical evidence to facilitate access, they require that data on the effectiveness and safety continues to be collected after approval. However, less than 10 percent of approved indications had one or more controlled studies after approval showing superior efficacy.[146] Patients are endangered while corporate interests are safeguarded.[147] The role of the FDA as an independent agency and arbiter is scaled down while manufacturers providing investigational drugs are protected from liability and data on negative clinical outcomes may be overlooked.[148]

It is remarkable that the ethical debate around the right to try has been so convincing for lawmakers and the public. The emphasis on hope and miracles presents an imaginative perspective that is hard to criticize. It is indicative of the transformation of patienthood in which patients play a more active role as health activists. At the same time, it indicates that the health context has changed into a global bioeconomy. A possible explanation for the narrowness of the debate is the focus on individual autonomy, while the broader context of the right-to-try debate is insufficiently analyzed. Narcyz Ghinea and colleagues, for example, observe that patient advocacy groups are demanding access but are silent about prices. One bioethical problem of today is that especially new cancer medication is extremely expensive, while the most expensive cancer medicines often provide only marginal benefits. Innovative cancer therapies generate hope, but they are less and less affordable for many people.[149]

A broader perspective should also analyze the cultural setting of the debate. It is evident that the right to try has an individualistic focus, but it was recognized decades ago that people suffering from life-threatening diseases should have the opportunity to use potentially beneficial treatments before they are formally approved. The FDA started its accelerated access programs in 1987. The main ethical argument was the principle of beneficence. The common name for such access was "compassionate use." Today the vocabulary has changed into rights language. The basic ethical argument now is that it is the patient herself who should determine whether to try a therapy and decide about possible risks and benefits. However, no such right exists, since patients can only ask while pharmaceutical companies decide. This individualistic focus is also reflected in the uneven balancing between individual and general interest. It is in the interest of everyone that there is a robust drug assessment system. All will benefit when reliable knowledge is acquired, and medical progress is only possible by rigorous scientific studies. It is in the public interest that we can be sure that new medications are safe and effective.

Given the commercialization of health care, it is also necessary to prevent conflicts of interests. The FDA can be criticized but at least it is expected that it operates as an independent agency that protects citizens when they use medication. Furthermore, comparatively, the agency is doing a good job. Average drug approval time for the FDA is 304 days; for the European Medicines Agency, 478 days; and for Australia's Therapeutic Goods Administration, 391 days. The United States has clearly defined regulations and mechanisms for expanded access.[150] It is an open question why accepted standards and policies need to be overridden and relaxed. The majority of Americans (6 out of 10 in 2016) oppose changing safety and effectiveness standards to allow for faster approval of drugs.[151]

The right-to-try movement is like stem cell tourism, another example of the conflict between science and imagination. Superficially the language of individual autonomy and freedom is used, emphasizing patient empowerment and suggesting that patients are in control. In practice the language of miracles and hope is more effectively used. Many patients are desperately ill; many of them do not have long to live. They are a vulnerable population that is sensitive to any suggestion that cures exist. Proposing that dying patients have a chance is providing false hope. Expanded access can harm patients. The new drug Avastin, for example, received accelerated approval for one type of breast cancer in 2008, but then approval was revoked in 2011 when follow-up studies showed no survival benefits and several side effects. Large numbers of patients had received ineffective and unsafe treatment.[152] This is not an isolated case: between 1993 and 2010, seventeen drugs were approved and later withdrawn, while being prescribed more than one hundred million times in the meantime.[153]

The imagery of miracles and hope is a major rhetorical move to promote the right to try. It suggests that the interests of patients are central. But in fact, it is doing very little for patients. It does not require manufacturers to provide investigational drugs. Instead of protecting patients, it provides liability protection to doctors, drug companies, and insurance providers. Many opposing the legislation therefore argued that it employs a false discourse of patient empowerment as well as a discourse of false hope.[154] It is important to observe that it also uses a distorted notion of hope as a product. Hope here is focused on a specific outcome: cure, remission, and recovery. Unproven products are labeled as "therapies." In the United States, a new program in 2012 was created to designate drugs as "breakthrough therapy." New products can be labeled as such if there is preliminary evidence of substantial improvement over existing therapies for serious and life-threatening conditions.

Against all expectations, hundreds of requests to designate drugs as breakthrough therapies have been made, and a growing proportion of approved drugs is now labeled as such (22 percent in 2014 and 37 percent in 2017). Breakthrough-designated drugs were approved in a median 1.7 months (against a median 8.0 years for nonexpedited drugs). Critics argue that the term "breakthrough" suggests claims that are not substantiated. Manufacturers advertise with that label while waiting for approval, which will only follow if there is substantial improvement in results compared to existing therapies. But it is not determined what will be appropriate comparators, and when newer or more similar medicines are excluded, it will be easier to designate drugs as "breakthrough." Based on experience thus far, breakthrough drugs may have only modest efficacy. The label does not imply superiority over the current standard of care and is thus misleading.[155]

Finally, the right-to-try debate illustrates another aspect of the social and cultural context in which bioethical debate arises. It reiterates the cultural ethos of a country. In his study of the birth of bioethics, Albert Jonsen argues that it could first emerge in the United States because of the typical American ethos, which is characterized by moralism, meliorism, and individualism.[156] Americans in general hold to the idea that the present can and should be made better. They will fight against disease, death, and disability. They believe in magic bullets, "a pill for every ill." Vulnerability is not a favorite topic. They have high regard for technology and have a strong belief in progress and individualism. Each person is free and responsible for his or her own life. Government should be minimized.

Still, this ethos is impregnated with the traditional imagery of miracles and hope rather than rational assessments. When signing the Medicare bill into law, President Lyndon B. Johnson stated that older Americans will no longer be denied the healing miracles of modern medicine.[157] Barack Obama, when he was a senator, declared in 2004, "That is the true genius of America: a faith in simple dreams, an insistence on small miracles."[158]

Conclusion

There is great power in the metaphors of hope and miracles. The present-day phenomenon of medical tourism with millions of people traveling around the globe in search of cures and health interventions reflects the ancient practice of pilgrimage. They are driven by similar motivations: desperation, hope, and a search for transformation and salvation. Miracle discourse in modern medicine and medical science is abundant. Rather than religious miracles, new scientific and technological developments promise scientific and medical miracles. But the dissemination of miracle language shows that for many people, medicine and religion are intimately connected. It also demonstrates how scientific and popular discourses are connected. Even religious miracles today are only acknowledged if they are vetted by medical experts. Miracle discourse presents a broader perspective than the prevailing worldview of science. It reiterates that there is a world of unexplained and unexplainable phenomena outside of the materialistic scientific view. The discourse also expresses hope and optimism.

However, in bioethics, miracles are generally rejected and often regarded as an obstruction of appropriate medical care. The fear is that people can have unrealistic expectations and ignore facts and evidence if they grasp for miracles. At the same time, and paradoxically, many health professionals and scientists believe that medicine can produce miracles. Today, especially in regard to genomics and biotechnologies, they have faith in the power of medical science to address and even solve many medical problems, now or in the near future. Miracles are frequently promoted by the medical and scientific establishment itself. Hope is actively managed, for example, in dissemination strategies for new medications. Using social media, patients create communities of hope. The right-to-try legislation recently adopted in the United States is also an example of how the imagery of hope and miracles is stronger than the scientific discourse.

The agenda of the ethical debate used to feature the latest medical advances and technologies, which evoked fascination because they promised hope and miracles for often-desperate patients. The metaphor of pilgrimage may help to explain why people are willing to undertake laborious travels in search of healing. The metaphor also broadens the bioethical imagination. It brings a perspective beyond the neoliberal images of the rational and individual decision-maker. Patients are not simply traveling around the globe as a result of individual decisions to receive the latest cures but are also driven by hope. Nonetheless, bioethical discourse does not frequently address this issue. Bioethical analysis does not consider the economic and cultural contexts in which expectations are created. It does not criticize the scientific imagination that fuels the popular search for salvation and healing.

CHAPTER SIX

Prophets

For a long time the essence of medicine has been located in its ability to predict the course and outcome of a disease. Hippocrates once stated that the best thing doctors can do is to practice forecasting. Prognosis is more important than diagnosis, and in his time, the practice to predict diseases was common. People consulted oracles, diviners, soothsayers, and fortune-tellers to know the future. For Hippocrates, medical prognosis was based on the symptoms of the illness and comprehensive knowledge of the past, current, and future condition of the patient. A physician must observe and analyze individual symptoms and combinations of symptoms because each have consequences for what therapeutic interventions might be used or are effective. It is impossible to restore every patient to health, but what is fundamental for the doctor-patient relationship is prognostication. It is a way to win respect and trust as a physician and it allows to practice preventative action.[1]

For a long time in the history of medicine, the burning question of patients was not "What disease do I have?" but instead "How am I doing?," "How long will I be ill?," "Will I recover?," "Do I need surgery or other treatment?," "When can I go back to work?," and "Do I need to make changes in my life?" And so, prognosis concerns these three dimensions: duration, recovery, and survival.[2] Predicting the outcome of an illness in a particular individual not only depends on the disease but also on many other factors, such as age, gender, history, and way of life. Prognostication has been considered an art.[3] It is connected to ethical considerations (how to inform patients about the future course of their illness) and influences patients' decisions. Providing information in a positive or negative way can modify such decisions. The prognosis also determines whether treatment would be valuable or not.

Making a correct prognosis demands that a physician is not focused on clarifying disease or declaring diagnosis. He should not explain what happens but rather what the impact of the illness will be on the patient's life. That requires careful observation of potential prognostic signs. A famous example is the *facies Hippocratica*. If the face of a patient is shrunken and pallid with nose sharp, eyes hollow, temples sunken, and ears cold and is contracted with a disease not improving within three days, this is (according to Hippocrates and still accepted today) a sure sign of death.[4] Galen, the second-century physician who developed Hippocrates's ideas into an encompassing medical system that would dominate medicine for centuries, built his reputation on correct predictions.[5] And one of the most influential medical texts in Medieval and Islamic medicine since the 11th century was the *Canon Medicinae* of Avicenna, which presents a list of prognostic signs that indicate the duration and stages of an illness and signs of death and recovery.

With the development of modern medical science in the 18th and 19th centuries, prognostic thinking became marginalized. The ideal of medicine as a rational and objective science became the identification and clarification of diseases as anatomical lesions and biological dysfunctions.[6] Medicine is now regarded as a branch of natural science, and it is characterized by a mechanical model. Since diseases are defects of bodily machinery, we need biological approaches to medical problems.[7] In this scientific view, prognostication can only be based on the disease entity and its characteristics. Knowledge is derived from empirical experience: data collected by observation, tested in experiments, and verified or falsified by others. Predictions of the future are therefore problematic since they are not based on observations but presuppose that the future will reflect the past. This is the logical problem of induction that states predictions are never certain.[8] Around 1970, medical textbooks hardly spent any space on prognostication.[9] In his well-known book *Rational Diagnosis and Treatment*, Henrik Wulff discusses a flow chart of the clinical decision process; from diagnosis to therapeutic decision is an immediate step.[10]

Despite its relative marginalization, prognostication is still very much alive today. Commercial sales of genetic testing kits are booming. In 2017, more than 7.8 million DNA test have been sold online in the United States.[11] The company 23andMe, founded in 2006, now has more than 5 million customers in forty-eight countries. This direct-to-consumer DNA testing is promoted as democratizing health. It provides people control over their destinies and rejects the traditional paternalism of experts. No need to see a doctor or to go to a laboratory. Within a few weeks, you can know your genetic profile and learn what future health risks you may have.

The attitude in Europe around this kind of testing is skeptical. Critics argue that it does not offer any potential cures and creates unnecessary fears. Driven by a commercial logic, companies want to sell as many tests as possible, but the belief in testing is in fact based on genetic superstition (i.e., attributing excessive influence to the genetic constitution of people). More controversial is the testing of children, and there is a lively debate in the United States over whether every baby should be genetically tested. Parents may want to know not only the health future of their offspring but also whether the child will be obese or athletic. Going even further, in Estonia, the government has set up a program to collect the DNA of 10 percent of the population to provide personal feedback about lifestyles.[12]

It is obvious that over the past few decades the Human Genome Project and the rise of genomics have promulgated an increasing emphasis on predicting health and diseases. New scientific areas have emerged—predictive medicine, personalized medicine, and precision medicine—all promising precise diagnoses and prognoses for individuals. The idea that genetics and genomics can offer more effective and personalized approaches to the prevention and treatment of diseases was strongly promoted decades ago, in fact. The introduction of life sciences led to individualized drug therapy. They could identify predispositions to diseases, provide a better understanding of aging, eliminate many diseases, and provide new cures. When the first draft of the human genome was launched in June 2000, Francis Collins, who has been in charge of the genome project at the National Institutes of Health, suggested in an interview that "in another 20, 25 years, we should be able to prevent or cure most cases of cancer, of diabetes, of heart diseases, of multiple sclerosis, of asthma."[13] We now know, however, that such predictions have not come true.

Mapping the genome has yielded few new cures. It has limited use in forecasting. For major diseases, the predictive capacity of personal genome sequencing is low.[14] Recent history has provided mostly promises and high expectations, but achievements have been disappointing. "The human genome is still largely a mystery," concluded one scholar in 2009.[15] How to explain, then, the continuing enthusiasm and optimism for predictive and personalized medicine? Here, the narrative of prophecy will be helpful. Prophets project a vision of the future. The idea that we will be able to predict diseases and their developments in the future, and that we can, on that basis, make decisions about our lifestyles and tailor treatment options, is an attractive vision, even if it is utopian.

Renewed interest in prognostication is furthermore stimulated by artificial intelligence. Earlier shortcomings in the ability to predict the future because of lack of data and insufficient knowledge can now be remediated. Gigantic quantities of data (Big Data) can be processed and analyzed with machine-learning methods that generate predictions about future genetic changes, for example, how tumors will evolve and spread.[16] Artificial intelligence is being rapidly introduced in clinical medicine. The supercomputer Watson, for example, evaluates cases, calculates prognoses, and recommends best treatments in oncology departments around the world.[17] It is predicted that in the near future, when everybody's genetic structure will be known and digitally accessible, robots will take over health care and provide personalized and precise care.[18]

Another reason why more and more people want to know the future has to do with rejuvenation. There is currently much excitement about the possibilities of life extension.[19] The expectation is that human aging can be slowed or even halted. The prospect is that human body parts can be replaced, not by artificial devices but by an individual's own stem cells transformed into tissues and organs. Forecasting is also significant for another new area that has emerged: regenerative medicine. This area exemplifies the connection between the power to explain and the power to change or engineer. It is important to know your future since you can design and redesign it into a better, healthier one. For example, parents can envision designer babies with the genetic makeup they prefer.

Some will argue that it is simply human to want to know the future. We cannot live without prediction, otherwise planning for tomorrow would be impossible. We are perfectly aware that predictions are not knowledge—they are inherently uncertain—but they present a vision of the possibilities to come. And we need such a vision to guide our lives, our aspirations, and our explorations.[20] What puts the issues of prediction and prognostication on the agenda of bioethical debate, then, is the ancient narrative of medical prophecy.[22] Humankind has a long tradition of medical forecasting using primitive or sophisticated mechanisms. Hippocratic physicians had to compete with oracles and diviners in predicting health and disease. Religious people relied on prophets who could foretell the future. The liver was used to read the future. The stars and planets were consulted in an elaborated art of astrology.[21] These phenomena of prophecy demonstrate that human beings have always been driven by the values of progress and perfectibility. And assuming that health and disease are determined primarily by genetics, genes become the augurs for our future.

Medical Prophecies and Prophets

Prophecy can be used in a broad sense to cover every instance of forecasting the future.[23] Traditionally, it is related to the activities of specific figures who have operated in the religious traditions of Judaism, Christianity, and Islam. In this context, prophecy has four characteristics. First, prophets have a strong imaginative faculty.[24] They have knowledge of the future, but more importantly they know the significance of this knowledge. They exhibit a form of knowledge that is creative and imaginative, but their personal experience is also significant since it has helped them to grasp the true nature of God.[25] Their predictions are not solely futuristic but also related to tradition. The eighth-century prophets in Israel, such as Amos and Isaiah, for example, are bound to the Mosaic laws.[26]

Second, prophets are intermediaries. They receive messages from the supernatural world. As inspired mediums, they convey these messages to a wider public. As intermediaries between the human and divine or supernatural world, their activities go through all the stages of ordinary thinking before becoming illuminations or revelations ready to be proclaimed. For Maimonides, a medieval Jewish physician and philosopher, this means that prophecy provides a higher degree of knowledge than philosophy.[27]

Third, prophets are visionaries, and their visions and dreams are the results of the combined actions of rational and imaginative faculties. In a religious context, prophecies are originating from divine inspiration. This allows prophets to speak with the "authority of inspiration."[28] They proclaim God's plan with the world and predict future events, aiming to edify, exhort, and console their audiences.[29] In light of the predicted future, they instigate people to take action now.

Finally, prophets promulgate discourses of doom and destruction but also of hope, promise, and restoration. The prophet Amos, for example, promised restoration: ruined cities would be rebuilt, and the land made fertile.[30] After the curses come the blessings. Prophetic literature has a vocabulary and image of hope.[31] Prophetic speech, therefore, can be negative, as in a threat, but also positive, as in a promise. It may predict an apocalyptic future in which the present world system will end but then be replaced with an ideal one, and God will inaugurate a golden age.[32]

As previously mentioned, there is a long tradition of speculation about the future in the field of medicine. While medical practitioners were powerless and ineffective for centuries, a future was imagined where disease was conquered and age unlimited. The Fountain of Youth, for example, was a legend that motivated Spanish conquerors in the 16th century, and other such medical utopias became more common as science and technology started to transform human life. The emergence of science fiction in the later part of the 19th century led to many predictions of future medical and technical progress. Worlds were imagined where technology would fundamentally transform society and even human nature. Such visions could be dystopias (e.g., Huxley's *Brave New World* and Orwell's *1984*) and there have been warnings (e.g., Mary Shelley's *Frankenstein*), but in general, science fiction is the prophecy of progress: it assumes that more technology will solve the problems of the world and that the benefits of technological advance outweigh the potential hazards and dangers.[33]

Lee Silver is a professor of molecular biology at Princeton University publishing widely in scientific fields like genetics but also in ethics, law, and religion. He can be regarded as a current medical prophet, providing a daily microdose of genius.[34] In his successful book *Remaking Eden* (1997), Silver predicts that future parents will be able to decide the genetic constitution of their children. A new discipline of "reprogenetics" will provide all possibilities for selecting and synthesizing genes so that parents can add and delete the genes they want their children to have in order to be happy, healthy, and successful. These selection processes will be determined by the market. In the long term, two species of human beings will exist: GenRich humans and Natural humans.[35]

Yuval Noah Harari is an even more well-known modern-day prophet. He is an historian at the Hebrew University of Jerusalem and bestselling author of *Sapiens* (2014) and *Homo Deus* (2016). Harari foresees that humanity's future will produce an elite category of upgraded superhumans. Science and technology will lead to species transformation. Our bodies will be fundamentally altered so that human capacities will be redefined. And as humans become more machine-like, they will become more god-like. Determined by algorithms, human beings will no longer have agency. Any belief in free-acting individuals will become fictional. The result will be a dystopian future of "unenhanced" human beings facing an upgraded elite of humans as gods.[36]

Eric Topol is a third example of a present-day prophet. His mission is to demonstrate that today's medicine is wasteful and archaic. He argues for the creative destruction of the current medical system since it does not take into account the advantages of information technology, social media, computers, apps, electronics, and robotics. Topol, a cardiologist and director of the Scripps Translational Science Institute in the United States, predicts the digitization of human beings. In the future, medical care will no longer be controlled by physicians. With the assistance of robots, patients will manage their own care and they will make decisions. Health care, therefore, will be personalized and democratized.[37]

Silver, Harari, and Topol are archetypes of a flourishing genre of scholars and authors forecasting the future by extrapolating the advances of science and technology. They profess contemporary prophecy, presenting a vision about what is going to happen with the world and with humanity. They connect the present state of science with imagination and inspiration to tell us more about the future. Unlike traditional prophets, their inspiration and revelation are not based on messages from a supernatural world. For most of them, religion is fiction that has been long surpassed by the development of science. Man himself has become God and can remake and perfect himself. Their prophecies, therefore, are secular, and their messages are derived from the natural world.

At the same time, it is not clear how they can be so sure about future developments. Are there laws of history and change or rational principles in nature that explain past events and predict imagined future? As secular visionaries, they discern "the deep currents under the surface of modern cultural and social trends."[38] They believe that the futures they predict are inescapable. The inspired message they have gives them the authority to admonish and exhort people to act now so that they can facilitate or slow down future trends. Generally, however, these prophets believe that progress is unavoidable. The superhuman upgrade will be here at some point.

Are these prophetic visions appealing? Prophecies certainly attract a lot of interest; people are curious, fascinated, astonished, and marveled . . . but are they really attracted to the futures being imagined? And why is there a trend toward the disappearance of what is typically human and the transition to something "posthuman"? Becoming a new kind of being seems only promising for a limited number of people. Most of us are predicted to end up in the underclass of unenhanced human beings. Nonetheless, the agenda of the bioethical debate is heavily influenced by these visions. While the number of publications on such topics as predictive medicine, life extension, and enhancement are rapidly growing, so do the number of publications analyzing the ethical dimensions and implications.[39]

The majority of current ethical studies focus on the moral queries and dilemmas that emerge within the framework of enhancement, predictive medicine, and life extension. They assume that the prophecies will materialize and that we must face the ethical challenges they present. That implies that the ethical discussion is already preformatted or predetermined by the future that is forecasted. It is rare that the prophetic activity itself is questioned. Predicting the future is usually based on the narratives of progress and perfectibility. It frequently assumes that human beings are essentially determined by their genetic makeup, that the social context is relatively powerless, that individual freedom is limited, and that technology is omnipotent. These underlying assumptions are, in fact, value judgments that are guiding the visions of medical prophets and shaping their conclusions and recommendations. For critical bioethical debate, it will be necessary to examine the value assumptions of medical prophecies.

Medical Utopia

There once was a young mother in the times of Buddha whose baby succumbed to a lethal disease. Exhausted by grief, the mother wandered through the streets and appealed to other people to help her. Finally, she meets Buddha himself. She puts down the corpse of her son at his feet and begs him to revive her child. Buddha whispers that there is only one remedy against her misery: she must go into town and return with a mustard seed from a family home in which nobody has ever died. The outcome of Buddha's request is clear, according to French philosopher Luc Ferry. Obtaining the seed is impossible since every family has experienced death.[40]

This type of story about death, dying, and grief is paramount in all major religions and spiritual traditions. It indicates that death and disease are intrinsic components of human life. No matter how much we strive to live healthily, disease will always overcome us. Philosophical schools, such as Stoicism, have underlined the transient and finite nature of the world. Philosophy is an exercise in how to die. Wisdom is the acceptance of finiteness. Human existence is characterized by the tension between love and death, or friendship and living together on the one hand and separation, loss, and extinction on the other hand. Those who ignore this fundamental tension are really exposing themselves to existential suffering.

However, in the new millennium, this old wisdom seems hopelessly antiquated. Modern human beings are destined to progress, to design the future, and to overcome our limitations. We are finally recognized, at least in moral theory, as autonomous individuals who attribute meaning to our existence and choose the projects we value. Science and technology are the most important instruments to making life more valuable; they provide individuals the means to liberate themselves from the tragic dimensions of nature, and particularly the absurdities of disease, aging, and death. In this perspective, it is obvious that medicine has emerged as the most influential and socially relevant science. The expectation is that medical science and technology will provide solutions for all kinds of daily problems found in human existence.

The development of modern medicine has been inspired by the long-held ideal that human ailments could be conquered by the progress of science. The philosopher René Descartes (1596–1650) argued that medicine could flourish if it was based on contemporary science, especially physics and chemistry.[41] He described all sciences as represented in the image of a tree: the roots are philosophy; the trunk, the natural sciences; and the branches, all other sciences. Ultimately, the tree will bear fruit, particularly in the domain of medicine. Although nothing in medicine at that time suggested this potential, Descartes hypothesized that human ailments could eventually be eradicated. He believed them to be unnecessary components of human existence as long as we attempt to improve medicine.

The belief in medical progress developed beyond Descartes's hypothetical approach during the Enlightenment. Self-confident assertions and predictions took the place of uncertain guesses and hypotheses. For example, Marquis de Condorcet (1743–1794), one of the ideologues of the French Revolution, expounded a far-reaching optimism. The means to improve human beings would not only become available through the progress of science, but it would also be obvious that there is a wide range of possibilities for improvement as long as human beings themselves are concerned. Condorcet pointed out that "the perfectibility of the human being is in reality indefinite."[42] Contrary to earlier utopian thinking, the new sciences will not merely help us overcome and eradicate the age-old ailments and suffering of humans, but they will enable the human person to perfect himself. Ultimately, medicine will be transformed from curing and preventing illness into promoting well-being and enhancing human existence.

Utopia Realized?

What Descartes and Condorcet regarded as an ideal that could be realized—if we focus on the further development of science—was for earlier generations nothing more than a dream, a phantasmagoric longing for Paradise lost.

This dream has been visualized over and over again in art and literature. A famous example is *Der Jungbrunnen*, a painting by 16th century artist Lucas Cranach of an ancient spa resort. From the left side, old, diseased, and crippled persons step into the thermal waters; being fully immersed, they leave the bath at the other side, completely rejuvenated, healed, vigorous, and able-bodied. The idea that humans will be able to improve nature, make artificial life, and fabricate human beings is as old as humankind.[43]

In our own times, we have a different situation. What started as a dream has become a realistic ideal and now grown into a situation where many people seem to believe that utopia has almost been realized. The general idea is that we are at the brink of a crucial transformation of our potentials. We are witnessing the emergence of a new medicine, one that is no longer focused on the treatment of the consequences of diseases but instead on the submicroscopic causes of diseases. These causes are considered to be the targets of elimination through therapeutic interventions that interrupt the processes leading from a gene to a symptom. The new interventions will furthermore use preventive strategies to identify the predisposition to develop diseases as well as the tests necessary to detect this predisposition at an early stage. The changes in medical science and practice under way at the moment prognosticate the enormous potential of molecular interventions, which may soon be attainable. The ideals that were imagined by Descartes and Condorcet appear to be approaching the stage of implementation.

The zeitgeist is demonstrated not only in the popular media but also in scholarly publications. Particularly during the time of the Human Genome Project (1990–2003) there was a multitude of very optimistic and hopeful messages promising that a revolutionary new medicine is around the corner.[44] The emergence of a totally new range of possibilities to enhance and perfect human existence is fascinating to the public imagination. The ancient ideal that we can finally be free from the limitations of the human body because diseases, at least in principle, can be eradicated, and age almost infinitely extended, is not far from being substantiated. The spirit of that time is well expressed by ethicist John Harris in his book *Wonderwoman and Superman*. He writes, "It would not be an exaggeration to say that humanity now stands at a crossroads. For the first time we can literally start to shape not only our own destiny in terms of what sort of world we wish to create and inhabit, but in terms of what we ourselves wish to be like. We can now, literally, change the nature of human beings."[45] In this optimistic, utopian discourse, two visions of the future are at work. One is a vision of health: life without disease. The other is a vision of immortality, or at least longevity: life without death.[46] Both visions are not new, but they are resuscitated within the specific social, cultural, and political context created by the Human Genome Project.

Geneticization

The moral debate concerning genetic technologies is usually focused on individual perspectives and practical applications. New discoveries and research findings in the biomolecular life sciences are rapidly presented and discussed in the media, the debate on the ethical implications generally tends to highlight the impact of genetics at the level of individuals, and the principle of respect for individual autonomy is often the starting point for considering these implications. Emphasis is on the proper management of information by individual citizens, informed consent, privacy regulations, the right to know and the right not to know, and the question whether to test.

The moral debate is also characterized by the immediate interest in translating the public fascination with new data, devices, and discoveries into practical applications. Often, very prematurely, benefits for medical advancement are pointed out. The purpose of the ethical debate is to develop guidelines and standards for the appropriate use of gene technology, assuming that the projected benefits will materialize. Various moral principles, rights, and rules have been developed to delineate what is regarded as appropriate use. Ethics literature aims, therefore, at reflecting upon and evaluating the rapid evolution of genetics. Ethicists try to analyze the potential effects of genetic information and determine the conditions for justified applications of gene technology. The basic assumption is that genetics provides new opportunities for individual consumers, who must then decide whether and how to use them.

Ethical analysis generally does not pay much attention to the social and cultural contexts within which genetic knowledge is promulgated, nor the social processes involved in the dissemination of genetic technologies. The ever-growing impact of genetic technologies on society as a whole and their diversified cultural manifestations do not often lead, for example, to a critical attitude toward moral statements that individual persons are free to choose among available genetic options exclusively following their desires and preferences, or that the development of unwanted scenarios involving others (for example, genetic discrimination) is unlikely. But social and economic pressures are actively influencing the use of genetic knowledge through, for example, the application of genetic testing in prenatal care and in various insurance arrangements, as well as indirectly through new imagery and concepts of health, disease, disorder, and abnormality.[47]

Since the 1980s, genetic explanations for all aspects of health have become more attractive. Many human behaviors, too, have been associated and "explained" by the existence of a specific gene. The well-known discussion on the "warrior gene" attributed to the Maori in New Zealand is no exception. The occurrence of the gene was used in a murder case in 2010 to argue for diminished responsibility, and it influenced the jury's decision ("not murder").[48] And what about the "gambling gene" (1996) and "adultery gene" (2010)? There is a search even for a genetic origin of human rights.[49] Through all this, popular news media generate imaginations about who we really are, our ancestry and destiny, but they also disseminate biofears and biofantasies.[50]

From an analysis of film, television, news reports, comic books, ads, and cartoons, it is shown that the gene is a very powerful image in popular culture. It is considered not only a unit of heredity but also a cultural icon, an entity crucial for understanding human identity, everyday behavior, interpersonal relations, and social problems. The growing impact of genetic imagery in this way has been related to "genetic essentialism," the belief that human beings in all their complexity are products of a molecular blueprint.[51] Moreover, the concept that genetics can be used to create postmodern human beings also has repercussions for health care and medicine, as well as science in general. Molecular biochemistry now has stronger claims to be the fundamental science in medicine and the life sciences than ever before.

The general conviction that future genetic possibilities will drastically change medical diagnosis, treatment, and prevention is increasingly being discussed, but the needs for critical moral discourse are often not clarified. In order to identify and analyze the various cultural processes related to the biomolecular life sciences, the concept of geneticization has been introduced to the scholarly debate.

"Geneticization" was used for the first time by Abby Lippman, professor at McGill University in Canada.[52] She defined it as "the ongoing process by which priority is given to differences between individuals based on their DNA codes, with most disorders, behaviors and physiological variations . . . structured as, at least in part, hereditary."[53] The concept does not simply articulate that genetics in current society is a hype or leading to exaggerated expectations. It points to a more fundamental question: How are science in general and genetics in particular influencing modern society and culture? Geneticization provides a way of thinking that can guide our attention toward these more fundamental processes. The concept aims to identify and analyze the interlocking and often unexplicit mechanisms of interaction between medicine, genetics, society, and culture. Not only human beings themselves but also health and disease, human behavior, and social interactions are viewed through the prism of biomolecular technology.[54]

Western culture currently is, supposedly, deeply involved in processes of geneticization. These processes imply a redefinition of individuals in terms of deoxyribonucleic acid (DNA) codes, a new language to describe and interpret human life and behavior in a genomic vocabulary of codes, blueprints, traits, dispositions, genetic mapping and a gene-technological approach to disease, health, and the body. The concept tries to make explicit what is often not well articulated. A growing number of studies these days explore our fascination with genetic technology. Genetics is regarded as more than a science, rather as a way of thinking, an ideology: "Whatever the question is, genetics is the answer."[55]

Lessons from Medicalization

The medicalization debate that occurred in the 1970s provides an analogy with the concept of geneticization. Lessons from that debate can be used to develop the analysis of the sociocultural impact of gene technology. Medicalization and geneticization are, in fact, examples of a more general, encompassing process: the philosophy of normalization advanced by Michel Foucault. Since the early 19th century, medicine created social order by polarizing the distinction between illness and health.[56] The value of medical knowledge in society was elevated when, in fact, it was more and more used to transform human beings into subjects. Biopolitics extended medical power. Within society, modes of power were developing that allowed some forms of individualization while, at the same time, denying other forms; the same movement that empowered individuals and liberated them from some forms of oppression resulted in other forms of domination. This is also the Janus face of medicalization: at the same time as it provides benefits, it also subjects patients to forms of discipline that regulate human behavior.

However, today's medicine is not simply "medicalizing." Instead of using domination and control, the field of medical power has been reformulated.[57] The locus of medical power is no longer the individual physician but instead is located in large, pervasive structures encompassing physician and patient alike. Medical power also is no longer exclusionary but has become inclusive; challenges from alternative health care, holism, bioethics, and the hospice movement are rapidly being incorporated into orthodox medical practice. The new field of medical power, therefore, is not so much dependent on domination and control as it is on monitoring and surveillance. Technologies of monitoring and surveillance incite specific discourse. They make the intimacies of the patient visible; they leave visible records. Everything must be noted, recorded, and subjected to analysis. And because of the dominant role of biosciences in present-day society, there is now new interest in what is called "biomedicalization."[58] Geneticization, therefore, can be regarded as the manifestation of medicalization today.

Medicalization is associated with several consequences. It is a mechanism of social control through the expansion of professional power over wider spheres of life, it locates the source of trouble in the individual body, it implies a specific allocation of responsibility and blame, and it produces dependency on professional and technological interventions.[59] Medicalization can also occur on different levels: conceptually, when medical vocabulary is used to define a problem; institutionally, when medical professionals confer legitimacy upon a problem; and relationally, when the actual diagnosis and treatment of a problem takes place between a doctor and her patient. These insights provide clues for studying the impact of geneticization. By analogy, the concept of geneticization can be studied on various levels as well: conceptually, when a genetic terminology is used to define problems; institutionally, when specific expertise is required to deal with problems; culturally, when genetic knowledge and technology lead to changing individual and social attitudes toward reproduction, health care, prevention and control of disease; and philosophically, when genetic imagery produces particular views on human identity, interpersonal relationships, and individual responsibility. In contrast to medicalization, the concept of geneticization seems to be broader because it also refers to developments and differences in the interaction between genetics and medicine. There is, for example, not simply an expansion of concepts of health and disease into everyday life, but a fundamental transformation of the concepts themselves. In medicine, there is also a tendency to use a genetic model of disease explanation as well as a growing influence of genetic technologies on medical practice.[60]

Reductionism Revived

Identifying and analyzing such fundamental processes in relation to the development of genetics will be complicated. The analysis should focus at least on two levels. The first is the surface level of the individual and doctor-patient interactions where genetic knowledge is offered as new possibilities for intervention, treatment, and prevention among which people can choose. This is usually the level at which mainstream bioethics is operating. The second level is the deeper, more hidden level of social and cultural expectations that guides the particular decisions in one or more specific directions depending on the discourse and images that have been created through genetic knowledge and technology. The understanding of processes of geneticalization will help patients to obtain a more nuanced and reflective comprehension of possibilities and impossibilities. This is also what Lippman had in mind when coining the concept. Asked in an interview about her hopes and fears for women facing the new genetics in the next decades, she replied, "I guess I hope people get a realistic understanding of the tremendous limits of genetics in explaining why people are the way they are so that they can assess this stuff really critically."[61]

Such critical examination is necessary since genetics has promoted a specific way of understanding that already prestructures or preformats how we are going to approach and interpret the advances of the new life sciences. This way of understanding is usually discussed as genetic reductionism. It is a new and more powerful edition of the mechanical model that has dominated medicine since its emergence as a natural science. As a scientific research strategy, reductionism assumes that complex phenomena can be explained in terms of simpler, more fundamental ones. In genetics, major aspects of life can be reduced to one chemical substance, DNA. The gene is the unit of heredity but also the basis of life.[62] Organisms are determined by their genes. However, this successful methodological approach has promoted a philosophical view that human existence and human beings are determined by genes. Phenotypes are regarded as mere survival machines for genes. This is the point of the geneticization thesis. The reductive notion of the gene is producing significant changes in human identity, redefining individuals in terms of DNA codes.[63] Postmodern society is using the language of the new life sciences to communicate about human life. Disease, health, and the body are explained in terms of molecular biology. It seems that for present-day human beings, the cultural meaning of DNA is remarkably similar to that of the immortal soul of Christian theology.[64]

The effect of reductionism is also noticeable in the tendency to reduce health and disease to genetic phenomena. The assumption is that genes basically explain the origin of health and disease, and there is only a limited role for psychological, social, and environmental factors. Changing this conceptual understanding of disease has significant implications for the process of diagnostics.[65] Genetic reductionism promotes a particular understanding of disease, based on an older ontological concept of disease. The disease is a reality within the body; it can be detected before it produces any symptoms. Diseases are entities in themselves, entities that exist in our body, like the cancer in our tissues or pathological body parts in general. When we remove them, the diseases are cured. Now we know that the more fundamental explanation is genetic; genes are the ultimate causal factors of disease. The phenotype can be reduced to the genotype. Therefore, we can speak of "genetic disease."

The same transformation takes place for the concept of health.[66] With the possibility of detecting mutations in the genome that might in the future produce serious diseases, one is only healthy if the genome is healthy. As soon as a mutation is discovered, one is no longer healthy but unhealthy. Even if one does not have a disease or any subjective complaint. With expanding genetic technologies, the number of healthy people will decline and most will be unhealthy. Even if we don't want to use the label "genetic disease," health will become exceptional and rare with the progress of genetics.[67]

The irony is that the advances of genetics and genomics have undermined the concept of the gene.[68] With the discovery of repetitive DNA with no coding function, as well as overlapping genes, the concept has become problematic.[69] An important lesson from the Human Genome Project (HGP), in fact, is the significance of environmental factors and the uncertainty of relating genotypic differences to phenotypic states. The image of the gene as determinant of traits has become much more complicated.[70] Living systems are characterized by complexity, development, and integrity. They cannot be explained in terms of the structure and function of the units that compose them.[71] The HGP findings that the number of human genes is relatively small and that humans are not so genetically different from animals (only 1 percent difference in DNA between humans and chimpanzees) show the limitations of genetic determinism.[72]

These new scientific insights have made reductionism untenable.[73] The suggestion that the new genetics can explain diseases but also human traits, such as violence and creativity, by locating them within individuals is false. It wrongly opposes nature versus nurture. The call to redefine medicine as "network medicine" goes in the same critical direction. The patient cannot be reduced to a collection of molecular sequences, disease cannot be separated from the sick person, and human suffering can only be understood as

the result of complex interactions between genomic, environmental, and social determinants.[74] This is, of course, nothing new. It reiterates the idea that
the patient should be regarded as a whole person and that patient care requires a biopsychosocial model. The problem is that in the public debate, the
popular image of the gene as determinative for future health and disease is
thriving.[75] One challenge for bioethical discourse is to reiterate that the concern with the individual genetic makeup is an oversimplification.

Critical analysis of genetic reductionism is important for bioethics. It scrutinizes how the agenda of the bioethical debate is often constituted. But it
also directs attention to the underlying value context of reductionism, promoting a specific perspective on human life, health and disease as determined
primarily by biology, and genetics in particular. This perspective narrows the
bioethical debate. The metaphor of prophecy helps to broaden the bioethical
imagination.

Widening the Bioethical Debate

The implications of geneticization on bioethics are now increasingly studied.[76]
Using the concept requires a critical analysis of the current ethical discourse
in genetics. In mainstream bioethics, the principle of respect for autonomy
plays a crucial role. Patients are not passive, docile bodies under the control
of medical power but instead articulate consumers and autonomous decision
makers. Hence, the moral requirements of nondirectiveness and respect for
individual choice in present-day clinical genetics. However, this emphasis
on autonomy tends to overlook that social arrangements frequently exist that
are predetermining the range of choices available to individuals in a particular
society. An example is the case of screening and counseling programs for beta-
thalassemia in Cyprus.[77] In this country, individuals can only marry if they have
a certificate showing their participation in genetic screening. Although a priest
performing a marriage is not allowed to inquire about the outcome of the
test, the question is what this knowledge will imply for the behavior of tested
young adults themselves. Autonomous choices generally take place within specific social and cultural settings and are therefore determined by the constraints
of this context. Patients' choices regarding the use of genetics are predetermined by the constraints of the health care system but more importantly by
the geneticized environment. This context is "consonant with the contemporary neoliberal emphases on individual responsibility, self-governance and a
prudential approach to controlling and transforming one's future."[78]

Assumptions of Prophetic Narratives

The current interest in genetic testing and predictive medicine is curious given the past decades of "gene talk." The Human Genome Project initiated a time of great expectations; it lauded the prospect that many quandaries of human existence could be solved. The powerful imagery of the gene and the intense faith in genetics as more than a biological science encouraged the construction of prophetic narratives. Genes are regarded as "internal oracle."[79] However, the promises of the new genetics have not materialized. There have been many stories and scandals over the years but not really any breakthroughs with significant consequences for medical practice and patient care.[80] A Nobel Prize sperm bank, set up in 1980 and closed in 1999; the cloning of Dolly the sheep in 1996; the death of Jesse Gelsinger in a clinical trial for gene therapy in 1999; celebrities announcing that they use surrogate mothers to have children; the story of DeCode Genetics in Iceland; legal interpretations of parenthood as a genetic rather than social relationship; the discovery of the CRISPR/Cas9 technology that allows genome editing; the introduction of noninvasive prenatal testing in 2011; a three-parent baby born in 2016 after mitochondrial replacement therapy; the creation of a human-pig hybrid in 2017; the announcement in 2018 that a Chinese scientist has successfully edited the genes of twin babies—these stories all dominated the news for some time. They raised ethical questions. But, in fact, they demonstrated that the popular imagination continues to be in the grip of gene talk.

In the same spirit of optimism, President Obama launched the Precision Medicine Initiative (PMI) in 2015. The promise is that future medicine will be unique and personalized. Health care providers can "customize" treatment and prevention strategies to the unique characteristics of people. The mission statement of the initiative is "To enable a new era of medicine through research, technology, and policies that empower patients, researchers, and providers to work together toward development of individualized care."[81] While the PMI website presents hallelujah stories that precision medicine is already saving lives, what is missing is information and research data concerning the interconnections between human biology, behavior, genetics, and environment. Since its inception, PMI has also been transformed into a research project (called All of Us) to collect genetic and health data from 1 million volunteers by 2022.

Why is the same prophetic belief in the promises of genetics continuing in one form or another? One reason is elaborated by Evelyn Fox Keller: gene talk is a powerful tool of persuasion.[82] It offers a relatively simple model of explanation. It suggests interventions to eradicate disease aimed at perfectibility and promises that human behavior can be predicted and controlled. For example, if alcoholism or addiction can be explained in genetic terms, we can identify who is at risk, apply preventive strategies, and perhaps develop therapies. Another reason is that emphasis on genetics allows us to abstract from the context. This is the ideological utility of genetic reductionism—problems are located within individuals[83]—and this emphasis is associated with a specific allocation of responsibility. The individual is to blame if he does not take appropriate steps to clarify his destiny, engage in testing, or change his lifestyle. Genetic explanations and interventions are therefore fitting perfectly well into the neoliberal ideology of individualism. They promise certainty, order, predictability, and control.

A third reason, mentioned earlier, is that gene talk exemplifies the mechanical model that dominates current medicine. Utopian expectations are culminating in today's acclamations of stem cell technology. We not only have the mechanistic vocabulary of "replacing failing parts," "neo-organs," "off-the-shelf organs," and "prefabricated spare parts," suggesting that all components of the body can be repaired in the body shop, but there is also the prediction that we can in principle overcome any obstacle presented by the human body and repair and regenerate components like we repair our cars and other equipment. From an ethical point of view, it is obvious that such promises and expectations are wrong, or at least one-sided. However, in bioethics they are often taken for granted.

Bioethical discourse seems to be seduced by the optimistic perspectives of science and technology and the utopian ideals presented. It does not deconstruct and criticize the prophecies. This is surprising since one of the sources for the emergence of bioethics has been the dissatisfaction with reductionism in health care. Contemporary medicine is focused on diseases, not on the illness and suffering of persons. It postulates that there is a mechanical defect that should be diagnosed, but it is often not clear how that impacts the person's well-being. This focus on diagnosis and disease ignores the subjective symptoms of the person. Knowledge of diseases is not the same as understanding the patient's suffering.[84] Reductionism, therefore, leads to a distorted view of human beings. Bioethics is supposed to bring in a broader perspective, but apparently it only reiterates the reductive views of the new genetics.

Critical analyses of prophetic narratives demonstrate that they are based on specific value assumptions that should not be taken for granted. Medical prophets are confident that the future is predictable and that they can deal with uncertainty. They assume that progress is inevitable while they disregard the context in which it happens. Furthermore, prophecies are based on an image of human beings as rational decision makers who want to know their future and take personal responsibility for their health. Finally, prophecies are sure that they can dismiss the legacy of eugenics that is darkening the history of genetics. All of these assumptions are questionable, which we'll detail here, but this may open an avenue for a different ethical assessment of present-day prophecies.

Predictability

In 1968, leading scientist Gunter Stent predicted in an article in *Science* the decline of molecular biology as an innovative science. The major discoveries have been made, he claimed. It was a "workaday field" in which one only needed to iron out the details.[85] Another prediction around the same time, wrongly attributed to the US Surgeon General, was that the book on infectious diseases could be closed because the war on microbes had been won.[86] Both anecdotes are notorious examples of how erroneous predictions can be.

The difficulty of predicting is well known in the practice of clinical medicine. Estimations of life expectancy in terminally ill patients are only accurate in 20 percent of cases. The majority of the predictions of survival are overly optimistic. In general, physicians are reticent to provide prognostic information, finding it too stressful or difficult to ascertain. They argue that patients expect too much certainty—something that cannot be provided.[87]

Nonetheless, despite inaccuracy, error, and bias, Nicholas Christakis argues that prognostication is desirable and important, especially in emergency and end-of-life care. His argument is a moral one. Patients need to know what choices are available and whether treatments or interventions have become futile so that they can prepare for death. The traditional assumption that the patient is a passive sufferer, not an active agent who will modify and interfere with his fate, can no longer be maintained.[88] Christakis articulates that against this background, doctors are in the role of prophets, whether they like it or not. They do not simply convey information but must deal with the emotions, wishes, hopes, and changing views of patients.[89] They help patients to understand the meaning of their condition; they assist in creating a future. Their predictions affect current behavior, and they also influence future outcomes, such that prophecies become self-fulfilling. Making a prediction changes behavior and attitudes and promotes a predicted outcome.[90]

Genetic testing technology is another reason why prognostics are important these days. Genes provide information that is inherently prophetic, but here the same problem arises: predictions are not reliable and often inaccurate. Direct-to-consumer (DTC) genetic tests, for example, produce 40 percent false positive results (i.e., they suggest that the individual is at risk of developing a condition while he or she is not). Often marketed online, these test results are deceptive and misleading.[91] DTC companies provide inaccurate information since they only predict disease risks on the basis of genetic factors and ignore the significant influence of environment and lifestyle. The usefulness of predictive information is therefore doubtful. Consumers do not know how to interpret the test results, they can become confused and anxious, and their uncertainty may lead to unnecessary medical visits and follow-up testing.

In contrast with the context of chronic and terminal illnesses, genetic testing mostly concerns healthy persons. Genetics thereby transforms the concept of health. The tests reveal that people are, in fact, presymptomatic carriers: there is a disease even though there is no subjective experience of it yet. This is an important difference with the previous context articulated by Christakis. Confronted with imminent death, prognostics will help patients to focus their remaining existence on the values they cherish. Genetic predictions, on the other hand, make healthy people into "pre-patients." It puts a burden on their future existence. It is not obvious that this is enhancing or restricting their personal autonomy. Is knowing your destiny liberating or oppressing? Is a prognosis empowering or not? Such questions are even more compelling since often no interventions are available to cure the detected condition or delay its onset. The new genetics does not bring cures or new medication, only prediction.

In 2010, more than two thousand genetic tests were available through clinical laboratories.[92] While the current surge of genetic testing, especially DTC testing, is severely criticized and calls for regulation are repeatedly issued, the general conclusion is not that predictions are impossible. As before, promises are projected into the future. It is expected that the accuracy, predictive value, and benefits of tests will greatly improve in the years to come. This prophetic ethos is clearly visible in the subsequent stages of predictive medicine: first personalized medicine, and now precision medicine.

Personalized medicine promises that medical treatments can be tailored to the specific characteristics of each patient. And not only will it be predictive but also personalized and participatory. It will empower individuals to manage their own health and illness, which is exactly the utopian future forecasted by medical prophets like Eric Topol. Who can be opposed to this? Every patient will receive the right drug at the right dose at the right time.[93]

If only that were true. The term "personalized" is a misnomer.[94] It is deceptive because the focus will not really be on the individual but on the average person.[95] Patients will be categorized in classes of genetic risk or in various subpopulations so that "stratified medicine" will be a more appropriate name.[96] Advocated over the past two decades, personalized medicine has not fulfilled expectations. Like earlier efforts to sell genetics and genomics, the real benefits still have yet to come. So far, personalized medicine has only limited applications in regular health services.[97]

Even more astonishing has been the launch of precision medicine, more than a decade after personalized medicine. Again, it forecasts future miracles. Not only are the hype and promises projected into the future, but it also repeats the dominating theory of reductionism.[98] The basic assumption is that genetics is the underlying factor in most health conditions. Disease is a genetic concept, and illness is not a subjective experience but the result of genetic makeup. The focus is on biology rather than on behavior, lifestyle, environment, or context. Even if the name has now changed and "personalization" is deemphasized, predictability continues to be founded on genetic reductionism. The very idea of the term "precision" suggests that there are facts to be discovered, "that there is a truth out there."[99] It furthermore revives the idea of a magic bullet: an effective drug that can be specifically targeted for one purpose. The genetic conception of the self is an unquestioned value. From these considerations it can be concluded that precision medicine is not politically or ethically neutral. It is driven by specific values that are at the core of medical prophecies.[100]

What can be learned from this repetition of history is that the possibility of prediction is not seriously questioned. Forecasting is possible because of a reductionist view of human beings. Humans are regarded as fundamentally determined by their genetics. This produces the idea that genes are augurs for our future; they can predict our health. This idea is also the uncritical assumption of many medical prophecies. However, predictability is not only questionable at the individual level, as argued so far. It is likewise problematic at a more general level.

Many prophecies predict the future of science, society, and culture. Philosophers have sometimes assumed that there are stages in human history and development. In the era of Enlightenment, humanistic scholar Giambattista Vico argued that human society was evolving through divine, heroic, and human stages. August Comte, one of the founders of sociology, distinguished three stages in the development of western civilization: the theological stage (marked by fictions and faith), the metaphysical stage (marked by reason and the abstract), and the positivist stage (marked by scientific knowledge and reasoning as decisive). In his sweeping reconstruction of human history, Harari distinguishes between various stages as well: the cognitive revolution, the agricultural revolution, and the scientific revolution. These stages will morph into a future stage where artificial intelligence makes humans superfluous and *Homo sapiens* will disappear.[101]

These efforts of prediction in the human and social sciences have been criticized by philosopher Karl Popper. The idea that the future course of history can be predicted is logically impossible because the growth of human knowledge is influencing the course of history. Human knowledge is growing, but "we cannot anticipate today what we shall know only tomorrow."[102] It follows that we cannot predict the future course of human history. There is no scientific predictor and no scientific method to predict the future results of science.

While natural sciences allow predictions, and prediction is regarded as one of the tasks of science—for example, in astronomy there is the possibility of long-term forecasts—the question is why this is not the case in the social sciences and humanities. If the methods of these sciences can be modeled on the natural sciences, perhaps the future of the human species can be foretold. Genetic reductionism, applying the methods of physics and chemistry in the area of medicine, precisely opens up the possibility of prediction. But the human being is not a mechanism. Perhaps it can be prognosticated what will happen in the mechanism, but this is not the same as predicting what human beings will do, how their existence will evolve, and what their future will look like. In order to make such kind of predictions, as is the aim of medical prophecies, one has to assume that there are historical laws or trends. This is what Popper calls the idea of historicism. Cultural and social phenomena are determined by history. There are certain laws and trends that underlie the evolution of history, and the nonnatural sciences intend to discover these laws and trends so that they can make predictions about how society and culture will develop.[103] This naturalistic approach usually implies that society does not significantly develop or change but instead follows a predetermined path through various stages.[104]

Popper distinguishes two different kind of predictions.[105] Both reveal the future of the human species but in different ways. One is historical prophecy. It forecasts that an event is coming but we can do nothing to prevent it. It is a warning, and it allows us to prepare for its impact. The other is technological prediction. We can do something if we want to achieve some result, for example, ameliorating the impact of the forecasted event or preventing major damage. This kind of prediction is the basis of engineering. It is fundamentally practical and typical for the natural and experimental sciences.

In short, predictions can have two aims: prophesying and engineering. For Popper, furthermore, there are two forms of engineering: piecemeal versus utopian.[106] Just as the engineer is designing machines, piecemeal social engineers design social institutions; they do not redesign society as a whole. Utopian or holistic social engineering aims at remodeling society in and of itself.

When reductionism is rejected, profound differences are articulated between natural sciences and human and social sciences. According to this antinaturalistic doctrine, social prediction is difficult, if not impossible. One main reason is the connection between prediction and predicted event. The act of predicting may lead to the occurrence or prevention of predicted events. This is the self-fulfilling prophecy that may occur in health care. Prognosticating imminent death may accelerate dying. Predicting genetic susceptibility to disease will make you feel ill and behave like a patient.

Medical prophets assume the predictability of the human future. It is not very clear why they have this confidence. They present themselves as expert mediums, and it seems as if nature or the state of the art in science and technology has revealed itself and given them inspiration and privileged access to knowledge. Lee Silver denies that he is imagining a utopian future; he claims he only explains things as they are. He simply presents the realities of science and technology. Yet he labels his book as "a book about imagined futures."[107]

Eric Topol presents himself as an involved and up-to-date expert in communication and information technology. He has a message to bring that will fundamentally change future health care. He envisions a better future since it will empower patients to facilitate personalized care and digitally define the essential characteristics of each individual. The only obstacle is the conservatism of the medical professional. But will the tsunami of data and information be beneficial? Digitizing humans will be the "ultimate life changer," but will it be progress?[108] Topol does not doubt that resistance is futile. There are certain laws of innovation and diffusion of new technologies. Even powerful professions will not be able to halt their dissemination. The future belongs to the emancipated consumer, he says.[109]

The present state of science and technology is also the platform from which Silver launches his prophecies, but he is in a double bind. He admits that it is impossible to predict the future with certainty. At the same time, he reiterates that "the use of reprogenetic technologies is inevitable."[110] Science fiction will be turned into reality. Why? It gives power to control the destiny of our species. Silver proceeds from a pronaturalistic position: the methods of the natural sciences can be used to foretell the future of society. Humans are bodies that are regulated by the same physical and chemical laws as in the natural world; they are part of nature. There is some law of progress that explains the tendency to perfect human nature and to enhance the conditions of life. As a prophet, Silver does not really predict completely new and unimaginable futures. He extrapolates from what is available now and forecasts expansion and diffusion, not elimination and retraction. His mission is to make the general public aware of the "incredible power" that is already here.[111]

Like Silver, Harari emphasizes that he is not a prophet. There are no determinate laws in history, but human history does have a pattern. Obviously, he is in the position to recognize this pattern so that we can be sure about "the general direction of history."[112] He admits that he is making predictions but, following Silver, this is not prophecy so much as merely discussing present choices.[113] The future cannot really be predicted. Nonetheless, Harari presents a grand narrative of human history and future. And again, this should be understood as "possibilities" rather than prophecies. The aim is to "broaden our horizons and make us aware of a much wider spectrum of options."[114]

Harari appeals to the imagination and encourages us to think and dream differently. This, however, is the typical mission of a prophet. It also supposes that there is some choice regarding the use of technologies. It seems that Harari supports Popper's technological predictions: through engineering we can change outcomes and envision an alternative future. At the same time, the possibilities for change in his own metanarrative are unclear. He proposes that humans are algorithms. They do not have free will. The idea that individuals are free to choose is a fiction. Human desires are based on genes, hormones, and electrochemical processes that follow the same physical and chemical laws as the rest of reality.[115] As a historical prophet, Harari is warning and admonishing us to explore our imagination.

Certainty

Lee Silver is convinced that it is with "almost certainty" that by the middle of the 21st century, virtual children and genetic choice will become feasible.[116]

Like Topol, he has no doubts that technology will determine a specific future. As a prophet, he is just diagnosing what is determinative and forecasting what are the effects of historical tendencies. There are mostly technical problems in realizing this future.

For Silver, ethical discourse is not very relevant: "Americans will not be hindered by ethical uncertainty."[117] It is in his view not possible to demarcate acceptable and unacceptable uses of technologies that combine genetics and reproduction. Ethics is often based on religious or personal concerns. Real objections against technology are in the realm of spirituality, not science.[118] All moral arguments are imaginary. The prophetic narrative apparently promotes the belief that the ends of development are not a matter of choice, moral decision, or deliberation but instead are "discovered" within the development of science and technology, assisted by the vision of the prophet.

Eric Topol also does not pay much attention to the ethical debate or the implications of digitalization for relationships of care, support, and assistance.[119] For Harari, too, it is certain that humans in the upcoming decades will strive for immortality, happiness, and divinity. There is no way to slow down technological advances.[120] Ethics, like politics, is built on subjective experiences.[121] Ethical discourse is fictional in that it belongs to the domain of religion and speaks about values rather than facts.[122]

However, fictions can be powerful. One of the continuing themes in Harari's book is this power of fictions. Nonetheless, ethical judgments for him are based on individual experiences, feelings, and sensations. Emotions and sensations are biochemical data-processing algorithms and natural sciences, therefore, can explain ethical judgments (for example, by referring to brain activities). What counts is not their validity or content but their social function and utility.

While medical prophecies do not admit uncertainty, predictions can only ever be certain in retrospect. Ahead of that, they have various degrees of probability. Doctors inherently face uncertainty since predictions move from the general/generic to the individual/specific as they attempt to foretell what the likely course of an illness will be in an individual. In clinical practice, prognostication requires the acceptance of uncertainty. This is also why physicians are hesitant to prognosticate. If they do, they are usually too optimistic.

Medical practice is characterized by several other sources of uncertainty as well: technical (due to insufficiency of adequate data), personal (due to characteristics of the physician-patient relationship), and conceptual (due to the application of general criteria to specific situations).[123] And advances in technology and an increase of information cannot eliminate the intrinsic uncertainty that exists in daily clinical care. It is a consequence of the crucial ambiguity of medical care. Uncertainty relates to the core features of health care. First, the practice of medicine will always have a subjective dimension. Second, the particular has priority over the general. The generalities of science must be applied to the illness of a specific individual.[124] Intrinsic uncertainty therefore cannot be eliminated from health care. The best practitioners can do is to deal with uncertainty, which means first acknowledging its inherent existence. It then provides room for alternative assessments. Most of all, it requires collaboration with the patient to understand what values should determine care and assistance.[125]

The inability to predict the future with certainty furthermore applies to genetic testing. The purpose of genetic screening is not an assessment of the present state of health but a probability statement about future risks. Prediction is the primary purpose.[126] However, most diseases are multifactorial. They result from the interactions of genetic and nongenetic factors. Tests indicate a risk of disease, not a certainty. For example, whole genome analysis reveals much information, but the significance of that information is uncertain. It is no problem to generate information. The question is whether it is helpful. Does testing actually lead to improved health outcomes? Probabilistic predictions also change the issue of responsibility. By emphasizing individual choices, the availability of testing individualizes prevention and decreases collective responsibility for health.[127]

The inherent uncertainties of genetic predictions seem to amplify the need for interpretation. What do test results mean for this patient and his stage in life? What will be the consequences; are further steps (tests and examinations) necessary or not? Contrary to what Topol assumes, more testing can lead to more expert assistance. The avalanche of data produced by new information technologies must be meaningful. Clinical experience will be an important tool to help sort out the differences between information, data, and knowledge.[128] How do we know that artificial intelligence will eliminate uncertainty by reducing care? Like personalized medicine, the virtual doctor will classify and standardize patients to facilitate faster and cheaper diagnosis and treatment, but is that what we want if the machine replaces our primary care physician?[129] There is also the risk of being overly optimistic. Promissory claims of test and drugs are deliberately created and promoted. Even if there is insufficient knowledge, low predictive value, and uncertainties about future applications, high expectations are promulgated by pharmaceutical and biotechnology industries. They disseminate prophecies that are performative and self-fulfilling. The future therefore is not fixed. It is the projected outcome of claims and expectations of powerful agencies.[130]

The awareness that uncertainty is intrinsic in health care and that medical prophecies therefore are not reliable is strengthened by the current movement to be more critical and skeptical toward modern medicine and its accomplishments. Critical studies emphasize that most published research findings are false. Many scientific studies make claims that are exaggerated and falsely positive.[131] In fact, an appeal is made to return to medical nihilism; that is, we should have little confidence in the effectiveness of medical interventions.[132] The methods of medical research are malleable, even in evidence-based medicine. They do not produce clear-cut results but instead are the outcome of choices at different stages: design, execution, analysis, interpretation, publication, and marketing. Most medical studies are sponsored by pharmaceutical industries so that financial incentives produce favorable results. Many studies show the link between the source of funding and the results of research. Instead of critical and independent scientists, many ghosts are at work to produce favorable outcomes and suggest effectiveness with the purpose to create new markets for medication (see chapter 3). Rigorous scientific assessment shows that many diseases are not treatable and that numerous medical interventions are ineffective. Critical analysis demonstrates that the benefits of medical interventions are systematically overestimated, and the harms systematically underestimated. The conclusion is that we should have low confidence in the effectiveness of medical interventions.

The benefits of medicine are often overestimated because of the belief in magic bullets. Penicillin and insulin are very specific and effective drugs. Precise medicine promises more of those drugs. But in fact, magic bullets are exceedingly rare in today's landscape. Nonetheless, belief in them is sustained by two factors. One is commercial: they are relatively easy to produce and distribute; they can be patented and be the source of huge profits. The other is political: they are much easier interventions than such alternatives as lifestyle changes and socioeconomic interventions. The bizarre irony is that although magic bullets are rare, they are the driving force for many grandiose projects and enormous financial investments.[133]

Inevitability

Biological determinism and reductionism imply the notion of inevitability. The course of the future is given by nature. Medical prophets are convinced that the predicted future is inevitable. For Silver, it is only a matter of when it will happen. Developments will not be controlled by governments or societies or even scientists. It will be controlled by the global marketplace. All attempts to limit technology are doomed to failure.[134] The same spirit is endorsed by John Harris, speaking about the new power to control human evolution and to manufacture new life forms: "The decision before us now is not whether or not to use this power but how and to what extent."[135]

How much choice do we have? Is this an autonomous process or an unstoppable progress? The picture is distorted. It projects enormous technological progress but in an unchanged context. New technologies will become available and be applied. The prediction assumes that the current context of a weakly regulated neoliberal economy will persist in future ages. In Silver's prophecy there is, in the year 2350, still a Health & Human Services Secretary who accepts a report from an academic commission on genetic enhancement, concluding that a new separate species of GenRich humans will have emerged.[136] In the scenario, by the year 2997, this species will have moved to other planets due to an overcrowded Earth. He writes nothing about climate change, environmental degradation, or biodiversity loss.

For Eric Topol, it is undeniable that the expanding use of information technology will produce democratization. In his books, he refers to the Arab Spring as an example of how the use of smartphones and social media has mobilized young people especially for democracy.[137] The example is awkward now that we know that countries, such as Egypt, have evolved into dictatorships where many young people are tortured and jailed. How sure can we be that information technology (IT) will empower citizens and not states or companies? A few vast corporations control IT today. They manufacture demand, and many IT practices favor the interests of corporations, innovators, and entrepreneurs.[138] The idealistic hopes concerning information technology have turned into disappointment, uncertainty, and disillusion. Technology companies are advertising miracles but creating ghosts. They search for profits and are not concerned about human rights, privacy, or the public good. Internet users learn every day that what is good for business is not good for everybody. Reliable use of cyberspace is based on trust, but confidence and trust are increasingly eroding from massive data breaches, hacking, and cybercrime. In many countries, networks for communication and knowledge are transformed into control systems.[139]

This broader context is also the reason why it is unlikely that in the future people will make major life decisions at the recommendation of Big Data algorithms, as predicted by Topol and Harari. What reason do they have to trust predictive algorithms? Companies running them have their own incentives, such as selling their products or peddling their applications. Making money, however, is commonly not what drives our major decisions in life. It is furthermore highly questionable whether algorithms can be developed for our life decisions since they are unique and local to each of us. The promise of precision medicine is simply repeating that information can be personalized while it is based on averaged and categorized data. This does not mean that algorithms are useless. They can optimize an "environment of choice," offering options and possibilities, but humans remain the agents of decision-making.[140]

Topol predicts that future doctorless patients will decide about their care.[141] They own their health care data, but he summarily discusses the current struggle about ownership of data, surveillance, and privacy. Can we assume that Big Data will be democratized given the enormous commercial and political interests they are generating? Harari carefully avoids the question of who owns the robots and artificial machines that will ultimately control humanity. His predictions for the next century first confirm Topol's dream: people will increasingly rely on Big Data algorithms to make important decisions about their lives. But Harari then adds the prediction of a gloomy next stage where these algorithms take over from humans and lead to the collapse of liberalism. Intelligent algorithms will be the monsters of tomorrow.

In such a perspective it is important who owns the robots because that determines how they will be used. The human species is not a collective agent. The idea to upgrade to godhood is usually promoted by governments and powerful corporations. It is particularly supported by the visionary billionaires of Silicon Valley. Harari is remarkable silent about the power structures manipulating and controlling human desires and determining who will belong to this new elite of upgraded superhumans. Despite multiple references to Google and Facebook, he does not elaborate but refers to the "system" that decides.[142]

Loss of Context

Medical prophecies claim that technologies will be applied regardless of the context in which they are developed and advanced. Their availability will necessarily lead to their application. This is a rather narrow view of innovation and diffusion of technologies. The fact that new technologies are on the market does not imply that they will penetrate medical practices. Health care is not like a commercial market. Using technologies is dependent on whether they are beneficial, or at least not harmful, to patients and whether they contribute to the goals of health care. Introducing new technologies often implies significant changes in the organization of practices and reimbursement of costs. Furthermore, health practitioners are not empty vessels that can easily absorb new knowledge and technology. The idea that new technologies will diffuse in medical practices since they are available and can potentially solve problems is naïve.[143] Technology transforms practices, and in order to do that they need to be examined within various contexts. These challenges explain why new ideas, such as personalized medicine or information technologies, have limited use in health services.[144]

Loss of context ignores how genetics itself is grounded in a political and economic environment.[145] Genetic reductionism serves as justification for a political ideology. Inequality and competitiveness in society are interpreted as flowing from the "scientific fact" that genes determine human nature. The cause of inequality can then be reattributed from the structure of society to the nature of individuals. The Human Genome Project and the Precision Medicine Initiative are the results of similar efforts at various levels. They are state-supported endeavours with enormous budgets available for a particular type of research, and they have dwarfed other possible priorities in health care.

Genetic knowledge and technologies flourish because they are deliberately promoted. Neoliberal policies have created a commercial context in which genetic materials are considered resources, commodities that can be patented and traded. Genetic reductionism is thus combined with economic reductionism.[146] In this perspective, genetic resources can be separated from their context in living nature and society. This separation promotes the idea of precision. For example, the claim that genetically modifying crops will make them more fertile, more nutritious, and resistant against pesticides is, again, a magic-bullet dream applied to human problems.

DTC genetic testing is another example of the impact of context. This type of testing is promoted by the present regulatory climate. In Europe there is at least some level of regulation. Countries such as France and Switzerland have legislation requiring access to genetic testing through individual medical supervision and counseling.[147] In the United States, free market ideology is restraining efforts of regulation. Prophets such as Lee Silver appreciate this context. They emphasize the importance of "personal liberty and personal fortune" as typical for American society.[148] They also assume that in the near and distant future this value context will not change.

Silver does not question the principles of market economics. Many things will change in human genetics and reproduction, but the political and economic context will not, he claims. This assumption is highly questionable today. Significant changes are, for example, happening in cyberspace. China is increasingly dominating the internet, promoting a top-down approach and a world of strictly controlled national networks. The bottom-up, private-sector-led model of internet governance is therefore weakening. In the future, the internet will be less global and less open. Its liberalizing potential prophesized by Topol is uncertain.[149]

The stability of the environment as assumed by medical prophets implies that current social inequalities will continue to exist as well. In the future there will be separate human species, like there are already major differences between social classes.[150] This view is the outcome of genetic reductionism. It suggests that genetics can focus on what is inherent and inside individuals as opposed to what is outside and environmental. Injustices are excused because they are determined by genetic makeup. As argued previously, this view is wrong. It denies that humans are actively constructing and modifying their environments. It also denies that most diseases are the result of influences from the environment and lifestyle; they are associated with age, diet, exercise, and stress, as well as socioeconomic conditions.

Loss of context furthermore implies that one can do nothing but accept the consequences of the advancement and application of new technologies. Predictive medicine is constructing a new kind of patient—somebody who has no symptoms yet but takes preemptive action as a "previvor," a preventive survivor who will submit to medical interventions in order to obtain preventive control. This expands the potential demand for drugs. It also creates a growing population at risk.[151] This tendency fits perfectly well with consumer genetics. But while the context in which this is happening is not articulated, the suggestion is that not much can be done to counter these trends. That the new genetics has many social consequences is clear, but they are often regarded as inevitable sequels that cannot be controlled.

Genetic testing is also redefining human relationships. Family members are now identified on the basis of genetic connectedness. Sperm donors are approached as fathers, and anonymity is no longer guaranteed. Genetic discrimination is a common and global phenomenon.[152] Precision medicine will lead to genetic stratification of disease categories and patient populations. This will create "molecularly unstratified patients"; that is, some patients will not be able to receive precision treatment. That means that certain forms of inequality and unequal treatment will be based on genetic differences.[153] Also, the aim of precision medicine to introduce genomic sequencing data into clinical practice will blur the line between research and care. In fact, all care practices for individuals will be turned into research domains. Genetic information will be applied before it is validated and when it is still uncertain. Risk and benefits are at that stage poorly defined.[154]

Finally, loss of context distorts priorities. If social inequality is based on genetic differences between individuals, there is little incentive to act on the wealth of evidence that health is strongly related to the inequalities in the conditions of daily life.[155] The nearly $3 billion that is spent globally each year to fund genomics has impacted other possible priorities of research and health care. It is time to reevaluate these funding priorities.[156] The United States spends more on health care than any other country, yet it has worse health than many. This is not a matter of genetics or insufficient application of technologies. It is due to ignoring the impact of the social, economic, and environmental context on human health and disease.

Desire to Know

Another argument promulgated by medical prophecies is that people want to know the future.

Since the new medicine will be able to predict the future, this information should be provided because people want to have it. An associated assumption is that this information will influence their behavior and their choices. Both assumptions are questionable.

A recent study indicates that most people do not want to know the future. Asked whether they want to know about future negative events, such as death and from what cause, or when loved ones will die, between 85 percent and 90 percent of people would not want to know. Between 40 percent and 70 percent do not want to know about positive events either. This study was done in Germany and Spain, but it indicates that the ideal of a rational decision maker who wants to foresee the future does not apply. Many people prefer deliberate ignorance. They willfully decide not to know, even if the costs of tests are low and they have no lack of personal interest. They choose to live with ambiguity and their decision cannot be dismissed as self-deception and moral weakness.[157] Therefore, the assumption that people want to know the future is contestable.

Perhaps there are differences in various parts of the world; the interest in the United States might be higher than elsewhere. Perhaps the urge to know will be higher as risks are greater and more imminent. In 1993, a genetic test for Huntington's disease became available. More than 70 percent of people at risk of this disease indicated an interest in the test. But actual testing rates are around 7 to 10 percent. This is not because the test is too expensive or inconvenient but because people conclude that it is in their interest not to know. They deliberately do not want to know their genetic fate to live a better life now. Uncertainty is regarded as a gift that allows them to live as they want.[158]

This data is remarkable since Huntington's disease is a serious inherited neurological disorder, caused by a single gene mutation. It is one of the relatively rare examples where genetic testing is perfectly predictive. From the beginning, patients were hesitant to undergo testing. What is the usefulness of a test if there is no treatment or cure? People were afraid of the consequences for health, life, insurance, and employment.[159] But more importantly, they were concerned that test outcomes would restrict their options and opportunities within their life span. They did not want to face and anticipate disability and death. They did not want to live a medicalized life.

The ideal of the rational decision maker assumes that more information is always better. This is why Topol is enthusiastic about the avalanche of data that can be obtained with the new information technologies. A rational individual wants to know information about his or her body, genetic makeup, and possible susceptibility. It ignores that knowing the future can be a source of suffering. Gathering information is only useful if it is actionable. In the case of Huntington's disease, there is no cure; nothing can prevent the onset of symptoms. In familiar breast cancer (5 to 10 percent of all cases), the option is regular screening or prophylactic breast removal. In familiar hypercholesterolemia, patients may use medication and change lifestyles. Such examples are relatively rare. In many cases it is not clear what should follow from predicted information.

Communicating disease risks from genetic testing has shown to have little effect on behavior. Tests that indicate higher susceptibility to lung cancer, for example, do not have an effect on smoking behavior. Claims that DNA testing and personalized risk prediction will motivate people to change their behaviors and lifestyles are therefore not supported by evidence.[160] This information should not be surprising. For patients with chronic illnesses, adherence to long-term therapy is only 50 percent. Even if there is a clear diagnosis and treatment is considered effective, many patients do not take medication as prescribed. In the United States, half of people diagnosed with hypertension are treated. Only 51 percent of this population adhere to the prescribed treatment. For patients treated for HIV/AIDS, adherence to antiretroviral agents varies between 37 percent and 83 percent.[161] The ideal of the rational decision maker often fails in health care practice.

Agency

The assumption that human beings are rational decision makers underlies most medical prophecies. The predicted future is the result of an accumulation of decisions by individuals who reflect on what is best for each of them and who want to maximize benefits and minimize harms. It is the neoliberal model of *Homo economicus* that is driving economic processes and market transactions. The economic human is primarily a rational, self-interested individual motivated by minimizing costs and maximizing gains for himself. The driving force of social life is rational egoism.[162] Promoting individual choice requires information. Some information is better than no information, and more information is always better. This is what prophecies do: they review the past and current state of knowledge and present information about trends so that rational persons can decide what will happen next.

The point is that this particular view of human beings is not just a model in economic discourse. It has been used in neoliberal policies and politics to reorganize societies and human relationship, especially at the global level. The market has become a metaphor for the organization of social life. It is assumed that it is the only framework within which individual liberty and freedom can flourish. Individuals will be self-actualizing when regulation and protection are removed. Constraints on free market competition should be eliminated and the role of the state restricted. This has been the purpose of national and international policies over the past few decades.

The domain of health care was not an obvious area of application for neoliberal policies, but the ideology that every area of human life should be open for market transactions so that individuals are free to choose what they want has significantly transformed health and health care. Patients are now increasingly considered to be consumers. Health has become a commodity that can be traded on the global marketplace. Predictive testing empowers people because it provides the tools to operate as an informed consumer. Personalized medicine, therefore, will realize the old ideal of Descartes: everyone can be his or her own doctor.

Predictive utopias proceed from this self-evident normative framework. The persistence of an economic view of human beings is often supported by the dominant bioethical discourse. Respect for individual autonomy as a fundamental principle of bioethics is reiterating the main assumptions of neoliberal ideology and applies them within the setting of health care. It translates the principle into specific guides for practical application: informed

consent, ownership of the body, and personal responsibility for health. Therefore, from its inception, bioethical discourse has promulgated the same image
of humans that has dominated the neoliberal context of policy-making. That
medical prophecies assume that this context will endure is therefore not a
surprise. What is amazing is that they completely ignore the dissatisfaction
with this image, as well as the possibilities for change.

One criticism is that the emphasis on individual decision-making is not
empowering at all. As discussed earlier, more information is not always better. Rather than being liberating it can restrict the possibilities to live a full
life. Addressing patients as consumers does not necessarily enhance personal
autonomy but primarily privileges consumption. Online advertising of genetic tests is first of all aimed at selling products; the goal is profit not health
promotion. The focus is on promoting the consumption of health information, health products, and health care by citizens who can afford it, have the
time, and hold the insurance. If attractive information is provided, people
will act as consumers and buy the products, especially when the products
enhance health. It would be difficult not to act. Who can ignore information
that empowers one to sustain or promote health?

Here, the idea of self-fulfilling prophecy shows its impact. Prophecies are
performative since they not only describe the expectations for the future but
also bring those futures into being. Predictive genetic information about disease susceptibilities changes the way in which individuals perceive themselves and their families. It also changes how other people see them.[163] The
fact that such information is available does not leave people much choice. They
are redefined as presymptomatic persons, carriers, and thus as a new category of patients. They redefine themselves, not as survivors of diseases, but
as previvors. The discourse of choice is not necessarily empowering. It is con-
venient for service providers and businesses regarding health care as just
another commercial context and marketing opportunity. It is one of the lessons
of medicalization that the soft normalizing power of biopolitics is disciplin-
ing citizens into responsible consumers. Rather than control and domination,
monitoring and surveillance are preferred with self-managing individuals as
the ideological cornerstone.[164]

Consumer language has been used in health care since the 1960s and 1970s.[165]
Especially in the United States, which is when health care came to be regarded
as a business. Patients were addressed as active, self-empowered health consumers with individual choice and personal responsibility as main values.

Consumerism promised to liberate patients from the paternalism of medicine. Policies suppose that citizens will be active consumers who not only search and check available information relevant for their health but who are also interested in a healthy lifestyle and demonstrate as much in the decisions they make. The assumption is that real empowerment will lead to the right choices.[166]

Another assumption is that consumerism requires the introduction of market competition and that this will constrain costs and optimize quality of care. Patients will shop for their insurance plans, doctors, hospitals, tests, and treatments, and they purchase what is best for them. It is clear, however, that in the domain of health care this consumer model is deficient. Adequate information as basis for choices and decisions is often lacking, does not exist, or is kept private. If information is available, it is less reliable and harder to evaluate than elsewhere. Doctors and patients do not regularly discuss costs.[167]

The major flaw of the consumer model, though, is that it is based on misconceptions about human nature. People do not always make decisions in search of control, shopping for the best combination of cost and quality. The continuous emphasis on people being consumers is exasperating. Citizens are concerned with issues affecting their neighborhoods, communities, or society at large. They are family members taking care for their children or disabled elders. They want to walk around and relax in a public park without being approached as customers.[168] Human beings do not want to be reduced to "aggregates of data" that are continuously monitored for detailed health information and how it can be used for profit.

Furthermore, the human context changes when people are ill, disabled, or suffering. While consumers are driven by desires and preferences, patients are in need of care, assistance, or treatment. Patients do not need a market with a diversity of providers and a range of offers from which they must choose. They do not have specific wants or preferences per se but simply care that is in their own best interest.[169] Patients, in distinction to consumers, are vulnerable. And it is precisely this aspect of vulnerability that is omitted in consumer discourse.

A second criticism is that the neoliberal model of humans has no grasp of its social consequences. In fact, these consequences cannot be envisioned since neoliberalism has a reduced notion of society; it is the collection of individual interests. Social intervention can only reduce the potential benefits of free individual decision-making. Emphasis should be on the autonomous individual rather than on the citizen who is connected to others and part of a larger community. And if the individual is the consumer in charge of the proper management of information and exploring its potential utility, then genetic reductionism is an attractive scientific approach. It underlines why lifestyle and environmental influences should not receive priority in health policies.

In reality, things are different when it comes to medical practice. Individuals cannot be isolated from their context. They are always in relationships and connected to other people, and the cumulative effect of individual decisions can lead to societal change. These effects are visible in the growing imbalance between the number of men and women in countries such as India and China. Selective testing and abortion turn individual preferences for having male children into a reality. A similar phenomenon is happening in Flanders and Iceland where the individual use of prenatal testing has almost eliminated the birth of children with Down Syndrome.[170]

A third criticism is that ethics is complex and dynamic, and existing normative frameworks can change. One of the effects of developing new technologies is that it calls into question the current preoccupation with personal autonomy and individual choice.[171] The two ideological departures from the individualistic ethos are (1) more focus on expert decision-making and away from patient empowerment, and (2) more emphasis on genomic profiling for the benefit of families, populations, and minority groups and away from individualizing treatment. This explains why it is an illusion to believe that precision medicine will empower patients. There is a deluge of data that can only lead to imprecision in the form of information overload that patients cannot interpret.

So what do we do with uninterpretable, uninformative, and clinically irrelevant information? It is suggested that we need "responsible coaching," thus, stronger professional gatekeeping.[172] Other bioethicists are bolder. They argue that choice should be combined with force. If people choose life extension, they should be obliged to have a limited number of children so that overpopulation, pollution, and resource shortages can be prevented.[173] Individual liberty is fine, but it is necessary that specific choices should be made. Choices that promote health and longevity are morally preferable and should therefore be encouraged. There have also been more arguments that enhancement should be imperative. Especially moral enhancement—that is, making people more altruistic and peaceful should be compulsory. Moral enhancement efforts should, however, also be kept secret from the population because it has a higher purpose that will be frustrated if openly applied.[174]

Shifting Responsibilities

Processes of geneticization imply a shift of responsibility from society to the individual. When human behavior is explained in genetic terms, not only is the social context put between brackets but personal responsibility also changes. When diseases are genetic, other interventions and treatments become more appropriate than changing lifestyle or socioeconomic context. Increasing individual control and discipline are needed in order to achieve health. The population must be educated and instigated to seek testing and counseling, and the range of conditions for which testing is feasible should be expanded to include predispositions and susceptibilities, making everyone aware that he or she is a potential patient. Genetic knowledge will not only enable people to foretell their individual fates from reading their genes but also to adapt their life plans in accordance with such predictive knowledge and, ultimately, to intervene and ameliorate the initial determinations.

Genetic interpretation carries with it the notion of possible intervention. If mutations are detected during pregnancy, the most "rational" approach is to abort. This is exactly the case when genetic testing is increasingly regarded as disease prevention.[175] If a technology can prevent suffering, distress, illness, and disability, destruction of embryos is irrelevant if healthy babies can be produced.[176] The overestimation of genetics may bring about a change from "right to know" to "duty to know"; parents are considered to have a moral obligation to have children free of harmful genetic mutations.

Today, a similar argument is used for predictive information. It may be prudent for individuals to use genetic diagnoses to predict future disabilities, and therefore appeals to (prospective) responsibility may be justified—but this argument in practice is often linked with the argument that individuals who deliberately have not used diagnostic options should be (retrospectively) responsible for any adverse consequences to themselves or their offspring. When, for example, a couple decides not to use prenatal diagnosis, or not to terminate pregnancy in case of diagnosed fetal disorders, it is argued that the couple then is responsible for the suffering of the child if indeed a child with handicaps is born. If suffering could have been avoided and a choice is made not to use predictive opportunities, parents should bear the consequences of their irresponsible choice. They can no longer argue that suffering has befallen them; they have themselves to blame.

This line of argument, if indeed taken seriously, will be a significant stimulus for individuals to obtain genetic information as much as possible, particularly when there is a threat that governments, insurance companies, and employers will work with a system of incentives and disincentives. When there is a cultural imagery that future diseases, disorders, and disabilities can be foretold by examining an individual's genome, persons can no longer claim that they are victims if they have deliberately decided not to use predictive diagnosis. It would be their voluntary choice to not know and to not eliminate potential disadvantages to their health. Fate would be replaced by choice. Personalized medicine is just the most recent effort to articulate that it is the patient's responsibility to implement disease prevention and comply with procedures for early intervention.[177]

The emphasis on personal responsibility for health is actually paradoxical. It is often combined with the belief in genetic reductionism, which sometimes assumes that free will is an illusion.[178] Yuval Harari advocates such a position. He proceeds from a dualistic worldview that holds the objective reality of genetic and electronic codes in the one hand and the imaginary realm of human fictions in the other. The first is the domain of facts; the second, of values. Religion, ethical judgements, and the idea that humans have free will or a soul—or even the idea that we are individuals with an authentic self—all belong to the second realm: they are empty fictions without meaning.[179] So much for the idea of personal autonomy and responsibility.

For Harari, human beings are biochemical data-processing algorithms. Separating the two worlds of science and humanities is therefore also an imaginary strategy. In the future, human fictions can be translated into genetic and electronic codes. Biology will then merge with history. We will be able to explain historical events in strictly biochemical terms.[180] Ethical judgments will be translated into factual claims. Finally, we will have a single overarching theory that explains everything. Harari's view, however, is not consistent, and it is also misleading. Considering living organisms as algorithms takes the model for the reality. Living organisms can be explained as algorithms (referring to genetic codes and electronic brain activity), but they are not algorithms themselves.[181] If people see themselves as a collection of biochemical mechanisms, why would they be interested in health or use wearable sensors and computers to monitor it? Why and how will some people be indispensable and therefore upgraded? Who are the masters of artificial intelligence and networked machines who are now racing to accumulate data? And why are they able to shape the future? If people lack free will, and choices are illusions, at least some are able to operate as conscious agents. In his reconstruction of human history, Harari emphasizes that humans differ from other species because of their ability of large-scale cooperation. It is not clear what happened to this ability to cooperate when humans are simply algorithms. It can make the difference between prophecies of doom and salvation.[182]

Bioethical discourse tries to avoid the dilemma between freedom and determination. The emphasis on personal autonomy underlines individual freedom, but bioethics also considers genetic determinants as the essence of human beings. Revealing that there is a genetic susceptibility shows that a particular disease or trait is within the genetic makeup. A person can therefore not be blamed for having this disease or trait. It is like having a certain inheritance, a built-in capital that needs good stewardship. What is in the DNA is unchangeable. What is inherited cannot be modified by the environment. This does not eliminate personal responsibility. Having a genetic disease or specific trait implies that the person will examine his or her condition and take steps to prevent harm: for example, frequent screening and testing, changing lifestyle, and preventive interventions. Individuals have little choice. Self-management is an ethical duty. If they do not take more responsibility for their health, they will demonstrate a moral failure, which will in turn show itself in the illness that will follow.

The discourse of responsibilization leads the ethical debate away from such issues as the right to health and collective duty. It ignores that major advances in health care have been made, not through personalization and individualization but through improvement of the environment: clean air and water, sewerage systems, better nutrition.[183] The affirmation of personal responsibility in current health policies and the transformation of public services into private markets is weakening the sense of collective responsibility, which may seriously affect the potential benefits of precision medicine. For example, one promise is that patients (or rather groups of patients) can be identified with a genetic profile that fits a specific drug so that treatment can be tailored. For small populations of patients, new and effective medication can be developed, but this new medication will be expensive.[184] Without a collective response and responsibility, few people will be able to afford new medications. Targeted drugs for a small subset of patients will be available but unaffordable, unless the individualized market perspective is balanced with some sense of social justice and social responsibility.[185]

Neglected Legacies

When the European Commission in 1988 submitted its proposal on predictive medicine to the European Parliament, it was confronted with severe criticism. The proposal was intended as the start of the European human genome research program, but it was rejected. Specifically, representatives from Germany and Denmark objected that the focus was primarily on the scientific research and not on the ethical implications, especially possible eugenic consequences. They argued that such ethical questions should be addressed before the research started. The language of the proposal was also a concern. It stated that predictive medicine "seeks to predict susceptibility to diseases with a view to their prevention and early diagnosis." It aimed to "identify high-risk individuals" and "to prevent the transmission of the genetic susceptibilities to the next generation."[186] Prediction thus suggested precision and prevention. It would target genetic susceptibilities and diseases, but diseases are not abstract entities—they are manifested in individuals who then become the target of medical interventions. The notion of prediction therefore raised the specter of eugenic policies. Based on such criticisms, the European Commission had to modify its proposal and drop the references to prediction. The revised proposal was adopted in 1990 as the Human Genome Analysis Programme.[187]

This episode shows how developments in biotechnology must define themselves in relation to history. The future cannot be determined without the past, and yet contemporary life sciences usually deny this history of discrimination, stigmatization, and eugenics. They reframe and relabel themselves as new: for example, the "new genetics." The argument is that classic eugenics should be rejected. It was a barbaric movement that resulted in the atrocities of Nazi Germany. It was bad science and was promoted and applied by states that aimed at coercive control of individual procreation. Finally, it violated human dignity since its focus was on the gene pool and not on the good of individuals.[188] The standard view is that these characteristics no longer apply for the "new" genetics. This is reflected in the shift of vocabulary. We no longer use terms such as "defective," and the vocabulary is more often emphasizing "healthy" rather than "abnormal."[189] Because of such differences, lessons from the past are no longer considered relevant for the present and future. The eugenics legacy is therefore generally neglected or discarded.

Nonetheless, the past never goes away. Eugenics and genetics are like conjoined twins.[190] A curious kind of social amnesia has repressed the memories about the enthusiasm for eugenics in the early 20th century in many western countries, not only in Nazi Germany. Eugenics was not an isolated movement of some barbaric scientists.[191] It was supported by highly respected scientists and politicians. Julian Huxley, who would become the first director-general of UNESCO, argued that eugenics would become part of the religion of the future. Huxley was an active eugenics campaigner, presenting his views as "humanism."[192] His activities illustrate that past eugenics cannot be dismissed as an unscientific activity. It was promoted by the scientific elite.

Eugenic practices were also associated with progressive reform and the expansion of the welfare state.[193] Practices such as sterilization programs continued until the 1970s in the Scandinavian countries where they assumed a model of responsible citizenship. Citizens were expected and encouraged to act in the interests of society, so they voluntarily participated in efforts to control reproduction and halt the transmission of defective genes. Rather than being compulsory, the state was enabling by inducing and encouraging individuals to take responsibility for reproductive decisions.[194] State involvement was often based on economic considerations as well: reducing the frequency of defective genes in the population assumedly curtails expenses for institutionalising the handicapped.

The picture of past eugenics as blunt racial hygiene is therefore false. It serves to demarcate the future of genetics from an unethical past. The new genetics is supposed to be ethical because it emphasizes respect for individual autonomy and human dignity, with practices based on informed consent and voluntary participation. References to class and race and talk about defective genes is avoided, and the focus is on the individual benefits that flow from genetic research.

However, old logic persists. Genetic testing today is regarded as efficient, effective, and cost saving.[195] Policy documents, such as the European Commission proposal, often start with constructing a specific problem: the challenge of genetic disease. Now that infectious diseases as the main cause of mortality can be controlled, many diseases are genetic. Luckily, the risks and susceptibility for such diseases can be predicted. If carriers can be detected and prenatal diagnosis provided, the birth of babies with abnormalities can be reduced. This line of reasoning formulates a specific problem, but more importantly it articulates solutions. An attractive policy perspective is presented. Predictive medicine will diminish suffering; it will decrease the prevalence of many diseases that are distressing for the patient and his family. At the same time, it will reduce costs and improve economic competitiveness.[196]

The idea that genetic determinants of illness and handicap can be eliminated, and that genetic knowledge should be used to control nature in order to avoid suffering and disease, is an appealing one. Although not framed in the language of eugenics, it expresses the same basic notion of genetic improvement. James Watson, the first director of the Human Genome Project in the United States, was obsessed with the notion of genetic perfection and purity. Joshua Lederberg, Nobel laureate, promoted the concept of orthobiosis, the perfection of man (sic), in 1969.[197] Joseph Fletcher, one of the pioneers in bioethics, argued that biological selection and control are ethical imperatives. The nonbirth of genetically handicapped individuals is in the interest of everybody. We should "increase the quality of the babies we make."[198]

The ethical framing of the new genetics in terms of individual responsibility will not prevent the persistence of eugenic thinking.[199] Obviously contemporary eugenics has a face that differs from the older one. It does not refer to a state ideology. It is operating at the microlevel of interactions between patients and professionals. Today it refers to a set of ideals concerning the future of healthy and perfected humans.[200] This is a consequence, as Barbara Katz Rothman has pointed out, of changes in the role of medicine. Medicine has been repositioned not only at the exit gates of life, where it decides about prolongation or termination, but also at life's entrance portals. It has the power to predict the future and can decide who will start life or not. Predictive technologies will decide who will be born.[201] Leaving such decisions to individuals and appealing to a respect for personal autonomy and individual reproductive freedom removes the fears for compulsion associated with the old eugenics. This approach creates what has been called "consumer eugenics" or "laissez-faire eugenics."[202]

However, individual choices are not made in a social vacuum. Although new mechanisms of decision-making are applied, the selective outcomes are often the same as in traditional eugenics. The intentions may be different, but the effects are the same.[203] Fletcher furthermore advocates compulsory screening. When individuals do not take responsibility for reproduction, they should be restricted from reproducing. Eugenics is introduced through the backdoor.[204]

The future paradises predicted by medical prophets such as Silver and Harari are populated with the results of eugenic practices and policies, even if they carefully avoid discussion of eugenics. These prophecies generally underestimate the impact of genetic discrimination, and practices of differential treatment of asymptomatic individuals and their relatives on the basis of genetic characteristics are now well documented in a range of countries.[205] Disabled people regard genetic progress as threatening to their very existence.[206] Because in medical prophecies the context is lost and the primary focus is on the benefits of testing, the social consequences cannot even be envisioned. The persistence of eugenic logics implies that a distinction between old and new eugenics cannot be made.

Neoliberal eugenics is not different from the old eugenics.[207] It is the specific mindset of enhancement and perfection that continues to inspire the advancement of the life sciences. The conclusion is that the new genetics is not a radical departure from previous selective practices. Historical legacies cannot be neglected in medical prophecies.

The Ultimate Prophecy: Immortality

A small group of specialized gurus believe that new technologies will help to liberate human beings from biology altogether. Their ultimate prophecy is a cure for mortality.[208]

Human life expectancy has been increasing considerably over the past century. It is now growing with two years every decennium. The process of aging is attracting intensive scientific interest and research with prolific funding. Studies of supercentenarians and longevity try to reveal the secrets of a long life. There is also a huge market for antiaging products. For medical prophets, these developments have a more radical message. They predict that aging is a curable condition, a disease that sooner or later can be remediated so that life can be extended. In which case, death will no longer be the end of life. Death will become a technical problem for which there will be a technical solution. No longer a natural event or a biological necessity, death will be optional.

The idea of extending life and overcoming death is not a new one. It has been a crucial component of many medical utopias. Since the 1960s, cryogenics has been promoted as a temporary solution, for example. Cryogenic facilities in the United States and Russia will preserve your body, or in most cases just your head, until future science can restore you or your head to life and cure your diseases.[209] There are many other longevity companies and start-ups as well. Banking stem cells, for example, has become insurance for a future of longer and healthier lives, and places like Forever Labs (https://foreverlabs.com) cryogenically freezes and stores adult stem cells for future therapeutic use.[210] And a growing range of new biotech companies aim at extending life span by targeting processes of aging, efforts that are lavishly supported by Silicon Valley billionaires.[211] There is even the Church of Perpetual Life (http://www.churchofperpetuallife.org) preaching that our ultimate destiny is indefinitely extended, healthy life spans.

Why is the prophecy of immortality so popular these days? One answer is that new technologies are promising redemption from our biological constitution. Especially embryonic, and later adult, stem cells have promulgated a miracle discourse, reinforcing the belief that medical technology is omnipotent and capable of repairing any affliction. Stem cell therapies are promising cures for age-related diseases and the prospect of rejuvenation. They have created the myth of regenerative medicine, suggesting that scientists will be able to replace, engineer, or regenerate human cells, tissues, or organs to establish, restore, or enhance normal functioning.[212]

Most arguments against and in favor of life extension are carefully examined and extensively debated.[213] Medical prophecies are essentially conservative. They assume that the present context will continue to exist. They do not question that the market will be the main determinant of the future while using the empowerment discourse of consumers and patients as active participants and emphasizing responsible conduct. They reiterate and extrapolate present developments into the near future. By doing so, they are not politically or ethically neutral. And by not questioning the prevailing value context, they implicitly endorse the assumptions that underlie medical prophecies in general, which we just examined in this chapter.

Life extension debates particularly combine two ethical discourses. One is the discourse of limits in biomedicine. There is no doubt that progress will be accomplished. The other discourse is the ethics of improvement or enhancement. Modern individuals have the duty of perfection. Combined, both discourses present a powerful ideology. As argued earlier, this ideology is not a statement of fact or a conclusion of scientific observation. It is a value judgment. It propounds a view of human beings that is highly questionable and that should be contested in bioethical analysis.

First, it reiterates the dualistic perspective that has pervaded medical science since its emergence as a natural science. Today, it formulates the self as an aggregate of data and codes driven by biochemical processes and algorithms. The self is imprisoned in a corruptible body. It is the body that produces frailties and makes life finite. New technologies promise redemption. Ultimately, we can overcome and redeem the limitations of human nature.[214] In the end, the autonomous subject will be able to obtain full mastery over the body.

Second, it rejects or denies human vulnerability. Prophecies of life extension predict that human beings can free themselves from corporeal restraints. Disease, decay, wear and tear, and decline can be halted and reversed. Human bodies are machines that can be repaired and upgraded. The artifact is often better than the natural. The possibilities for human improvement should therefore be rigorously explored and the notion of vulnerability should be discarded. This view of vulnerability is commonly articulated in mainstream bioethics discourse, assuming the priority of the notion of individual autonomy. It affirms the neoliberal worldview in which each person decides and identifies his or her interests. Vulnerability means that in some circumstances this autonomy is diminished or affected, thus requiring special protection. However, this view of vulnerability is biased. In real life, vulnerability is the general predicament of human beings and autonomy is the exception.

Vulnerability has become a universal phenomenon due to processes of globalization. Medical prophets are wrong to assume that vulnerability is only a biological phenomenon due to our imprisonment in a body. On the contrary, vulnerability is an anthropological phenomenon, a defining characteristic of the human species. Being human means being vulnerable, and this inherent vulnerability of the human condition has different explanations. It is argued that humans are biologically imperfect compared to other species, that the body is the source of decay, that vulnerability is an existential experience since it is not located in the bodily machinery but at the level of personal existence where humans are open to the world and thus susceptible to harm, that humans are always situated and interacting with the world, and finally, that human beings are not only related to other people but fundamentally dependent on others. These views explain why human life is necessarily precarious. But it is also the case that vulnerability can be exacerbated by social, political, and economic circumstances. Human existence is precarious because we live together with other human beings. This observation highlights that vulnerability is not a negative condition. It is a precondition for specifically human activities because it implies that humans are first of all social beings. Being vulnerable promotes solidarity, cooperation, assistance and care. It is a resource to create community and relationships.[215] It is this social dimension of human existence that is missed in the interpretation of vulnerability as frailty, deficiency, and weakness. Medical prophets promoting invulnerabilism suggest that the sources of vulnerability can be controlled. Perhaps we can control our body, but we cannot control our relations to the world. Prophecies of immortality cannot accept that some things are beyond our ability to control and master.

Third, life extension discourse is fundamentally ambivalent since human aging itself is an ambiguous process. People know that aging is unavoidable and associated with decline. Dying after a completed life is not necessarily regarded as tragic. Many people prefer to have a long life, but in a healthy state without a long trajectory of suffering, and then to die rather quickly. What is important is extending the human health span, not the life span as such.[216] The ambivalence is reflected in public support for life extension. Most people would support life-extension technologies but only a minority will actually use them.[217]

Conclusion

In health care, prediction plays an important role. Individual patients want to know the prognosis of their diseases. Healthy people use predictive genetic testing to know their future. There is also a lively genre of predictive medical literature combining artificial intelligence and biotechnology to forecast the future. Regenerative medicine is supposed to conquer human aging and postpone death. This chapter examined some fascinating scenarios for the future and the role of bioethics.

Frequently, bioethics debate is concentrated on the promises and expectations that are mostly projected in the times to come. The concern with prognosis and prediction can be best explained with the narrative of prophecy. This narrative articulates imagination and vision rather than facts and observations. Prophecies are founded on the values of progress and perfectibility and are influencing the agenda of bioethical debate. They are attractive for policy makers as well as the general public, but guiding the prophecies are underlying value judgments. It is these underlying values that should be scrutinized by bioethics.

Prediction is at the core of medical prophecies. It is also the main promise of such major scientific projects as the Human Genome Project and the Precision Medicine Initiative. The assumption of such projects is that progress will continue as it has done in the past. It is taken for granted that the future is predictable and that uncertainties can be eliminated. This confidence is based on a reductionist view of human beings as biochemical complexes and organic algorithms. Reductionism maintains that precise biomedical interventions can target some of the components of this basic human constitution with foreseeable results.

Another value assumption is that human beings can be abstracted from their context. In fact, most prophecies suppose that the present context of neoliberal policy-making and free market ideology will continue to exist. Contemporary inequalities and injustices will therefore persist or even deepen, making most medical prophecies a conservative exercise. Medical prophecies are often a new form of Whig history in that there is inevitable progress from the past to a better future that justifies the dominance of ruling elites as the outcome of natural forces or as the consequence of scientific evolution.

Furthermore, predictions and prophecies assume that individuals are rational decision makers and enlightened consumers. They only need adequate and reliable information (the more, the better) to make responsible choices. In the domain of health care, as argued in this chapter, this model of the human being is deficient.

Finally, medical prophecies assume that new developments can break with history. This is paradoxical since prophecies are based on the past. They identify trends and patterns in history and forecast on this basis what will happen in the decades to come. Analyses show that the legacy of eugenics in the modern history of the life sciences cannot be ignored. Many prophecies predict clearly eugenic futures.

The value assumptions of medical prophecies are based on the fundamental ideas of progress and perfectibility. For Karl Popper, science is a "spiritual adventure" but aimed at solving practical problems.[218] It is not primarily an objective, distanced, and engineered activity but is instead inspired by values and ideals. The motor of the scientific endeavor is the idea of progress. Science can make the world a better place. In his critique of contemporary pessimism, Raymond Tallis argues that we need the utopian hope of progress: it is "an essential idea for collective human morality."[219] Humans cannot live without utopian ideals. Belief in progress is especially strong in the United States. Everything in life can be taken as problem and solution. The solution is the application of technology. Technology, sooner or later, can alleviate suffering and enhance human capacities. Scientific advances will redeem us from disease, disability, and death.[220]

The ideal of progress is not a neutral statement or program. It goes together with a reductionistic and materialistic philosophy. The narrative of genetic reductionism exemplifies that progress is possible because humans are part of nature. They can be regarded as mechanisms that can repaired and improved. There is no immortal soul, no essential human self, no individual freedom; thoughts are the product of electrochemical impulses, as are our emotions. The human future can be determined by technical expertise, computer science, artificial intelligence, and biotechnology. For medical prophets such as Harari, there is no escape from biology. Even human imagination is the product of biochemical algorithms.[221] Everything can be reduced to physical explanations.

There is another price to be paid for the ideal of progress: ethics is transformed into an obstacle. Medical prophecies celebrate the conviction that technological advance and application is inevitable. The simple fact that knowledge is available will determine that it will be used. There shall be ethical debate, but it will be useless, ineffective, and futile. Debate is often cumbersome since it only delays the applications of science and technology. It obstructs progress. At the same time, prophets advance a rather simple but normative argument: there is objective, value-free science that should rationally be applied, and health improvements will logically follow. This argument also reiterates a conservative position. Prophecies reinforce the existing value context. They do not question that the commercial market will be the main determinant of the future. They use empowerment discourse and frame patients as active consumers to implement neoliberal policies. This position is not politically or ethically neutral, which is contrary to what prophets may lead us to believe. Prophets furthermore argue that it is impossible to control technologies, especially in the context of globalization. They forget how many technologies have been controlled, and that in many countries today successful efforts of control are introduced and expanded.[222]

So what is the future as presented by contemporary prophecies? The future predicted after the fusion of bioengineering, genetics, and artificial intelligence is, in fact, more dystopian than utopian. Robots are the new monsters. Enhanced superhumans populate the universe. It is an article of faith to assume that progress will result in paradise or heaven on Earth. Has progress been made? Tallis argues that within the domain of health care, progress is undeniable: life expectancy has increased, infant mortality has decreased, and there is better nourishment and less famine.[223] A survey in 2017 asking people in thirty-eight countries whether life today is better than fifty years ago, however, presents a mixed picture. In countries such as India, Indonesia, and Japan, but also Germany and Spain, a large majority replies that life is better today. But in the United States, 41 percent say that life is worse (with 37 percent saying that it is better). For major developing countries such as Nigeria, Kenya, and Mexico, a majority replies that life is worse than in the past.[224]

Percy Bysshe Shelley, the husband of Mary Shelley, pointed out two centuries ago that progress and advances in science, economics, and political knowledge had not resulted in a more equitable world; they only served to expand human misery. Progress is not the same as moral advancement. In his view, this is due to the excessive development of the "calculating principle" (mechanistic and reductive thinking) at the expense of the "creative principle." For Shelley, humanity can only advance through the strengthening of the imagination: "The great instrument of moral good is the imagination."[225] Creative impulses should therefore oppose the calculating ones. In his analysis of cultural pessimism, Oliver Bennett adds to this view the impact of globalization in the present century. The global economy has promoted widespread vulnerability and insecurity, growing inequality, instability, and precariousness in social life, creating conditions under which people have less and less control over their lives.[226]

The relentless and uncritical push of the ideal of progress presents a serious challenge to bioethics. As long as it frames ethical issues in terms of reductionistic and individualistic models applied in the sciences, it can never question the dreams of progress. In order to argue that availability of a technology should not dictate its applications, bioethics should interrogate the quest for control. It should also use its imagination to go beyond a medicalized life with screening, testing, enrollment in wellness programs, annual physicals, and waiting for the next colonoscopy. The metaphor of prophecy clarifies that prognosis and prediction are normative activities that are articulating values rather than facts.

CHAPTER SEVEN

Relics

Suitbert Mollinger was born in Maastricht, Netherlands, in 1828. He studied medicine in Naples, Rome, and Genoa. Later, he entered the Catholic seminary in Gent to become a priest. Here he met an American bishop looking for missionaries. In 1854, he arrived in New York to be ordained as priest a few years later. In 1868, he became pastor of a parish in a neighborhood of Pittsburgh with many German and Italian immigrants. Mollinger organized healing sessions that were attended by thousands of people and reported to have had many miracle healings. He became famous because he built the largest collection of relics outside the Vatican. With his own money he constructed Saint Anthony Chapel, which now displays more than five thousand relics, such as particles of the true Cross, the skull of St. Theodore, and a molar tooth of St. Anthony.[1]

In 1988, Dick van Velzen was appointed professor of fetal and infant pathology at the University of Liverpool, based at Alder Hey Hospital, the largest children's hospital in Europe. He was invited to expand the hospital's heart collection that had systematically removed and stored more than 2,500 hearts from deceased children since 1948. Van Velzen removed many more organs from more than eight hundred babies during post-mortem examination and stockpiled them. In 1999, an inquiry into the higher than average mortality rates of pediatric heart surgery at the Bristol Royal Infirmary revealed that hearts and other organs of 170 deceased children had been removed and kept without informing the parents. Shortly after, similar news at Alder Hey broke out after groups of affected parents went public. A nationwide inquiry demonstrated that more than 105,000 organs were retained at hospitals and medical schools in the United Kingdom, and a special inquiry into Alder Hey practices in particular primarily blamed Van Velzen as a rogue doctor, accusing him of malpractice, unethical conduct, falsification, lying, and deception. Years later, the General Medical Council banned Van Velzen from practicing medicine in the United Kingdom because of serious professional misconduct. In 2013, he lost his medical license in the Netherlands. Van Velzen never acknowledged that he did anything wrong. He wanted to do research, dedicating his life to helping children. As a foreigner he was scapegoated for practices that existed everywhere.[2]

Finally, a more recent phenomena in health care is the establishment of repositories of biomaterials in many countries across the world. In 2009, *Time* magazine listed biobanks as one of the ten ideas that will change the world.[3] Two-thirds of now existing biobanks have been established in the last two decades. They can store any type of tissue for research but also for therapeutic purposes (adult stem cells or umbilical cord cells). One of the largest clinical biobanks in Europe is the Biobank in Graz, Austria. It has approximately 20 million human samples. A commercial biobank in Shanghai in China has 10 million human samples. The UK Biobank has recruited 500,000 people to participate.[4] The All of Us research program in the United States aims at collecting materials from 1 million volunteers.

Collecting and storing human biomaterials is not new. From relics to organ collection to biobanking, the practice has only shifted the effort to new and unprecedented levels of gathering data and samples and bringing them together in centralized facilities. When the discipline of anatomy was first developed in the 17th century, professors and surgeons made private collections, and these later merged into medical museums. Pathology became a separate discipline in the 19th century, which required post-mortem examinations of major organs, such as the brain, heart, lungs, liver, intestines, and kidneys. They would be removed, sliced, examined, and often stored. The argument was that they were necessary for medical research, medical education, and therapeutic purposes, even if in many cases they were simply stored and not used. And it seems that these bodily materials, like relics, were attributed with special meaning and endowed with some magical power.

In the 19th century, collections of human remains were also initiated and expanded for anthropological and archeological purposes. The Musée du Quai Branly in Paris, inaugurated in 2006 and uniting different collections of non-European civilizations, has 370,000 objects, most of which include human remains.[5] At the same time, there are many stories of fascination with bodies and body parts, for example the theft of Albert Einstein's brain. And one of the most popular public exhibitions over the last few years is *Body Worlds*, which features dissected and plastinated human bodies. People are also eager to preserve their bodies or specific parts such as stem cells for possible future use through cryogenics. Finally, there are immortal body parts, such as HeLa cells that are essential tools for scientific research.

This chapter will examine the current trends in collecting and storing increasing amounts of biodata and biosamples. What is the relationship between a person and their body parts? Historian Rina Knoeff has suggested that it is useful to think about anatomical collections and preparations in terms of relics,[6] and not just those specifically related to religion. There are many secular relics that are collected for different reasons, too. One is to connect the past and the present. Collecting and exhibiting the physical remains of a person is an act of remembrance. Anatomical and pathological preparations "provide a material mnemonic link to the past, to stories of lost lives."[7] This is also an argument used by Kenneth Woodward in his study on making saints.[8] Today, saints, and therefore relics, do matter because they provide connections. They create communities of memory. Furthermore, they remind us of dependency. Human beings interact with and are receptive to other people. Human lives do not only show mastery, autonomy, and control but also reciprocity, vulnerability, and sensitivity. Finally, relics indicate particularity. They underline the significance of the local, the circumscribed, and the particular. They have materiality, and emphasize the concrete, the physical, and the tactile.

Body parts are sacred objects because they refer to fellow human beings. This is why they have a particular power. They affect people here and now. As discussed in chapter 5, religious relics were associated with miracles and can be the destination of pilgrimages. In health care, body parts as relics refers to the symbolic value of human remains. They still seem to be connected with salvation, which is not surprising since the words "health," "well-being," and "salvation" are etymologically related. "Health" is derived from the old English *hal*, meaning whole. The same root has produced words like "heal," "hail," "holy," "whole," and "well."[9] These characteristics of relics explain why collecting and displaying human remains is a sensitive issue. Many museums no longer exhibit human remains, and they are also confronted with requests of indigenous communities to repatriate remains so that they can be properly buried. This has revived the issue of the dignity of the dead. Why do human remains deserve respect? Can educational benefit and research with human corpses and body parts outweigh the respect for the dead?

Traditional medical collections have lost some of their significance since research has become more focused on the cellular and molecular levels. Collections of frozen tissue samples have become more important. Blood samples, too, have become the focus of new collections. The introduction of newborn screening for metabolic diseases since the early 1980s, for example, means that in some countries, such as Denmark, body materials from every person born since 1982 are available for research. The Danish National Biobank, opened in 2012, incorporated these extensive older collections and is a testament to how medical museums and biobanks are closely related institutions.[10] Contemporary biobanking continues a much older tradition of collecting. It shares similar practices as anatomical and pathological collections. Both provide a bridge between specific cases in the clinic and medical research.

The Cult of Relics

A relic in a general sense is a reminder, a memento of an earlier time. The term is derived from the Latin word *reliquiae*, which means "remains," "remnants," or "leftovers." Relics are important in many religious traditions, but they also play a role outside of any religious context. The Science Museum in Florence, for example, displays the middle finger of Galileo, removed ninety-five years after Galileo's death. At an auction in 2009, two more fingers and a tooth were for sale. They are now together in a jar with other relics of the scientist. When Albert Einstein died in 1955, the pathologist doing the autopsy stole his brain for scientific research. For a long time, the brain was considered lost, until the first research paper based on it was published in 1985.[11] An even more bizarre story concerns the skull of Emanuel Swedenborg, a scientist and theologian who founded a spiritual movement in the 18th century. He was buried in London in 1772 in three overlapping coffins. Nonetheless, his skull was stolen in 1816. The thief was persuaded by Swedenborg's disciples to return it, but he exchanged it for another skull. In 1906, the Swedish Royal Academy of Sciences wanted to repatriate the remains to Sweden and bury them in Uppsala Cathedral. At that time, a message was received that the skull in the remains was false. After an intensive search, the real skull was discovered in Swansea and finally bought by the academy at an auction in 1978.[12]

In a religious context, relics are primarily the physical remains or ashes of a dead human being, specifically a saint, or items directly associated with the events of the life of Christ. These are first-class relics, to be distinguished from other relics, like items used by a saint (e.g., fragments of clothing) or items touched by a first-class relic. Veneration of relics is common in many religions. As the remains of a holy person, relics are often associated with special powers and miracles.[13] In Christianity, the cult of relics started at the tombs of martyrs buried in cemeteries outside of cities. Disturbance of human remains was not allowed so the dead and living were physically separated. Basilica were built on the tombs, and later the dead were brought into cities and located in crypts. Christians were accused of too much intimacy with the dead, of disturbing human remains, and breaking the taboo of separating the living and the dead.[14]

In the fourth century new saints emerged, such as ascetics and monks. They were venerated because of their exemplary lives. At that time, the cult of relics was widespread and supply became insufficient. Relics were needed to consecrate churches, newly founded abbeys, and monasteries. There were also enormous private collections of relics by emperors and royalties. Many "new" relics were discovered, but there was often the problem of authenticity. In the fifth century, for example, two heads of John the Baptist were venerated in Constantinople while a third head was discovered in France.[15] There was a lively trade and a black market of false relics. Relics were also looted and stolen.[16] The body of Mark the Evangelist was smuggled out of Egypt in the 820s and transported to Venice, while the corpse of St. Nicholas was stolen from Myra in Anatolia and transported to Bari in Italy in 1087 where it gave rise to a growing sect and pilgrimage.[17] These practices were not condemned; they were regarded as acts of veneration and respect. Their aim was to preserve the remains of holy persons when their cult was no longer active or threatened.[18]

It was also common practice to remove organs and parts from the bodies of saints. The belief was that the whole of the saint was in the smallest fragments of the body. Fingers, nails, and bones were sliced off and given away as acts of friendship, diplomacy, or patronage. Body parts were transferred from one place to another and used to start cults in other places.[19] St. Anthony died in 1231, and his body was transferred to a new basilica in Padua in 1263. At that occasion his sarcophagus was opened. It was discovered that his tongue was uncorrupted. This was taken as a powerful symbol since the saint had been famous as an eloquent and inspiring preacher. The tongue is now on permanent display in a golden reliquary in the basilica. (In 1981, the sarcophagus was opened again. It turned out that the vocal cords were perfectly preserved.)

When St. Francis de Sales, founder of the Order of the Visitation, died in 1622 in Lyon, his heart was extracted and given to a monastery in Italy.

Thomas Aquinas argued that there are three reasons to venerate relics. First, they are physical reminders of the saints. Second, bodily relics have an intrinsic merit; they are not mere objects but connected to the soul of the saint. Third, the saints are working miracles at their tombs.[20] To further explain their power, two notions are important: intercession and liminality. Intercession means that saints can intervene on behalf of another person. A saint was an ordinary human being who led an exemplary life. They succeeded in overcoming evil and their holiness induces to imagine a world that is much better than the present one. Saints can therefore be invoked for help; they are a medium through which heaven can be reached.[21] They are ambassadors, patrons that can mediate and advocate so that diseases can be healed and disasters averted.

Relics also show that in a certain way, though in heaven, saints are still with us. Relics define the liminal space between heaven and Earth, between the secular and the divine. They have sacred power and are associated with miracles because they are at once real and spiritual entities. As far as the body of a saint is concerned, there is no absolute barrier between life and death. The physical body can be horribly mutilated (as in martyrdom) but transformed into a spiritual body that transcends the world of mortal flesh.[22] The corpse of a saint can be recognized because of its wholeness, lack of corruption, and wonderful fragrance. Remains therefore triumph over putrefaction and decay. The mortal remains are transformed into spiritual flesh.

During ritual activities like the veneration of relics, participants pass through a threshold (limen) where they go from one stage to another.[23] This transition involves ambiguity and disorientation. One's sense of identity can change, and new perspectives will emerge. Common distinctions dissolve and boundaries are blurred. Relics show that there is communion of the living and the dead. Everyone who has existed and who exists is interconnected.[24]

The veneration of saints and the cult of relics demonstrate the significance of narratives. Saints exist in and through their stories.[25] They present human behavior and virtues that inspire and motivate, that call for emulation. The cult of saints, therefore, has three basic elements: a name, a body, and a text. Relics refer to the bodily remains of the saint, and they highlight the positive role of dead bodies in particular. But it is also necessary to have a narrative, documenting and exemplifying the life of the saint even if historical evidence is scarce, deficient, or missing, as well as a book of miracles. The emphasis on narratives means that different models of holiness can be presented. Such models are produced locally and can be critical of the models of the institutional Church. For a long time, the cult of saints and relics was primarily driven by popular religious practices. The making of saints was a spontaneous act of the local Christian community.[26] Efforts to bring the cult under control of the Church were not very successful.[27] The sale of relics, for example, was prohibited in 1215 but the trade continued to exist. Until the eleventh century there was no formal procedure to recognize sainthood. In fact, there are more than ten thousand known saints while only a few hundred have been canonized by popes.[28] From a popular perspective saints are important for intercession. They may provide help and protection against evil. For the Church, saints are important as virtuous exemplars rather than wonderworkers.

The cult of relics substantially declined after the Reformation, especially in Northern European countries, as Protestants rejected the idea of intercession. They believed veneration of relics to be idolatry, superstition, and pagan. Calvin, in his *Treatise on Relics* (1543), argued that most relics were false inventions for deceiving silly people.[29] Catholics, such as Erasmus of Rotterdam, became more skeptical as well. Erasmus mocked relics and thought they were unnecessary to achieve salvation.[30] Despite all this, however, the collecting of relics did not cease.[31] And Mollinger's efforts are a good example. As political struggles against the Catholic Church in Germany, the unification of Italy, and the loss of Papal States destroyed churches and monasteries across Europe over the next centuries, many feared that relics might be lost. Mollinger, though, somehow procured many of them through his networks between 1868 and 1892 and brought them to Pittsburgh.

Medical Museums

Before Mollinger and his relics, there were others growing medical collections. In 1797, for example, Gerardus Vrolik was appointed as professor of anatomy and botany at the Athenaeum Illustre in Amsterdam. He was also made responsible for the maternity ward in the hospital at the center of town. Since he was the first professor in the Netherlands with his own obstetrics clinic, he had access to embryonic and fetal materials that would otherwise have been difficult to obtain. Thus, Vrolik started a private collection of anatomical and pathological specimens, particularly of congenital malformations, and was celebrated for his studies in comparative anatomy.[32] Vrolik was succeeded as professor by his son Willem in 1831 who expanded the personal collection. In 1869, the collection was purchased by the city of Amsterdam. The Museum Vrolik is now located in the Academic Medical Center of Amsterdam.

Collections of body materials started as private initiatives motivated by curiosity (they were often called "curiosity cabinets") but also as an occasion to admire the diversity of nature and to marvel at God's creation. In medicine, the anatomical theater was the place to display body materials, especially skeletons. And university professors often explained that collecting has higher purposes, such as research and education. Pieter Pauw, the first professor of anatomy in Leiden in 1577, saw his collection as instruction material for teaching. The same purpose motivated three generations of the Monro family, chairs of anatomy at Edinburgh since 1705.[33]

Like producing relics, gathering and displaying human remains was not a clean business. Corpses were taken from burial grounds and charnel houses. A body had to be boiled for some time to separate the flesh from the bones, and for the skeleton to become a scientific object in the 16th century, the human body needed to be seriously violated.[34] Beyond this, not much is known about how body parts were taken, but it is presumed that most times they were simply removed without informing anybody and materials were stored anonymously. John Hunter collected 17,000 samples for his museum in London. He assembled these by dissections and making the preparations himself. He received some as gifts from friends, but he also traded specimens, sometimes bribing family and undertakers (as in the case of the Irish giant discussed in chapter 4). Embryos, fetuses, and infants were a different story. The Amsterdam anatomist Frederik Ruysch reported in the early 18th century how he negotiated with the parents of a conjoined twin, born after an eight-month pregnancy. He wanted to embalm the babies and include them in his famous anatomical cabinet. In return, the parents would have free access to his cabinet; and indeed, the parents frequently visited their children.[35]

Collecting was encouraged because this was a time of explorations across the globe. Anthropological and biological treasures were brought back to Europe, and the enormous influx of materials and specimens stimulated the building of collections as a way to create order and to impose systematic classification on these rapidly increasing empirical resources. And it was not long before private collections began to merge into medical museums. They were especially important in the late 18th and early 20th century. The role of museums as public institutions was to disseminate knowledge, and with the rise of clinical medicine there was a need to verify diagnoses with post-mortal examination. Autopsies could associate symptoms during life with bodily lesions so that disease processes and symptomatology could be explained. Collecting organs and specimens was necessary to further research, not only to specify and classify diseases but also for educating future physicians. And with the advance of pathology, more and more smaller body parts—such as tissues, blood, and histological slides—were frozen, archived, and stored. In 1999, it was estimated that in England alone 408,600 samples of tissues have been stored in museums and other archives.[36] The majority of museums in England have human remains as well. An estimate in 2003 found 61,000 human remains, most of which were acquired through archaeological activity. Approximately 25 percent is from overseas. Some or all of these remains were on public display in more than two-thirds of English museums.[37]

While human remains have been collected, stored, and displayed for centuries, they have become increasingly problematic.[38] Since the 1990s, in fact, thousands of human remains have been repatriated from museums in Europe and the United States. In 2011, the medical museum at the Charité in Berlin returned twenty skulls to Namibia. The decapitated heads were in fact trophies of colonial repression more than one hundred years ago. In 2009, the Dutch government returned the head of a Ghanaian king killed in 1838. The head was discovered by accident in the Leiden anatomical collection and was returned to members of the Ahanta tribe who had requested it as a relic. And the repatriation of Sarah Baartman in 2002 led to her eventual burial in South Africa (see chapter 4).

Many museums no longer display human remains. The fundamental argument for halting these activities is that these remains should be treated differently than other museum objects. Human bones are not just bones; they have special significance. Compared to artifacts, they have a unique status. However, the argument is often not elaborated. Why exactly do human remains require special consideration? This is where the analogy with relics can be helpful. Human remains deserve respect because they connect the past and the present. They are not specimens but ancestors. They once were individuals, and they are still part of human communities.

Respecting human remains, including body parts, is an act of remembrance. Like relics, dead human bodies are also symbolic objects. Their status is ambiguous because of liminality. Corpses are definitely "things" that can be displayed, but simultaneously they refer to persons. They are real and not real at the same time. As material things they can be located and treated as physical objects but are also associated with values. That turns a dead body into more than a complicated mechanism and transforms it into a symbolic object. This is why its storage and display may provoke distress and offense.

A purely scientific approach dehumanizes human remains; it removes what makes them special.[39]

The debate about the status of human bodies and body parts in museums since the 1990s coincided with the Alder Hey scandal. Repatriation and re- burial of remains was advocated as a way to repair injustices of the past. Also, the therapeutic effect of return was highlighted. It was pointed out that in- digenous communities and parents of dissected children experienced simi- lar distress and trauma because of disrespect for the mortal remains of their relatives.[40] Scientific arguments that human remains are research material and invaluable scientific resources quickly lost support. In fact, materials were rarely studied. Van Velzen, for example, did not do substantial research with his collections. Materials were simply stored and not used. Convincing cases of future benefits were rare. At the same time, researchers and policy makers were continuously confronted with the immediate needs of commu- nities and parents that should be addressed.

I argue that these problems cannot be resolved by emphasizing consent. Articulating ownership raises the question of whether human remains *can* be owned.[41] Who can claim human remains, especially if they have been col- lected long ago? And even if consent is obtained, there is still the question of the dignity of the dead. What can be done to human remains even if permis- sion is granted to use them? Property discourse regards the human body as simply an object that can be used for scientific purposes. It does not fully explain why the collection, storage, and display of bodies and body parts is controversial, even if permission has been granted. The discourse of respect goes beyond the issues of ownership.

Medicine as Collection Science

The role of collections for medical progress is often underestimated. The rise of the life sciences is commonly attributed to the importance of experimen- tation rather than observation and collection. Histories of medicine, partic- ularly older ones, used to present a picture of medical progress as the result of medicine redefining itself as a natural science and applying measurement and experiment as well as instrumentation.[42]

The mechanical view of the human body and the systematic application of physics and chemistry produced exceptional results in the 19th century. Experimental physiology, cell biology, and later molecular biology not only explained the functions of the body but also clarified disease mechanisms and provided the tools to intervene. The laboratory became the most important resource for medical science. Gifted experimenters, such as Johannes Müller in Germany and Claude Bernard in France, turned medicine away from natural history and toward experimentation.

The distinction between observation and experimentation is not as clear-cut as generally assumed, however. Collecting and observing materials was not a passive enterprise necessarily. Anatomical and pathological specimens were not simply admired or venerated like relics. They were actively examined. They facilitated comparisons so that relationships, similarities, and patterns could be determined. Private collectors allowed visitors to handle preparations, to inspect them from nearby, and to even touch them. Preparations were often regarded as working material that needed adjustment and improvement.[43] They were, therefore, not simply the object of observation but were actively used.

Studying human remains and body samples elicited admiration for the wonderful machinery of a living being but at the same time inspired ideas about the structure and function of bodies and body parts. Importantly, preparations affected visitors and observers. Aware that these are not simply scientific objects but were once fellow human beings, a distanced or detached view was not fully possible. The observed preparations show facts, but at the same time they have a special meaning and value.

Nonetheless, due to the central role of experimentation in the production of medical knowledge, medical collections and museums lost much of their allure. Accumulating specimens and body parts was no longer regarded as encouraging science or leading to new knowledge. Instead, the museum was contrasted with the laboratory, losing even more significance because medical research was becoming more focused on the cellular and molecular levels. Since the 1980s, there was also a substantial decline in dissections.[44] Many teachers argued that using plastic models, computer simulation, and virtual imaging were more effective ways of educating medical students.[45]

Remarkably, the current growth of databases and biobanks has raised serious doubts about the emphasis on experimentation as the primary mode of producing scientific knowledge. Scholars now argue that practices of collecting, describing, naming, comparing, categorizing, and classifying objects are essential for experimental knowledge.[46] Contrary to the predictions of Bernard, natural history is not dead. Collections are tools for producing knowledge. Biomedical research today combines experimentation and collection in producing knowledge. On this basis, Karin Tybjerg argues that "the medical museum and the biobank are closely related institutions."[47] They share a range of practices. They provide a bridge between specific cases in the clinical setting and medical research. Medical collections as well as biobanks proceed from historical understanding: past diagnoses are used to comprehend present diseases. Similar museum practices are used at biobanks for preservation, registration, and storage of human materials. Body materials now travel from hospitals to research centers. Following the "translation" of relics from one place to another, we now have "translational" science, turning observations into interventions.

The parallels between the practices of medical collections and biobanks demonstrate that medicine still is a collection science. The body has not disappeared but is divided into smaller units. The focus is on materiality, physical body materials. Collecting and comparing is necessary for producing knowledge since it brings order in the plethora and diversity of information. It saves us from drowning in the tsunami of data and helps to make sense of the growing multitude of empirical materials.

Another parallel with the cult of relics is in the value of the materials. The relics of a saint are not intrinsically valuable. They can be just a bone, for example, that looks exactly like any other bone—not rare, abnormal, or curious. But the bone is unique because of its source, its provenance. It is a sample of a specific history of a specific individual.[48] Though databases deal with information, and biobank collections with biomaterials, the source of the data and samples is equally important. It is their history that makes these materials valuable. But, like in the case of relics, the past needs to be connected to the present. While narratives used to do that in the case of relics, biobanks and databases connect their samples and data to health records that report on medical history, family, and lifestyle factors. Provenance of body materials must be related to the information about the current status of the person in order to be useful for further research.

Biobanks

It is estimated that in 1999 there were more than 307 million human biospecimens from more than 178 million individual people stored in the United States. Since then accumulation of samples has rapidly increased. Some estimates are that currently more than 1 billion biosamples have been collected worldwide.[49] A relatively new establishment, biobanks remain diverse in their individual features, policies, and practices. There is not an agreed upon definition of a biobank. And since a national or international registry of biobanks does not exist, the actual number of banks and stored specimens is unknown. A distinction is made between population-wide biobanks, for example set up in Iceland, the United Kingdom, Estonia, and the United States, and disease-specific biobanks, particularly in the area of cancer. Most biobanks have been established to facilitate research on a particular disease. Serum and plasma are most commonly stored, with solid tissues a close second place.[50]

Biobanks do not carry out research themselves. They are research platforms, providing a generic infrastructure with resources for research projects. Biospecimens are necessary for translational research, with up to 45 percent of this research using biomaterial.[51] In the United States, most biobanks are affiliated with academic institutions, hospitals, and research centers, and the federal government is the largest funding source. The US Congress has authorized $1.455 billion over ten years for the All of Us project, for example.

What is new, compared to previous collecting efforts, is that biobanks are not just amassing biological samples but instead combining them with clinical patient data (e.g., demographics, diagnoses, treatments, and outcomes). The use of information technology facilitates the study of populations throughout their life spans, and because biobanks do rely on patient participation, public trust is an important issue. Biobanks need to recruit numerous individuals to provide biospecimens as well as relevant health information to research efforts. Scandals like in Alder Hey can seriously jeopardize public trust and engagement. At the same time, biobanks play a crucial role in the promises of personalized and precision medicine. They are a kind of investment in the future, delivering new diagnostic tests and targeted treatments for specific diseases. Unfortunately, reporting in news media on biobanks heavily emphasizes the benefits and underreports the risks.[52]

The development of biobanks has followed the same pattern as medical collections in the past. Many researchers have created individual banks that have merged into institutional banks and later incorporated into national biobanks. This tendency to create gigantic biobanks has been criticized, though. The fear is that the emphasis is too much on increasing the number of specimens and serious concerns have been raised about underutilization of stored biospecimens.[53] Perhaps a different collection model would be more sustainable and focused on specific needs. Collection networks may be more flexible than centralized facilities, for example.[54] One important issue in this regard, too, is the overestimation of the commercial value of biobanks. They have been heralded by neoliberal policies as a way to commercialize medical progress, but in general they do not generate profits. They are only sustainable with public support and subsidies.[55] Involving commercial entities in biobanking activities affects the willingness of persons to participate and erodes public confidence. A series of data breaches in health care has also increased concerns about insufficient protection of personal health information.[56]

More than practical and scientific issues, biobanking has generated ethical concerns. The main focus has been on questions of consent. Biobanks have applied consent strategies that are different from the usual ones in research and clinical practice.[57] Supporters generally assume that there is consensus about consent but in fact there is not. There is a lot of ambiguity and uncertainty regarding ownership of samples and data. Again, the narrative of relics provides a way to go beyond the usual preoccupation with consent and ownership. Some argue that collections of body materials have become more valuable since the rise of molecular biology.[58] DNA has become a magical force that facilitates the prediction of the future. Future healing powers are ascribed to the gene, and so genes are regarded as sacred entities. Biotechnology company StarGene, in fact, produces cards with DNA of celebrities so that one can be close to the revered person.[59] These genetic relics are not just biological entities but have a symbolic meaning. The language of properties and commodities is not sufficient to articulate their value. The concept of intercession, although completely secularized, also seems to apply. Molecular relics are situated between the past and the future. They determine our destiny and mediate our fate, especially when personalized approaches to remove diseases and target treatments are developed in the near future. Like the veneration of relics in the past, this is primarily popular belief exaggerating the benefits, while scientific orthodoxy is more skeptical.

Questions concerning the relationship between persons and bodies received a central place in ethical debates because of scandals such as Alder Hey and disputes about display and repatriation of human remains. Until the late 1990s, organs and tissues were routinely taken during autopsy without informing relatives and parents or asking them for permission. The common assumption was that relatives would not object to the removal and use of body parts for research, therapeutic purposes, or education. Scientific and medical arguments were assumed to prevail. Building a collection of human organs provides knowledge, leads to major developments in the treatment of diseases, and provides an indispensable training tool for students. Scientists and medical professionals could simply not imagine that anybody would object to these uses of human remains. They were convinced that their view of human bodies as objects, products, or resources could not be contested from an ethical point of view. And this is reflected in the language used: body materials are "taken," "retained," "collected," and "stored."

In the area of biobanking, the case of DeCode Genetics provided a similar impetus to the ethical debate as previous collection scandals. It was the first large-scale biobank, established in the United States but operating in Iceland. The private company constructed the false narrative that the population of Iceland was genetically homogeneous and therefore attractive for genomic research. The idea was to combine three databases: genotypes, genealogies, and health data. In reality, the last database needed to be created. Because Iceland has a universal health system, medical records were available for almost every citizen dating back to 1915. The government was ready to sell the medical records of an entire population, including DNA samples, to a private company and give it monopoly access. Draft legislation was quickly processed without significant public consultation. The government claimed that they owned all the data. Universal consent was presumed, and there was no possibility for opt-out or withdrawing data.

In December 1998 the law on a health sector database was adopted, this time allowing the possibility to opt-out. It was strongly resisted and boycotted by the medical profession. Almost 7 percent of the population opted out. The National Bioethics Committee that opposed the legislation and proposed ethical regulations was simply dismissed and replaced in August 1999. However, in 2003, the Supreme Court ruled that the law was unconstitutional. Data of deceased persons could not be included, and individual consent of all participants was necessary. In the end, the company never built the database. In 2010, DeCode filed for bankruptcy. In 2015, it was bought by a Chinese company. All in all, the Iceland biobank case is an instructive example of how not to proceed. It was a fiasco because of the authoritarian and greedy conduct of some scientists and politicians who merely regarded data and biosamples as unexploited resources to be mined. They simply extended the tradition of lack of respect for persons and their bodies.[60]

Bodies as Spectacle

When I participated in a conference on utilitarian philosophy in University College in London in 1983, Professor Herbert Hart, a famous legal philosopher and chair of the meeting, made a lunch speech. He first addressed a box placed at the head of the table. The box contained the body of philosopher Jeremy Bentham, the founder of utilitarianism. In his last publication, Bentham had argued that the remains of a person could be used as his own statue. He requested that his body be preserved as an autoicon and be present when friends and disciples were gathering to discuss his philosophy. Bentham thus made himself into a relic. As other relics, too, it was not undisturbed. The head was stolen several times, and Bentham is still displayed in University College, dressed in his usual clothes, walking staff, and hat, but he has a wax head.[61]

Display of human bodies became especially controversial with *Body Worlds*, the successful public exhibition of dissected bodies. Over the past two decades the exhibition has toured the world and attracted almost 50 million visitors. Using the technique of plastination, body fluids are replaced with silicon rubber so that most tissues are transformed into moldable plastic. The bodies are then exposed in lifelike and active positions. Most visitors have been fascinated but some experienced a mixture of curiosity and discomfort. While the purpose of the exhibition was articulated as pedagogical, it combines commerce and theater. Bioethical responses have been critical, especially in Europe. The National Bioethics Committee in France produced a negative advice in 2010 arguing that the exhibition is dehumanizing, reducing persons to their corpses and the display of human remains to entertainment. They claimed it continues the tradition of freak shows and encourages voyeurism and sensationalism.[62]

While questions have been raised about the provenance of the bodies and whether consent has been given for the use of the bodies, more fundamental questions have focused on the issue of respect. Is displaying plastic bodies as a spectacle showing proper respect for human remains? In most anatomical collections, the story of the body is strictly separated from the story of the person.[63] Also, in dissections medical students learn how to detach from the personal story of the corpse. What is striking in *Body Worlds* is the same ambiguity that has been experienced with relics. The bodies are not simply objects. They are not plastic models but real human bodies. Dead bodies are presented as if they are alive. The narrative of the body (how it was acquired, preserved, and molded into a specific pose) cannot be fully separated from the personal narrative, although efforts are undertaken to eliminate personal identifiers. Cadavers are first depersonalized by removing the skin and then repersonalized in specific lifelike poses, suggesting living and moving bodies.[64] In this way, bodies are perceived as objects of spectacle, commerce, and art. The dignity of the dead person is therefore offended, even if people have consented to be plastinized and even if the dead cannot experience harm.

Bodies on Ice

In 1953, newspapers reported a daring experiment with hearts of chicken embryos that were first quickly frozen and then brought to life again when thawed. Basile Luyet, who carried out the experiment, wanted to show that it is possible to arrest life with deep freeze and restart it after some time. Luyet, the founder of cryobiology, was captivated by latency, the phase of life between active life and death.[65]

In 1701, Antonie van Leeuwenhoek had observed that Rotifera (microscopic aquatic animals) can be completely desiccated but revive even after thirteen years. This phenomenon was interpreted differently.[66] If it was anabiosis, it was assumed that life in this condition had arrested and that it returned after rehydration. As latent life, however, it was assumed that life under drought conditions very slowly continued. The controversy was addressed by the Société de Biologie in Paris in 1860. Repeating the experiment, the society decided in favor of anabiosis. This conclusion was supported by experiments of Dombrowski who could revive bacteria that were "frozen" in salt deposits 200 to 500 million years ago. Anabiosis is a state of suspended animation; life has stopped but can return when appropriate conditions are restored. For Luyet, various techniques of cryopreservation produced such a state in between life and death. It is possible, then, to create a kind of immortality that is different from plastic bodies. Plastination seems to overcome human mortality, but in fact the exhibits are dead. Cryopreservation, on the other hand, is extending life indefinitely; death is only apparent. For Luyet, frozen specimens were potentially alive: "like a watch that has unwound."[67]

Luyet's techniques were initially applied to preserve human tissues, particularly blood, and they allowed the setting up of tissue banks in the early 1950s. But they were also used in the cattle industry for freezing sperm for artificial insemination. These applications inspired Robert Ettinger in 1964 to publish his book *The Prospect of Immortality*, advocating cryonics. Ettinger argued that immortality, interpreted as indefinitely extended life, was technically attainable, practically feasible, and desirable.[68] In 1966, the first body of a person after death was frozen (and it is still in cryopreservation today). The American Cryonics Society was established in 1969.[69] Since then, cryopreservation techniques have facilitated the application of in vitro fertilization, creating the possibility of cryobanking oocytes and embryos.

The narrative of relics in this process was never far away. Between 1960 and 1970, major efforts were undertaken, supported by the World Health Organization, to set up a global infrastructure of frozen human tissues, particularly blood sera, of many populations across the world. The motivation was to preserve these samples as relics from the past and as resources for the future. These freezers filled with biospecimens have been called "a secular reliquary of latent life,"[70] thus connecting the notion of latency with the religious idea of liminality, or the limbo between life and death. The idea of salvage, too, was especially applied for indigenous populations. Large tissue and blood collections have included materials from Brazil and the Pacific islands. Although in the 1990s, the status of these materials became questionable and many blood samples, some frozen for decades, have been returned recently to indigenous populations.[71]

Immortal Body Parts

In the past, many efforts have been made to keep tissue alive outside of the body. Tissue banking became possible because human tissues could be archived at low temperatures for indefinite periods of time, as just discussed. Then in 2010, the story broke of the HeLa cell line. These cells are regarded as immortal because as long as they are cultured under appropriate conditions they continue to divide. Parts removed from the body of a person can continue to exist in the laboratory, which demonstrates that life can be extracted from the body and exist apart from it. The HeLa cell line in particular, which originated as tumor cells taken from a patient named Henrietta Lacks who died in 1951, have been growing in tissue cultures and used in thousands of experiments across the world. They have been crucial for medical progress (e.g., used to test the safety of the first polio vaccine), and more than 100,000 publications have resulted from research with HeLa cells.

While this is a "technical form of existence," the notion of immortality is only restricted to cells.[72] The Lacks family, however, regards the cells as relics from the mother. For many scientists, the cells were never just biological materials either; there was a personification of the cells associating them with the woman from whom they were taken. There was an assumption of continuity between the person and the cell line, even if they were physically separated and considered discontinuous entities from a legal point of view so that the cell line could be regarded as an invention.[73] In this experiential narrative, Henrietta herself gained immortality through her cells.[74]

Neither Mrs. Lacks nor her family were initially told that cells were taken from her. And although the HeLa cells were not patented, they were commercialized and the Lacks family never received anything in return. At the same time, the courts in the United States have justified that cell lines can be created from patient materials, that body parts taken from individuals for research purposes are no longer their property, and that they cannot claim to participate in any profits resulting from the use of these materials. Only in 2013 did the National Institutes of Health (NIH) announce that they had an agreement with the descendants of Henrietta Lacks concerning the HeLa cell line. The agreement followed a controversy earlier in March that year when a team of European researchers had published the genome sequence of the cell line. To protect the privacy of the Lacks family, the NIH agreement required review of requests to use the genomic data. The role of Henrietta Lacks and her family was finally acknowledged and respected. However, the agreement about data sharing is an exceptional case. There are many other human samples and data that are accessible to researchers or even publicly available but have been obtained and collected without any consent.[75]

Ethical Debate around Consent and Bodies as Property

It was after the Alder Hey scandal in the 1990s that the rules around obtaining and using human remains changed. In the United Kingdom, the Human Tissue Act, initially from 1961, was revised in 2004. It strongly emphasized that human remains should be treated with appropriate respect and dignity. Human bodies cannot be used as objects. The wishes of the dead, and in the case of children, the parents, should be taken into account. Retained organs should be returned to the parents or relatives to complete the burial.[76] A change in language was initiated. Body parts can no longer be "taken" or "retained" but are instead "donated." In this approach, the notion of respect was first of all translated in the notion of consent. Curiously, the Vatican, issuing new rules for the use of relics, also emphasized consent. Before unearthing the remains of candidates for sainthood, consent must be obtained from surviving family members or heirs. They are invited to donate the mortal remains to the Church.[77]

The emphasis on consent, however, is problematic. In reality, practices of biobanking are widely varied. Very different types of consent are used while several biobanks do not apply consent at all.[78] And the question remains about what to do with previously collected samples. In most cases, it is impossible to retrospectively ask for consent. The more fundamental question is whether the notion of consent that is now commonly used in research ethics and clinical ethics can be applied in the context of medicine as a collection science. It is argued that biobanking requires new ethical approaches and several reasons are advanced. One is that individual consent does not only concern the individual but has implications for family members. Genetic information about Henrietta Lacks, for example, also concerns her family members. Another argument is that biobanks are a resource for many projects. While consent can be asked for a specific project, future projects are not known and promising use in the future of biospecimens cannot be articulated now. There is a difference between storing and using. Since biobanking is future-oriented, consent cannot be fully informed. Furthermore, there is always a risk that data are disclosed. Confidentiality cannot be guaranteed even if anonymization has been done. Finally, consent will imply the right to withdraw data. But when specimens are anonymized, they cannot be removed from the collection.[79] In practice, biobanks are using various modes of consent. For example, UK Biobank uses blanket or open consent.

While the emphasis on consent is contested, it is considered some degree of guarantee to prevent the paternalistic behavior of the recent past as well as the greedy strategies of commercial businesses, such as DeCode Genetics, that transferred personal medical records to a private corporation without informed consent of individuals or serious public debate. Professional paternalism is less of a risk today than complicity of politics, media, and private business to gain monopolistic control over medical data.[80] Most scholars agree that such guarantee is essential for the success of biobanking. Gathering an adequate collection of samples and data will take time. Potential benefits will only emerge in the future. Individual participants will generally not benefit since the population at large and future generations will be beneficiaries. It will therefore require a long-term effort to recruit people to contribute. This will require a climate of public trust. Perhaps the most important ethical concern is ensuring that trust.[81]

Life as Mechanism

Because of the fundamental role of trust for the long-term flourishing of biobanking, a broader ethical perspective is advocated beyond the articulation of consent. Such perspective necessitates a critical review of some of the basic tenets of mainstream bioethics. Consent is regarded as a powerful form of control since it is the expression of respect for autonomy, one of the basic principles of bioethics. It is associated with the idea that one's body is one's property. Body parts, tissues, and cells can only be used when they are given away or donated by their owners.

Property language is used to designate the locus of decision-making authority. It is regarded as the best way to protect the interests of people concerning their body and body parts. The individual person as owner is in control over his or her private body. Without informed consent, the body might not be touched.[82] Property language is furthermore related to the issue of patenting of body materials and to commodification; body parts are items that can be exchanged, traded, bought, and sold as long as the private owner agrees. Referring to the body as property presents a moral vocabulary that promotes a particular view of human life and human existence and channels moral experience in specific directions, foreclosing possible other perspectives.

The basic elements of this moral vocabulary dominating current bioethical debates have been identified in previous chapters. Chapter 4 argued that science is not a value-free enterprise. Influenced by the ideology of reductionism and materialism it promotes a view of human beings that is disconnected from moral and subjective experiences. Chapter 6 showed how medical prophecies and the bioethics agenda are governed by a reductionistic worldview, particularly genetic determinism. Human life is regarded as a mechanism. Human beings are complexes of basic components with causal relations between these components that can be examined with basic science research and experiments. Disease is a biological dysfunction. The patient is a rational decision-maker, an autonomous subject, and owner of her body. Medical intervention is aimed at controlling disorders, eliminating diseases, and prolonging life. Bioethical discourse in general is endorsing this scientific understanding of human life and the world.

However, also argued earlier, doing so means that the experiences and subjectivity of individual persons are not taken seriously. After the exorcism of theology and philosophy from bioethics, it is difficult to do justice to moral experiences. Dissatisfaction with property language and problems with consent perhaps point to the need for a different imaginative vision of the world and ourselves.

Life as Narrative

The concept of body ownership is morally problematic. Separating the person and the body is contrary to human self-experience as embodied self. We are embodied beings. We *are* our body but we *have* it at the same time.[83] Respecting our dignity requires acknowledging our embodied existence. The body is an integral part of the person.

Philosopher Rom Harré advanced the notion of metaphysical ownership to articulate the internal relationship between person and body.[84] The human body is constitutive of our individuality and identity. We use our bodies to fix our identity and to determine and express the kind of person we are. The human body should therefore be morally and legally protected because it is the material vehicle of personhood. The human body is not just one item among other items of privileged private property. Legal ownership presupposes that there already is somebody. We would not be here without this body.

In this perspective, the idea of ownership of the human body should be rejected. It presupposes a dualistic anthropology, postulating a nonbodily person as owner of the body. Being in possession of a body does not imply that it is property; therefore a distinction should be made between possessive and nonpossessive uses of the pronoun. Calling somebody my wife indicates that we belong together, not that she is my property.[85] Nonetheless, not all parts of the body are the same. Some body parts are more like ownable objects than others. Perhaps limited ownership in regard to body parts should be allowed. Hair, urine, blood, and even one kidney can be alienated and transferred; removal of these parts does not threaten the integrity of the body. Once separated from the body, they acquire a different moral status. They do not simply become things that can be owned, however. They continue to refer to the person from which they have been separated. The genetic information they contain identifies the person from which they are separated.

Criticizing property language is associated with a different moral vocabulary.[86] It rejects the suggestion that person and body can be separated. The human being is a coherent whole (a narrative). Even if he or she is composed of various physical, psychological, and spiritual elements, these components have relations of meaning that are important for the identity and subjectivity of the person. They can be analyzed separately but need to be brought together in a holistic perspective to understand what they mean for the person. In this perspective, the patient is regarded as a subject, as the author of his or her life. This subject is autonomous but always embedded in a community and related to others.

This is especially clear in health care since it is a field based on communication and care. Medicine is not merely a natural science but a science of the human person. Causal thinking, analysis, technical approaches, control, and manipulation are valuable but not sufficient to grasp what is essential to human beings. Medicine is different from physics and chemistry; there is always a need to overcome disconnections and abstraction. This inherent tension was appropriately framed long ago by Frederik Buytendijk, one of the representatives of the anthropological tradition in medicine: If medicine is not objective, it is impossible; if medicine is only an objective science, it is inhuman.[87]

Neoliberal Context

Property language has emerged in the context of neoliberal policies. Imagining the body and its parts as property invites market thinking when collections and banks are created. The idea of property encourages the view that bodies and body parts are commodities that can be bought and sold. Since the 1980s, in fact, patenting life forms and human body parts has been legally permitted in the United States.[88]

Applying the term "bank" to amassing biospecimens and data has inserted these collections into the domain of commodities and commerce. The human body and its parts are thus regarded as resources that can be harvested, mined, exploited, patented, and traded. And it assumes that there are activities of exchange now that are different from in the past. Anatomical and pathological collections were historically static, built slowly and preserved, there to be admired and studied, and also benefiting science and education. Many of these collections were incorporated into museums. Biobanks, in contrast, are dynamic—they can change in purpose and are not just collecting but also operating to accomplish certain goals. Banks are focused on exchange; samples and data are stored in order to be used.

The expectation of banks is that stored materials are assets with economic potential; researching them will deliver new and profitable diagnostic tools and cures.[89] Thus, applying the metaphor of banking is not a neutral move. It transforms the value of bodies and body parts. No longer simply clinical waste that can be discarded, biosamples become valuable for future gains. Collecting them is a kind of investment. The traditional values are still there—connection to loved ones, advancement of scientific knowledge, essential for medical education—but now they have added value as well. They can be sold to companies to make products that can deliver profits. They also have informational value: combining biospecimens with health and genetic data will allow predictions for the future. Though mostly speculative as discussed in chapter 6, these genetic predictions are based on promises, and perhaps hopes and hypes, of beneficial returns in the decades to come.[90]

Effects of Commodification

Property language is not innocent. It presupposes the ideological context of capitalism and neoliberalism. One concern is that it displaces other forms of exchange. Turning blood into a commodity, for example, and commercializing it as a market item fundamentally changes the nature of blood donation. While the language of commodity is the language of the market economy, the vocabulary of giving is the language of solidarity.[91] Blood as a commodity is no longer the expression of solidarity between members of a human community.

Another concern is that commodification changes moral experience. It is perfectly possible to establish a market for awards (the highest bidder will receive the award), but this undermines the notion of merit that make awards valuable. It is equally possible to commodify voting, for example. Wealthy politicians who can buy the most votes will get elected, but the process disqualifies the idea of citizenship and democracy. These examples show that market thinking threatens the realization of particular kinds of human interests.[92]

There are moral boundaries to commodification because noneconomic values are disregarded and transactions on the market transform the good that is exchanged. Many people have feelings of disgust and moral repugnance if things are for sale that should not be regarded as commodities (e.g., organs or dead bodies). For them, it is disrespectful and degrading to buy and sell certain items. This is not merely an emotional response but represents legitimate rational objections. It indicates that the scope of the market should be restricted. The concern is that the market promotes some values at the expense of others. Commercial blood donation displaces a practice that promotes solidarity and a sense of community. The meaning of blood donation is transformed by paying for it.

Such transformations are also visible in the area of biobanking. People are willing to donate materials to biobanks for the benefit of future patients and the progress of medical science. However, they feel discomfort when materials are used for other purposes. Donors of materials have concerns about how their donations will be used. They may have moral concerns that materials will be used for more effective methods of abortion or developing biological weapons. The willingness to donate will decline if such future uses are known.[93]

These concerns show that providing materials is not simply a transaction of transferring ownership that is morally justified with the consent of the provider. Furthermore, studies show that lack of benefit-sharing is also a major issue. Biobanks are regarded as a public good. Public trust declines significantly when commercial entities get involved. People provide materials altruistically and as gifts. When it is noticed that self-interest or profit motives are driving the biobanking, interactions with the public change. People contributing usually do not expect any immediate rewards or benefits for themselves, but they consider it unfair when commercial entities are profiting from their voluntary and altruistic contributions.[94]

Theoretical and practical considerations of these sorts have led to serious doubts about the appropriateness of property language in the context of current medicine. Ownership, patenting, and consent primarily emphasize control and mastery. Patenting body parts and genes have created monopolies. Rather than facilitating the production of knowledge and sharing social benefits, they often impede invention, innovation, and health care.[95] If trust is more important than control, however, contributors to biobanks would not be considered as mere sources of materials and information but as participants. They would be engaged in a joint effort to advance medicine for the benefit of future generations. The emphasis, therefore, should be on cooperation and shared effort.

Instead of using property language, a different vocabulary might be more useful. For example, "library" could be a more appropriate metaphor than "bank." It represents another model of exchange but one that refers to what people have in common and what they can share. The collection is taken care of as a collective good, a common pool of resources, rather than an asset to be exploited for future gains. The objects in the collection may generate knowledge and research and stimulate creativity but the value they have is not projected in a speculative future. Libraries also are public spaces, accessible for all citizens.[96] An alternative metaphor is the notion of stewardship, which expresses the idea that humans are guardians, not owners, of their bodies. And what follows are the notions of solidarity, justice, and responsibility.[97] If biobanks are considered as social rather than commercial institutions, an individualistic ethical framework is inadequate.[98]

The Relics Narrative

The suggestion to regard human remains, biospecimens, and body parts as relics might explain some of the ambiguities and discomfort in regard to collecting, storing, and displaying human materials. The Nuffield Council on Bioethics in an early advice on human tissue recommended that human body parts should not be displayed in connection to public entertainment or art. Some uses of human remains are unacceptable, even if consent has been obtained. Consent, therefore, is not the primary consideration according to the council. Moral notions of respect and dignity come before the notions of property and ownership.[99] But why? Why do human bodies and their parts demand respect even after death?

Relics have special value because of intercession. They mediate between the past, present, and future. Intercession is the work of the dead. The corpse and human remains are not viewed as living agents but also not as an inert collection of chemicals. They have "quasi-human status"[100] and symbolic value: they are endowed with power since they promise a better future. The language of property and commodity is not adequate to express this symbolic value. It is, instead, a matter of faith: faith in a future where new knowledge and innovative medical applications will be developed.

Value is generated in an interaction between materiality and abstraction. Like relics, human remains are part of the material world, nothing but bones and dust. At the same time, they have moral value as symbolic capital. Genomic information becomes virtuous since it is linked to potential therapy. It is a precursor of therapy,[101] albeit one based on speculation. Body samples are valuable because they have the potential to change the future, and as such they impact the present. These forecasted benefits are intangible abstractions promising salvation in the future. The point is not whether the promises are true or false but whether they are credible or not. The current discourse is promissory, just as was the discourse of saints. It uses the imaginary of salvation as well as new miracle drugs.[102] Biospecimens, like relics, have mystical force.

The concept of liminality is also relevant. As discussed earlier, relics and the human body itself are in a liminal state. They are part of the material world but simultaneously part of a person, even if deceased. They have been integrated into an individual subject and still refer to this subject after he or she is gone. The human cadaver is a thing but at the same time it is also a symbol of the human person who had been present in the world. The subject is clearly no longer alive but also not like other objects or things in the world. That is why mutilating a body is a moral offence. It also explains why commercializing the human body evokes moral repugnance.

This liminal state also applies to the pilgrims venerating relics. They break away from their ordinary lives and pass through a transitional period before being reintegrated and incorporated into a community of believers. The concept of liminality further applies to medical collections. Body materials bring us from one state to another. They are in between the present and the future. And the banking metaphor particularly proposes that biological materials and information are currencies that will only increase in value.

Moreover, liminality applies to the donors of biomaterials. We should break with the idea that they are simply the sources of biospecimens and transition them to a new state of biological or scientific citizen. Contributing to biobanks should be considered as an act of responsible citizenship. Reconfigured as a special type of subject, donors become citizens that are concerned about research as a public good and active participants engaged in shared projects that will enhance the health and well-being of future generations.[103]

In the past, the cult of saints included three elements: name, body, and text. In the present, the same elements are required to make collected body samples valuable. Human tissues need to be connected with the disease history and lifestyle of the persons donating them. In order to have future value, a narrative must be constructed that connects biology and biography. It demonstrates that the embodied person has diachronic relations (history, tradition, and biography) as well as synchronic relations (community and culture). This emphasis on narratives clarifies two issues that are important for building and maintaining public trust.

First is that people are part of a community, not simply contributors or donors but participants acting out of a sense of solidarity with future patients. Individual autonomy is interpreted as a social responsibility. An individual as a citizen is more than *Homo economicus*, concerned only with his own self-interest. Second is that the body is never an abstraction; it is always somebody's body. The lesson from the history of relics is that separating cadaver and person will not work. In medicine, however, the story of the person is often suppressed, or at least separated from the story of the body. Anatomical and pathological exhibits are usually depersonalized. Museums and anatomical theaters put a lot of effort into make dead bodies pedagogically useful by distancing them from their original living context.[104] *Body Worlds* has erased personal history and all markers of identity from the plastic bodies that are nonetheless presented as active and interactive. But human remains are not simply objects.

Respect and Dignity

In ordinary life, an ontological distinction is made between persons and things. This distinction is relevant for ethics. Persons need to be treated with respect, while things are objects that can be used for various purposes. Things can be exchanged in transactions; they have use value and can become commodities that are saleable for money or things with equivalent value. Simple shorthand says, things have a price and persons have dignity. The debate about the moral status of the cadaver draws on this distinction. Death means that the person is no longer there; what remains is the dead body that has become a thing like other objects in the world. However, the narrative of relics indicates that the distinction is too rigid. Relics can be both; they are objects that are persons as well as things. In fact, they are objects in-between persons and things; a kind of sacred entities that have special value.[105]

The intermediary status of human remains between subject and object, or person and thing, exemplifies the interconnection between the two perspectives on human life just discussed: life as mechanism and life as narrative. The first highlights life in the biological sense: life is a biological phenomenon. The second focuses on the story in which the subjectivity of the person is formed: life is biography, a fact of experience. Over the last few decades these two perspectives have been more and more opposed. Life as biology is often prevailing over life as biography. As a fact of nature, human life can be examined and explained by reducing it to its most fundamental material units, namely genes, cells, and molecules. Nonetheless, in ethical discussions, human life is a fact of experience—it is not a matter of science but refers to the existence of a singular subject with a unique biography.[106]

The distinction between person and object, which seems simple from a scientific perspective, is problematic and ambiguous from an ethical perspective. Ethical ambiguity explains why in most cultures, human bodies and human remains are awarded special respect. Dead people should be buried. Human corpses should not be used to feed animals in the zoo. Burial places should not be desecrated. The human body is worthy of special protection regardless of the wishes of the person incorporating the body. Respect for the dead is therefore regarded as a human universal.[107] Anthropologists today identify remarkable similarities in this regard among most cultures and civilizations. These universals are rooted in a common human biology, but they are also related to basic human needs, which elicit often comparable cultural responses.

Explaining the respect awarded to human remains appeals to the notion of dignity. The moral injunction not to mutilate or desecrate the corpse is difficult to explain with the language of property. In such dualistic vision in which bodies are perceived as property, it is arduous to identify who exactly is the owner (since the initial person is no longer there). For body samples and human remains collected long ago it is difficult to identify heirs or communities that can claim them. But even if property rights can be established, and thus consent from the "owners" obtained, it does not justify all possible uses of human remains. The notion of dignity operates at a more fundamental level than the usual ethical considerations of rights, liberties, harms, and consent.

In regard to persons as embodied human beings, dignity articulates that they should be treated with respect in and of themselves. The notion is used to protect subjects against objectification and degradation. Respecting dignity is not dependent on an individual decision or interpretation but is a collective value. Whether or not a subject has consented to be used in a specific manner does not remove the indignity of some practices. One example is the "freak show," discussed in chapter 4. For some people, exhibiting themselves at traveling fairs was the only way to earn money. Some were exploited but others made a voluntary choice to display their disabilities and monstrosities. But regardless of possible harm and consent, the display was regarded as an affront to human dignity. Another example is "midget-tossing," a contest to see how far one can throw a dwarf. It is still practiced in some places as a form of entertainment, although banned in many others. The argument for the practice is that the participating dwarf has consented, is paid, and wears a helmet and protective gear. The counterargument is that the game is degrading—it does not show respect to little people in general. Finally, a commonly used example is cannibalism. Britta van Beers discusses a show that involved live cannibalism on Dutch television in 2011 when two presenters agreed to eat each other's flesh.[108] Nobody was murdered (the flesh was removed by a surgeon to minimalized physical harm) and mutual consent was given. Nonetheless, the show was considered disgusting, but for what reason?

Within the framework of neoliberal ethics, nothing is wrong with freak shows, dwarf-tossing, or cannibalism if informed consent is provided and no harm is done. People are autonomous persons in charge of their own bodies and can voluntarily decide what to do with them as long as nobody else is hurt. These practices are regarded as offensive, however. They provoke unease, disquiet, revulsion, and repugnance. Displaying or exploiting human bodies strikes against some core value that requires respect for human bodies since they have dignity as embodied persons. Using them as objects signifies denigration of the human person.

In bioethics discourse, moral repugnance is often downplayed as merely an emotional response. It is interpreted as "visceral reactions" or "rhetorical claims."[109] It is not recognized that revulsion and disgust have a moral value or that there is a level of ethical concern beyond the standard considerations of consent and harm. However, these visceral reactions are not simply irrational tendencies but instead "unarticulated wisdom from which we can learn."[110] They voice the worry that unrestricted interventions may not always be beneficial, that respecting some limits may be better for the long term and for humanity as a whole, and that ethics is about more than respecting individual autonomy and preventing harm. Like Frankenstein's metaphor indicates, the concern is that the availability of technology should not dictate its application. Perhaps humans should be more cautious in applying technology and new knowledge because not every intervention will be an improvement. As philosopher Michael Sandel argues, some activities are objectionable because they express "a stance of mastery and domination that fails to appreciate the gifted character of human powers and achievements."[111] It makes it more difficult, then, to explain the relevancy of the notion of dignity in regard to entities between persons and things. If human remains fall somewhere in between, or like relics are sacred entities that are both, can it be argued that displaying and exploiting dead bodies and body parts is violating dignity?

In the second part of the 19th century, there was a movement in England to close public anatomy museums because they were regarded as obscene, and in the 1870s the last of these museums was closed. Physicians argued that display of anatomical models was indecent for the general public and should be the domain of medical teaching and research. This movement was associated with the general cultural Victorian climate at the time,[112] but it illustrates that the argument around respect for human remains relates to potential uses. The Nuffield Council on Bioethics, for example, outlines research, education, diagnostics, therapeutics, and forensic purposes as ethically acceptable. Other uses (including display for entertainment) are unacceptable even if consent has been obtained. These uses "injure human beings by treating them as things, as less than human, as object for use."[113] According to the Nuffield Council, injury is the result of destruction (irreparable), damage (reversible; impairing function or causing pain; constituting serious harm) or degradation (depending on cultural and personal differences). Examples include using human tissue as food or other manufactured products (e.g., soap).

Focusing on injury or harm is problematic since it presupposes that human remains or body parts can be harmed. The claim is not that disrespectful treatment of the dead is harmful for the survivors—though this can be the case, it is not the reason respect for the dead is required. So if the dead can no longer be physically harmed, what interests do they have that can be affected? One postmortem interest is the fulfilment of the wishes of a decedent, justified as "prospective autonomy."[114] People are autonomous decision makers while they are alive, and their premortem choices need to be respected. Still, not all choices are equally respectable (e.g., they cannot violate the dignified treatment of the dead, such as that of decent sepulcher and quiet repose). Another postmortem interest might be a "memory picture."[115] People often cultivate a narrative, a lifetime image of the kind of person they are or wanted to be during their life. They want to retain that legacy for their family and survivors, and there is moral harm done if it is not. Both arguments apply the idea of extended personal autonomy and assume that self-determination continues at least some time after the person has disappeared.

Another type of argument extends the notion of human dignity and assumes "postmortem human dignity."[116] The corpse is associated with the former person; it has some quasihuman status. It is the human origin of the dead body that continues to determine its future. What is important here is the history of the person. The difference with the previous argument is that the emphasis on dignity rather than on autonomy transfers the issue of respect beyond the individual level. Respect for the dignity of human remains is required from survivors; it gives them peace of mind to notice that the corpse is not mistreated. But it also appeals to all members of a society to ensure the decent disposal of human remains, whether or not there are survivors who can be injured. What is regarded as degradation and indignity is determined at the level of communities and cultures.

The question of why the dead have moral claims against the living is difficult to answer with the argument that there are posthumous interests or potential harm. It can be argued, however, that the moral commitments extend beyond one's lifetime. As a moral personality, one can, as Ernst Partridge points out, "transcend, through imagination, the bounds of one's immediate time and place."[117] We can draw up formal contracts, such as wills, and make promises affecting events beyond our death. But these provisions for the future demand posthumous respect because it is in the interest of the living—they want to be sure that their wills and concerns will be respected in turn.

The narrative of relics illustrates how the human cadaver is not regarded as somebody's property but, nonetheless, sometimes commodified.[118] Already in his time, St. Augustine complained about the commercial trade in relics. Although officially prohibited by the Church later, and repeated in recent Vatican rules for the use of relics,[119] there always has been a commercial market for relics. The large collection of relics in Pittsburgh, for example, grew as Father Mollinger bought relics in Europe that would otherwise be destroyed. He also combined his healing services with an adjacent pharmacy that sold his patented herbal medicines. Mollinger Medicine Company was incorporated in Pennsylvania in 1925 and his herbal pills and tinctures were advertised for decades in Pittsburgh newspapers whether or not combined with the veneration of relics.[120]

Biobanks often have an ambiguous attitude toward commercialization. Donated body materials are not considered commodities but gifts. But as soon as they are donated, they become the property of the bank. Donors no longer have any claims on the potential benefits that result from research with their body materials. As soon as they are included in the bank collection, the samples do move from gifts to commodities. This reflects the position of the Nuffield Council on Bioethics: procurement of tissue should be noncommercial, but the use of human tissue can be commercially organized (e.g., therapeutic products may be derived from tissue). The biobank can therefore operate as an intermediary institution connecting nonmarket and market structures.[121]

Conclusion

Parts of the human body have been gathered for centuries in anatomical and pathological collections, first in private cabinets and later in medical museums. For most of that time, it was assumed that people would not have any objections when organs and tissues were removed from their bodies during surgery or after death. Body parts were simply taken without informing the dying or their relatives. Historically, these parts and remains have been preserved, stored, and exhibited for various reasons: medical research, education, historical reconstruction, archaeological exploration, and public enlightenment. Over the last few decades, however, public display of human remains has become a sensitive issue even as huge efforts are being made to collect biomaterials and store them in biorepositories. In this chapter, the historical narrative of relics clarifies the contemporary significance of collecting body materials.

The cult of relics, which is not necessarily connected to religious perspectives, demonstrates that dead human materials may have intercessionary power. They can bring future salvation because they transcend the world of present mortality. Relics furthermore show the liminal state between life and death. Human remains are no longer embodied persons, but they are also not just inconsequential objects. Finally, relics demand narratives. Their materiality has special value because it is linked to personal stories. These characteristics of relics are recognizable in present-day attitudes to human remains and body parts, and they have led to such scenarios as museums being asked to repatriate remains to be buried within their communities and parents learning that organs retrieved from their children during autopsies led to incomplete burials.

Until now, ethical debate concerning body materials has focused on the issue of consent, assuming that property language is adequate to determine what will happen with biospecimens. This focus reflects the perspective of mainstream bioethics that takes respect for individual autonomy as the most fundamental ethical principle. This chapter argues that a broader concern is at stake: the space between "person" and "thing." Human remains are more than just objects abandoned by autonomous persons. This is why experiences with exhibitions and displays of human remains are often uncomfortable and disquieting. It also explains why cryogenic body parts and immortalized cells are not considered merely material products. It may also clarify why people feel repugnance when they witness the maltreatment of human bodies. Apparently, there is a level of moral sensitivity beyond the usual level of rational discourse.

Put another way, the basic image of mainstream bioethics assumes that the individual person should be in control of his life and body. Respect means that decisions of the autonomous person are accepted and recognized since they are based on personal values. If a body is property, consent of the owner is determinative for its appropriate uses. This image has been very productive for medical science but, as argued here, is deficient from the perspective of human experience. Persons and bodies cannot be separated. The narrative of relics reminds us that life itself is a narrative with persons embedded in communities, histories, and traditions. Individuals feel connected to other people. They are aware they cannot fully master their fate and destiny alone and are dependent on social and community support.

Processes of globalization have further highlighted these features of life as narrative, but when viewed through the lens of health and medicine, globalization illustrates that concerns with individual autonomy, consent, body property, and personal control are important but insufficient to ensure they will be beneficial for everyone. Bioethics needs to extend to another level of concern: universal human dignity.

CHAPTER EIGHT

Critical Bioethics

In November and December 2018, protesters paralyzed France. So-called yellow vests (*gilets jaunes*) blocked roads and roundabouts throughout the country to object to a rise in gasoline taxes. The complaint was not one of ecological measures but instead against another tax that favored the rich at the expense of the majority. The movement emerged spontaneously from a Facebook post without any connections to trade unions, political parties, or established institutions and was supported by 75 percent of the French population. It united individuals with different political orientations and appealed primarily to ordinary people from suburban and rural areas who had difficulties in making ends meet. All activities were coordinated online through *groups colères* ("angry groups") with hundreds of thousands of members, which made the movement difficult to characterize since it had no specific leaders or agenda. In general, it represented average lower-middle-class citizens who did not feel represented, and even ignored, by the political elites. They felt victim to growing inequalities and injustices. One of their common slogans, in fact, was "Dignity for all," which communicates that people felt humiliated and were acting out of a sense of social justice and human dignity.[1]

Similar movements around the world have articulated the same concerns. In 2011, *los indignados* ("the outraged") protested in Spain and Greece against austerity measures of their governments. They rejected the subjugation of politics and denounced the impotence of government to protect citizens from economic injustice and hardship. In opposing political and economic elites, the Spanish protesters promoted the image of an occupied square to emphasize the public space as a global icon that represents the value of active and direct democracy.[2] After all, neoliberal policies are implemented by socialists, conservatives, liberals and Christian-democrats alike, and yet they subsume all values under the economic one.

The parents of Charlie Gard (chapter 1) launched a social movement in 2017 with several hundred thousand supporters across the world. They contested decisions of medical doctors and courts concerning their severely handicapped child and created a global controversy that many people got on board with. In a short time they raised $1.67 million for possible treatment, mobilizing support for their case with the use of social media and the internet. More than 60,000 people joined Charlie's Army on Facebook to support the family. Charlie's parents set up a foundation to help other babies with mitochondrial diseases, and they also initiated legal proposals (Charlie's law) requesting independent mediators to be involved in disagreements. One of their main arguments has been that the focus in cases such as theirs should not only be on the best interest of the child but also on the question of whether treatment would cause significant harm to the child.[3]

Bioethics witnessed another outbreak of moral outrage at the end of November 2018 when He Jiankui, a Chinese researcher, announced that he had produced two genetically edited babies.[4] He used gene-editing technology called CRISPR to alter the genome and make the twin babies resistant to future HIV infection. The scientific community was shocked. There was a global outcry of anger and outrage, not only about what he had done but also how he did it, seemingly working in secret outside established protocols. The responses of scientists, authorities, and the general public in Chinese and Western media condemned his work as "unethical," "disgusting," and "monstrous."[5]

The above events illustrate two points. First, they show the power of social media, particularly virtual social networks. Interconnectivity and new information technologies provide people with the means to quickly share views, denounce practices, and mobilize for protests. Second, the events demonstrate that in general, but for bioethics in particular, debate and protest is motivated by anger and outrage. In fact, moral outrage is frequently expressed on social media and blogs when our sense of fairness is violated. This social, political, and economic context of contemporary human life has changed with processes of globalization.

Critics of globalization regard the phenomenon as an exploitation of trust by an elite minority that reaps most of the benefits at the cost of the majority. Inequality is growing between and within countries. The driving force of global policies seems to be maximizing profits and no longer reflect the interests of average citizens. This is especially visible in bioethics today, and a critical analysis of the driving forces of globalization leads us to neoliberalism as the fundamental ideology, the view of the market as the primary model for human life and society, and the role of bioethics as a supporting force for this ideology and model. It is furthermore argued that other critical approaches—those based on moral imagination that can propose different ethical discourses—are needed.

And so we refer to the role of metaphors and imaginative visions that have been discussed in previous chapters. Bioethics can learn from philosophical and theological traditions to develop a broader approach that scrutinizes the interpretive frameworks and imaginative views that determine our understanding of the world and human existence. This broader approach is especially needed now that academic bioethics is increasingly confronted with social media and popular movements.

Moral Outrage, the Internet, and Bioethics

Moral outrage is anger over a violated moral standard. It is a powerful emotion and a source of motivation to engage in action to reaffirm or reestablish the moral standard. It should be distinguished from personal anger (which is not a moral emotion but related to the affection of one's own interests) and empathic anger (which is also not a moral emotion but triggered because somebody else's interests are thwarted, not because a moral standard is violated).[6] Not all forms of anger are emotional responses to violations of moral transgressions, and only 5 percent of people report that in daily experiences they witness immoral acts.[7] It is argued, however, that social media has transformed the expression of moral outrage. The internet exposes people to a vast range of misdeeds, corruption, trafficking, dishonesty, and fraud. Most people learn about immoral acts online these days, and this online content is being widely shared.

There is evidence that immoral acts encountered online incite stronger moral outrage than immoral acts faced in person or via traditional media.[8] Digital media seems to alter the subjective experience of outrage. If there is a continuous online experience of immoral conduct, a person can also experience outrage fatigue. On the other hand, it could also become easier to express outrage since the threshold for consequences of expressing it is lower. There is not much chance of encountering the wrongdoer and little risk of physical retaliation. The costs of shaming others is lower for the shamer, even as the damaging effect on the perpetrator's life can increase with massive online retaliations. Anger and moral outrage can be destructive. They can lead to the perception that offenders are less than human (i.e., monsters) and lack core human qualities.[9]

Anger and outrage can also have a positive meaning. They signal moral quality to others and publicly clarify when behavior is not socially acceptable. It may even enhance one's reputation by expressing moral outrage. This more positive view was endorsed in a recent publication by Charles Duhigg. He argues that anger has always been an important emotion. The main question is how to deal with anger and indignation. Anger can make bad situations better because people are listening, speak more honestly, and are more accommodating to complaints and grievances. Moral indignation can be a force for the good as it transforms practical and economic complaints into emotional and moral issues. The protest of the yellow vests, for example, was not simply about higher taxes but rather about justice and equality. People did not merely want higher wages or better working conditions; their movement demanded righting an injustice. Moral indignation therefore locates their protest and discontent within a broader fight about rights and wrongs; it reframes complaints into moral offences. Duhigg concludes that we "need moral outrage that motivates citizens to push for a more just society."[10]

Looking again at He Jiankui and his gene-edited babies, the case illustrates a major issue: science cannot regulate or police itself. Science and technology apparently have been caught in the same processes that could endanger humanity. In a climate of global competition, science is driven by performances and outcomes.[11] Even when there is broad consensus that gene editing should not be applied, for example, how can an individual scientist or groups of scientists be prevented from unilaterally deciding to go ahead with experiments? On the other hand, those experiments have potentially global consequences so why do those scientists think they can decide for humanity? The real moral outrage, therefore, is against the arrogance of scientists who think they can decide what is ethically appropriate.

Because of such arrogance there is a need for global standards going beyond self-regulation. Scientific advancement should be subordinated to social debate and control. Is the fundamental question whether altering genes of future children should be done at all? Or maybe the focus should be on how to proceed more cautiously and not on moratoria or prohibitions. The dismay surrounding He was produced by the rogue nature of his process, not by outrage over the ethics of gene-editing per se. His critics assume that it is up to the scientific community itself to determine when and how to apply new technologies. Against all odds, they reiterate that scientific academies should develop criteria and standards, while the example of He clearly demonstrates that this will be insufficient.[12]

This language of moral outrage is common in bioethics debates, as seen in each chapter herein, but often it is not taken seriously. It employs such concepts as revulsion and repugnance as well as lack of decency, dignity, respect, and humanity, rather than just anger. And the focus is on offensive practices (e.g., selling organs) or inappropriate transactions (e.g., prostitution, human trafficking, or dwarf-tossing).[13] In order to understand why this language is more than an emotional or irrational response, we must explore the changing context of human existence that impacts the current bioethical debates.

Globalization and Vulnerability as a Global Phenomenon

One of the remarkable features of contemporary bioethics is the emergence and expansion of the notion of vulnerability.[14] It was first used in the context of research ethics but later expanded to other areas of bioethical debate. The emphasis on vulnerability articulates that the human person is not only an autonomous subject but also a body that is susceptible and fragile. The notion also calls attention to the social context in which persons are exposed to threats and harms. However, in mainstream bioethics vulnerability is construed as an individual deficit.

Vulnerability is a central topic in a variety of discourses, such as nursing science, public health, and social sciences. It is also used in new fields of study concerning pandemics, disasters, environmental degradation, climate change, bioterrorism, and human security. The fact that the world has become increasingly interconnected and interdependent has created a sense of mutual vulnerability. Being vulnerable is often the result of a range of social, economic, and political conditions, and processes of globalization have resulted in a world that not only creates more and new threats but also undermines the traditional protection mechanisms (social security and welfare systems, family support systems). The abilities of individuals and communities to cope with threats are being eroded. Entire categories of people are disenfranchised, powerless, and voiceless. It is clear that this interpretation of vulnerability as global phenomenon—and therefore beyond the power and control of individuals—is at odds with mainstream bioethics framing it as an individual affair.

Wasted Lives and Inequal Opportunities

Although vulnerability is manifested at the individual level, its roots are in structural conditions and circumstances. Over the past few decades, neoliberal policies have made life more precarious for most human beings. They have created a context of structural violence and multiplied opportunities for injustice and exploitation. Emphasizing efficiency, private responsibility, and reduction of social safety nets, these policies have instituted a "politics of disposability."[15]

A growing number of individuals and groups are becoming redundant, superfluous, and unworthy of care and protection; they are the "undesirables" and "deplorables." Entire populations are no longer relevant for economic development—for instance, slum dwellers in Abidjan, beggars in Mumbai, or the poor population of Flint, Michigan. They are groups of people most susceptible to exploitation, exclusion, and state violence. Public services to sustain their basic needs (health, food, and water) are regarded as unnecessary so that they can be cut. Polish sociologist Zymunt Bauman has called these excluded and discarded populations "wasted lives."[16] In present-day societies, precariousness is not an exception; it has become the rule of modern life,[17] and these populations form a new "precariat" class who are the losers of global processes. In fact, they are not even necessary for development and economic growth.

According to the neoliberal ideology, poor people are lazy. Unemployment and poverty are matters of individual responsibility. Lives are precarious because of our own choices, not because of exploitation, power differences, disease, and poverty. Ignoring structural forms of injustice, neoliberal discourse celebrates diversity and difference because people are not equal, and by focusing on individual differences, it ignores the inequalities produced by economic and political systems.

More and more, though, inequality is being recognized as a major social, political, and economic problem.[18] Even International Monetary Fund experts have concluded that at least some neoliberal policies have overplayed growth and increased income inequality, thereby jeopardizing the economic system.[19] Inequality is blamed for political and social instability, which includes the undermining of democracy, a lack of social cohesion, a decrease of solidarity, an erosion of a sense of community, and the destruction of a spirit of fair play and justice.

For a long time, inequality and vulnerability have been ignored and deprioritized as the inevitable consequence of globalization, with the argument being that they will diminish in the long run as priority is given to the market ideology. Today it is no longer possible to deny that inequality exists and is growing. The market is not solving this problem but amplifying it. Recent polls in the United States show that two-thirds of the population is dissatisfied with the distribution of wealth and income.[20] Many people feel that something is fundamentally wrong and unjust; the economic system has brought them into an undignified situation.

The Bioethical Response

While it can be argued that inequality is one of the basic problems of globalization today, the bioethical response is rather muted. Although more attention is paid to the issue of vulnerability, the bioethical perspective on it is not much different from the neoliberal perspective. According to market thinking, the human person is primarily *Homo economicus*: a rational individual motivated by minimizing costs and maximizing gains for himself. In this perspective, humans relate primarily to others through market exchanges. Citizenship, the public sphere, and social networks erode since there are only individuals and commodities that can be traded.[21]

This economic discourse is similar to the dominant discourse of contemporary bioethics that considers human beings primarily as autonomous individuals. Being ill, receiving treatment and care, and participating in research are first of all individual affairs; involving others requires consent and individual decision-making. However, what does the dream of individual autonomy mean when you go to bed hungry, drink dirty water, and do not have a home, basic health care, medication, or a reliable job? Unequal power structures and structural violence are simply ignored in many ethical and political discourses.[22]

The point is that focusing vulnerability primarily on the individual person ignores why it has become a bioethical problem in the first place. It has surfaced as a problem because of globalization. The discourse of vulnerability has particularly expanded in the context of global phenomena such as natural disasters, pandemics, poverty, hunger, and environmental degradation. Vulnerability is the result of the damaging impact of neoliberal logic. Take for example the loss of biodiversity. It is associated with economic growth and development, but the impacts are unequal; poor and traditional societies are most vulnerable since they rely directly on the benefits of the ecosystem.[23] When natural medicinal products are lost, for instance, indigenous populations cannot afford to purchase the drugs developed from their traditional knowledge. Loss of biodiversity exponentially exposes vulnerable groups, and further underlines inequality.

Vulnerability is, therefore, not an individual concern but a social one since society itself is affected. In this perspective, ethical analysis should focus on the broader context that produces vulnerability.²⁴ This is only possible in an ethical framework that goes beyond the individual perspective to include justice, solidarity, care, and social responsibility. The principle of respect for human vulnerability, in fact, is closely related to the principle of justice. Global health is frequently regarded as an ethical commitment to the most vulnerable people and populations, for example. The idea that everybody should have an equal opportunity to have equal health status implies that efforts are specifically directed to the most disadvantaged persons and groups. Those vulnerable groups with the poorest health should receive priority.

Reframing Bioethics

More and more bioethicists today argue that bioethical discourse needs to change. It should rethink the agenda that is currently focused on issues arising in wealthy countries while common global issues are ignored.²⁵ Bioethics should develop into a critical discipline that examines the social processes that determine bioethical problems.²⁶ It is too distanced from the values of ordinary people and too far from the social context in which problems arise. Ethics should be "resocialized" (i.e., located into specific contexts; for example, considering the setting of poverty with the lack of access to treatment). The environment of injustice and inequality frequently denies the fruits of science and medicine to many people.²⁷

A recent concern is how bioethics should respond to populism. Bioethical analysis often has a narrow perspective and does not engage with "the deeper underlying values" that are important for democratic deliberation and for the dignity of individuals and groups. Particularly, it has neglected the common good, interdependence, solidarity, and structural justice.²⁸ However, recognizing that mainstream bioethics is deficient in the face of new challenges is one thing, and providing a remedy is another. It will not be sufficient to add more topics to the bioethics agenda or integrate empirical and sociological data into its analytical framework. More will be required than just expanding the scope of bioethics.

Solomon and Jennings hint in a certain direction (clarifying "underlying values") without elaborating what this entails.²⁹ Bioethics should emphasize the common good, develop trustworthy information, ensure the integrity of agencies, and defend civil rights. Bioethicists are in a silo and have become a professional elite. Many will agree with the authors that this comfortable exclusiveness should not continue. But, again, a broader approach will not be sufficient without an analysis of why bioethics so far has not taken these larger issues and deeper values into account. The authors do not refer to the economic and social value contexts in which bioethics has emerged and materialized as an established discipline; they do not discuss how bioethics often implicitly endorses and reinforces the neoliberal value policies that promote scientific and technological advances; they do not articulate how the emphasis on individual autonomy, self-interest, and competition have undermined such values as mutual respect, trust, and dignity. Without attending to such contextual issues, expanding bioethics will be inadequate; another kind of bioethics will be needed.

This book has followed the suggestion of Solomon Benatar: what is necessary is a greater moral imagination, enabling us to completely alter our outlook and actions.[30] Current bioethical debate is bizarre because it concentrates on exceptional cases while not paying attention to underlying value perspectives that determine the agenda of the debate. Often triggered by issues such as experimental drugs, gene-edited babies, and life extension, bioethical agenda-setting is currently driven by two main forces: neoliberal market ideology and faith in scientific advancement. Both forces bring their own value context that is usually not critically addressed in bioethical analysis.

Neoliberal Ideology

As a main driving force of globalization, neoliberalism has significantly changed the environment of science and health care. Commercialization, publication pressure, promotion of drugs, and lack of accessibility to medication have discredited the objectivity of scientific knowledge, the reliability of scientific facts, and the independency of scientists (all part of the ghost phenomena discussed in chapter 3). Leading medical scholars, such as the dean of Yale's medical school and the new president of the American Society of Clinical Oncology, are not reporting financial relationships with pharmaceutical and health care companies.[31] Despite all discussions and codes of conduct, these trends are not receding. The corrupting role of money in science and health care has been known about for decades, but not much is changing because the roots of the current system are not being scrutinized.

The promotion of genetic testing and precision medicine (chapter 6) are also examples of basic neoliberal values that are assumed without serious questioning. These scientific advancements are regarded as tools and resources available on the market to assist individuals in predicting future health and realizing life without disease, and possibly without death. The underlying values are that individuals are rational, self-interested decision makers; progress and growth are inevitable; and socioeconomic context is irrelevant. The same assumptions are maintained in the context of present-day biobanking (chapter 7). Individuals are the owners of their bodies, and body materials are resources that can be commodified and patented.

These dominating value perspectives significantly restrict moral imagination. Considering health care as a market and patients as consumers predetermines the moral perception before an ethical debate has even started. Issues of vulnerability, altruism, or empathy are not relevant. Moral agency is instrumental rationality, aimed at maximizing benefits and minimizing harms.

Scientific Ideology

Earlier chapters have pointed out that science presents itself as an objective and value-neutral enterprise, even as it is haunted by ghosts, monsters, pilgrims, and prophets. The idea still is that scientific facts are neutral and reliable. Scientists, however, do not limit themselves to observation, experiments, careful studies, and tentative conclusions. They often offer predictions about the future, make promises for future advances, and persuade policy makers to support their work because it will deliver miracles. They assume they can do this on the basis of a scientific worldview grounded on reductionism, materialism, and progress. This view is in fact an ideology inspired by a specific value perspective that regards a human life as an explainable mechanism that can be controlled by appropriate knowledge and techniques.

This is a powerful image of human beings that has delivered impressive results. However, it is a limited image that does not fully recognize the human subject and the world in which she exists. Human beings are not simply algorithms or biomolecular engines. They are not merely subjects in charge of a physical body. The insufficiency of this view of human life is now acknowledged in science itself (but not promoted in public discourse). Genetic reductionism is no longer a tenable position, and the significance of genes is greatly diminished. But what is more remarkable is that scientific discourse is often invoking metaphors and narratives that go beyond its own ideological worldview. In the area of prediction and prognosis, for example, science often employs imaginative visions of progress and perfectibility, projecting promises and expectations into the future. Similar projections of hope and miracles are made in regard to the search for cures across the globe. These futures are promoted by science itself, though not warranted by the worldview of scientism.

The Challenge for Bioethics

It is the hallmark of an ideology as a system of normative ideas and ideals that it applies itself to other areas of theory and policy outside of the domain in which it has originated.

Neoliberalism proclaims liberalization—individual liberty and human well-being will be fostered if constraints on free market competition are removed; self-interest and competition are the driving forces of social organization and advancement. What initially emerged as an economic view has developed into a broader view of human beings and human life. The "market" has become a metaphor for how social life should be organized. Every domain of society should be arranged according to market thinking: education, social services, prison systems, and health care.[32]

A similar expansion can be observed in scientific ideology. Science proclaims itself as an objective, neutral, and value-free endeavor. It is founded on a reductionist methodology that has produced many advances but has expanded into a broader worldview of scientism, assuming that it can fully explain social and cultural events, historical developments, and human beings. In the new life sciences, and applied in health care, individuals are treated as biomolecular engines. Organisms are determined by their genes. Health and disease can be reduced to genetic phenomena. Social and cultural phenomena can be explained by the biological and physical sciences.

For bioethics, two challenges are on the table. First is the challenge of the worldviews that are presented by neoliberalism and scientism. They are restricted value perspectives that reduce other perspectives to either the market or the mechanism of life. If it is required that bioethics will engage with other values, it should first of all be more critical about the values endorsed by neoliberal and scientific ideologies. The second challenge is a practice already visible in contemporary discourses. This book shows that multiple interpretive frameworks are at work in scientific and bioethical discussions. Traditional metaphors are operative in these debates, not only in popular responses but also in scientific and ethical debates. Health care professionals, scientists, ethicists, and lawyers themselves appeal to interpretive frameworks by calling patients monsters or vegetables, or by promoting miracle drugs and cures. Imaginative visions are frequently used in science and health care, and bioethics can build on these visions to propose alternative approaches.

The Role of Moral Imagination

Perhaps philosophy and theology can help to revitalize bioethics. Critiques of the dominant scientific and economic worldviews that have moved out of mainstream bioethics should move back in. This is another way of saying that ghosts, monsters, pilgrims, prophets, and relics have a place in the current debate.

How Can Moral Imagination Help?

Metaphors and imaginary visions are cognitive ways of framing and acting on the world. Bruce Jennings and Angus Dawson describe moral imagination as "the capacity to take a critical distance from the given, to think reality otherwise."[33] Imaginary visions are not just rhetorical devices. Many metaphors today are mechanistic: the organism as a machine and genes as building blocks. The emphasis on precision is connected to "targeting." Metaphors do not provide explanations but furnish ways of thinking, give structure to ideas. The concept of solidarity, for example, may broaden the moral imagination of bioethics by shaping sensibility to go beyond particular acts and individual agents.[34] Imaginary visions can reflect negative as well as positive experiences: deception and trust, horror and fascination, hope and despair, disgust and admiration, and compassion and self-interest.

In her final book, philosopher Mary Midgley focuses on patterns, world-pictures, and frameworks as ways of thinking to explore the constantly changing world. She argues that there are always different perspectives available. In her view, imaginative visions of how the world is "are the necessary background of all our living. They are likely to be much more important to us, much more influential than our factual knowledge."[35] There is a close relation between how we think and how we live. We should be looking at life as a whole, and thus take into account different perspectives. For example, there are always two aspects of human health: the physical aspect appropriate for medicine and science, and the imaginative or sympathetic social aspect that reflects the point of view of the patient and the subject.[36] Philosophy suggests new ways of thinking that call for different ways of living.[37]

The danger is that one worldview will become dominant. In the current era, this danger is manifested in scientism. Midgley contends that the scientific worldview is a myth and that human behavior cannot be understood within it. It articulates a world of objects without subjects.[38] Physical sciences are regarded as the metaphysical source of all our knowledge. Scientism is the new sedative,[39] and it promotes a reductive program: that there is only physical matter and the universe is an immense machine. However, Midgley argues, this is not real science but instead a vision, an image, "an enormous act of faith."[40] Science itself is the product of the human mind, and it illustrates the immense power of the human imagination. Nonetheless, this reductionist outlook is sold to the wider public and is mostly uncritically adopted in bioethics discussions.

Religious scholar Courtney Campbell has advanced a similar plea for moral imagination. He argues that metaphors and symbols play a vital role in communicating meaning. Symbolic narratives express values, and they articulate a worldview. They represent, in his view, a different type of rationality than the instrumental rationality that prevails in scientific and neoliberal ideologies. This symbolic rationality can help to overcome the reduced moral worldview of mainstream bioethics that attends to "a world that few people actually live in."[41] Paying attention to these background worldviews expressed in metaphors and symbols can make bioethics richer and more real.

Different imaginary visions have played different roles in health care for a long time. The metaphor of monster, for example, has played a role throughout the history of medicine. It has been used to objectify people so that they could be displayed, collected, and traded, which reinforced the tendency of modern medicine to reduce people to their physical bodies and deny their personhood. At the same time, the metaphor is employed as a warning against hubris, suggesting limits to what science should accomplish. It is therefore an imaginative instrument used to implement some restrictions on scientific interventions. If the metaphor is not allowed in bioethical discourse, however—being passed off instead as a purely emotional response of disgust that should play no role in rational debate—the consequence is that scientific advancement cannot be criticized as morally transgressive. The ideology of scientism based on reductionism, materialism, and progress is taken for granted. The technological imperative cannot be doubted.

Another area of globalization that shows the limitations of scientific and bioethical perspectives is the rise of medical tourism, which reflects the ancient narrative of pilgrimage. People want to use scientific and technological opportunities even if they have not been proven safe or effective. Patients are motivated by hope and the search for health and salvation. The paradox is that this narrative is promoted by science while at the same time miracles, especially religious ones, are formally rejected and hope is not a serious consideration in health care. Miracle discourse nonetheless is actively promoted in science and politics, for example, in right-to-try legislation in the United States. Medical tourism is regarded as a symptom of the global market in health care; it offers a wider range of choices to individuals who can now receive the latest cures everywhere in the world. The context of hope that is fueled by the promises of scientific advancement beyond the restricted value perspective of science as objective, neutral, and evidence-based is not addressed.

That the ideology of scientism itself is not restricted and often goes beyond its self-imposed limitations is evident in the increased interest in prognosis and prediction. However, in studying how science is operating in the area of predictive medicine, it is clear that scientific discourse does not restrict itself to its limited worldview. Often discourses are used that are inspired by the traditional narratives of prophecy. They employ imaginative visions of progress and perfectibility. Medicine has a long tradition of utopias, and today there is not a lack of medical prophets. These prophecies are based on the narrow worldview of scientism, assuming genetic determinism and reductionism. But in predicting the future, they quickly transcend the value perspective of scientism and appeal to others, assuming predictability, inevitable progress, the unchanging context of neoliberal market priorities, and faith that mortality is curable. Rather than evidence and facts, promises and expectations are projected into the future. Bioethics is not debunking such visions of progress and perfectibility.

The current trend to collect and store human biomaterials is taken as another example of the confluence of neoliberal and scientific discourses. Body materials are regarded as commodities that can be exchanged and sold. From a bioethical perspective, the major questions are therefore related to property and consent. All discourses assume that human life is a mechanism so that the person can be separated from the body. But this separation is contrary to human self-experience. Humans are embodied beings. As persons, and through their bodies, they are autonomous subjects but at the same time are embedded in communities with other persons, and therefore vulnerable. Medicine as a modern collection science—eager to gather human remains, from bodies to cells and genes—is now confronted with a growing sentiment against collecting, displaying, and exploiting human remains. This is not so much related to the lack of consent (body parts used to be taken not donated) but to the neglect of the relationship between person and body. Property language is inadequate since human remains have a symbolic value; they have an intermediary status between persons and things. Moral notions like respect and dignity are therefore more appropriate than the vocabulary of property. This is in fact a lesson from the traditional narrative of relics.

Critical Approaches

Recently, bioethicist Devan Stahl, reflecting on her own illness, described how science is focused on technical control of life and not on issues of meaning, purpose, or goals of life. For patients, however, the latter are important concerns and usually delegated to theology and philosophy. What medicine can offer to these topics is necessarily limited but at the same time includes many promises and predictions. Medicine represents a specific way of seeing the world, making the patient as a subject nearly invisible. This predicament should not be accepted in bioethical discourse. Ethics should appeal to moral imagination in order to broaden perspectives.[42]

This book argues that a range of imaginary visions is already operative in current scientific and bioethical approaches that present a broader view of human beings and the human world than are usually engaged in. The human being is not merely an object of scientific analysis and manipulation, nor solely an individual subject that rationally chooses and decides. These visions articulate that human beings are embedded in relationships and contexts. They emphasize that other conceptual frameworks are necessary for reframing the bioethical debate, including focusing on the common good, the role of the social context, and the importance of the environment.

Common Good

In his extensive study of the idea of common humanity, Siep Stuurman explains how regarding human beings as fellow persons and not as strangers or barbarians is an idea that was invented over history in multiple places and cultures.[43] The concept of common humanity illustrates the power of ideas. It is the result of critical reflection on prevailing notions of otherness and inequality, but also of imagination, of envisioning an imagined world that is different from the world that is empirically given. It employs the image of the widening circle of moral concern, gradually including others beyond the free white man: women, foreigners, slaves. It was the effect of cross-cultural encounters and exchanges that produced dialectic processes between universalism and particularism. Humanity manifested itself not in isolated individuals but in interpersonal relationships.

Stuurman's analysis exemplifies the role of ethics in the struggle against barbarism, violence, and arbitrariness. More importantly, it demonstrates that moral thinking and acting relates to communities of human beings. We are moral actors, not just as persons but as members of a community. Moral life with all its ambiguities and uncertainties is a common life.[44] The presupposition is that there are shared values and ideals, that there is a common good. Such common good is about inclusion; it is essential for social trust and obligations to other people; it defines what we owe each other as members of society.

It is indisputable that today the idea of a common good is compromised.[45] The economic and political elites have altered the rules to the game of life to their own advantage. Governments are not run for the benefit of all. No wonder people are angry and distrust the basic institutions of society. Ethics, if appealed to at all, is about avoiding legal troubles or public relations disasters rather than rethinking the idea of common good.

However, this situation is changing, at least at the level of ideas and reflection. Economist and Nobel laureate Jean Tirole published a call for an economics for the common good, for example. The challenge was to reconcile individual and general interests. He argued that economics so far has developed on the basis of the fiction of the *Homo economicus* despite the reality that human behavior is not driven by self-interest. For a flourishing society, it is indispensable to have arrangements and institutions that are independent from market forces.[46] Similar trends are visible in bioethics, too, where notions such as "common heritage" and "global commons" have been receiving more attention.[47] Other examples include the opposition to biopiracy and patenting and to privatization of water resources on the one hand, and on the other hand an emphasis on data sharing in research, redefining science as an open and shared activity, and open access publications.

Social Context

Some time ago, it was pointed out that bioethics often represents a world of moral fiction that is abstracted from genuine moral experience. This is because ethics is not seen as "an intrinsic part of a society, but as some sort of abstract system to be imposed upon, chosen by, or rejected by a society."[48] On the contrary, our moral values are the result of social and cultural factors. Moral concepts are embedded in a way of life. However, these factors are not within our reach or under our control, which complicates our desire for action and intervention.

Furthermore, the abstract features of bioethics are related to the dominating influence of the notion of personal autonomy. Other people are regarded as competitors, not as "fellow searchers for civilization."[49] The ghost phenomena of modern science, for example, are interpreted as individual cases of aberration. Criticism is usually not directed at the context in which phenomena emerge, although they are based on deception and erode the public trust that is necessary for the social credibility of science. The same is true for the focus on vulnerability. It requires a transformation from individual to social ethics, a concern with moral issues beyond the level of individual autonomy, like solidarity, justice, social responsibility, and protection of future generations. Vulnerability requires different types of action—not simply protection but care, creating a social environment that is more hospitable, and connecting so that people can overcome and compensate for vulnerability.

The significance of the social context for health has been understood for ages. The emphasis on individualistic approaches in health care and bioethics, however, is not favoring to address the structural drivers of health inequities. The World Health Organization recommended in 2008 three principles of action: (1) improve the conditions of daily life; (2) tackle the inequitable distribution of power, money, and resources; and (3) measure and understand the problem, evaluate action, and raise public awareness about the social determinants of health.[50] However, after fourteen years, the gap in global health care is not closed, and it is contested how much progress has been made. A recent report of the Pan American Health Organization clarifies that inequalities between and within countries continue to be widespread. Many people are not able to lead a dignified life or enjoy the highest attainable standard of health because of poverty, the legacies of colonialism, structural racism, and displacement from land.[51]

Role of the Environment

The significance of environmental issues is now undeniable. Climate change, biodiversity loss, and environmental degradation are seriously impacting health. It is clear, regardless of one's opinions and views, that human health is being degraded and health care systems are affected. Many advantages accomplished by public health over the last centuries will be undone. Climate change, biodiversity, and health care can no longer be approached as separate domains. This environmental nexus must be taken into account in bioethical approaches.[52]

Much damage to the planet is already done and irreversible. Data of the World Wildlife Fund show that wildlife populations have globally declined by 60 percent over the past forty years. Almost all seabirds in the world have fragments of plastic in their stomachs.[53] And systematic tracking of the impacts of climate change on human health can only lead to the conclusion that there is a public health emergency.[54] Direct risks to human health are increasing. For example, Europe's significant elderly population is vulnerable to a rise in heat waves. In other parts of the world, communities are exposed to extreme weather events (droughts and extreme rainfall) and natural disasters. Rising temperatures are associated with a growing spread of such diseases as dengue, cholera, and malaria, and the mortality rates for malignant melanoma have significantly increased.

Health effects of climate change are also indirect. Global economic losses have tripled since 2016. Billions of labor hours have been lost. Public health infrastructures are overburdened and risk collapse in many countries. Yields in agriculture have declined, threatening food security. Air pollution has worsened in 70 percent of cities in the world since 2010. The social consequences of increased temperatures are visible for everyone (as numbers of climate refugees, displaced people, and disasters grow), and existing health disparities are deteriorating since vulnerable populations are most affected.

Though the evidence of an expanding health crisis is overwhelming, attention to these threats to humanity is relatively low. For populist reasons, policy makers are employing counterproductive tactics and denialist arguments. The contribution of bioethics to the analysis and debate on environmental degradation is limited. Since 1973, 27,720 journal articles with the keyword "bioethics" have been published. Only 3 percent of those have also included the keyword "environment."[55]

Conclusion

With the many challenges outlined throughout this book, bioethics may seem rather bizarre to many people. It does not pay attention to their daily worries and concerns. Their vulnerabilities, discontent, and anxieties are not acknowledged. Instead, bioethical discourse celebrates the promises of scientific progress that will be available for a limited population—the same elites that are reaping most of the benefits of globalization. Bioethics also operates with a distorted image of human beings as rational and individual consumers who are free to choose what they need and desire. With such an image, many feel abandoned and blamed for their conditions, while what they need is cooperation, solidarity, and support.

It is important to emphasize again that bioethics is not a homogeneous discourse. It comprises various discursive practices. It moves from cases and specific situations to more elaborate and abstract arguments. Imaginary processes play a role in this movement so that a richer and broader conceptual and analytical apparatus may emerge, giving voice to discourses that are easily silenced. Moral imagination, as a creative faculty of the mind, is not fantasy. It activates value systems that are not dominant and provide alternative styles of thinking that can create new norms and resist the imposition of current norms. That ideas can be powerful and have the potential to change the status quo is illustrated in field of disease diplomacy. States have been redefining their interests and responsibilities in regard to emerging infectious diseases. Now that local outbreaks of diseases quickly develop into global threats, national approaches and the traditional norms of containment (quarantine and border control) are no longer sufficient. The need for collective action, sharing information and transparency to secure global health, has changed the normative behavior of states. A new set of expectations concerning responsibilities of states, a "new package of norms," has been created based on a greater sense of global solidarity.[56]

These possibilities for change through broader normative discourse problematize the usual stance of bioethics as a rational and objective discourse. Bioethics identifies itself as an academic discipline that is neutral. It assumes that there are clear boundaries between academic analysis and political engagement. Bioethics provides a "detached authority" that overrides the responses of patients, parents, families, and communities that are often emotional and not well informed, as demonstrated in the Charlie Gard case. Ethical activism in this view will compromise the credibility of bioethics.[57] Angus Dawson and colleagues have argued that this neutrality is a myth. They are right. Mainstream bioethics is not a neutral discourse. It endorses the ideologies of neoliberalism and scientism, neglects issues of vulnerability and solidarity, and treats justice merely as the distribution of resources. Confronted with human rights violations—for example, the inhuman treatment of asylum seekers in Australia and elsewhere—bioethicists should criticize policies that directly harm the most vulnerable people and seriously affect their health.[58]

Finally, bioethics is not only crucial for democracy but also essential for civilization.[59] Moral imagination locates us within the stories of undesirables, disposables, and victims. It interrogates our relationships to other people and how they are affected by scientism and neoliberalism. It questions the rules of civilization and standards of appropriateness. The inhumane treatment of other persons threatens "society's conception of civilized life."[60] Concepts such as respect, recognition, and dignity must be reconfigured from a moral perspective that primarily applies to individuals into a perspective that is fundamental for society. Human dignity, for example, has motivated the search for shared humanity, and therefore human rights.[61] It is the basis for mutual respect in decent societies.

As argued throughout, to understand patient cases and medical problems, interpretation of moral experiences is unavoidable. Imagination is important to facilitate interpretation; it generates the worldviews, ideals, and values that guide moral perception. Imagination is reshaping and reconstructing our experiences. Philosopher John Dewey has argued that we all have the capacity to imagine.[62] It is the creative ability to make the absent become present. Imagination projects ideals and values, offers possibilities for thinking and acting, helps us to bring new realities into existence, and conceives alternatives to problematic situations. It also makes use of past experiences to suggest alternative possibilities. It furthermore has the advantage that it brings us together as human beings. Imagination is the extension of experience beyond our limited and familiar realm of everyday life. It fosters sympathy and dialogue because it involves taking the perspective of others. Without imagination we cannot see situations from the viewpoint of other persons and cannot understand the experiences of others. Imagination is therefore a crucial activity for ethics.

Chapter 1 • Questioning the Paradigm of Bioethics

1. Lakoff and Johnson, *Metaphors we live by.*

2. For the Charlie Gard case, see Savulescu, "Is it in Charlie Gard's best interest to die?"; Dyer, "Law, ethics, and emotion"; Sokol, "Charlie Gard case"; Rimmer, "Charlie Gard's parents end legal fight to keep son alive"; Hurley, "How a fight for Charlie Gard became a fight against the state"; Wilkinson and Savulescu, "After Charlie Gard,"; Truog, "The United Kingdom sets limits on experimental treatment"; Lantos, "The tragic case of Charlie Gard"; Shah, Rosenberg, and Diekema, "Charlie Gard and the limits of best interests"; Wilkinson, "Restoring the balance to 'best interests' disputes in children: Editorial,"; Wilkinson and Savulescu, "Hard lessons"; Caplan and McBride Folkers, "Charlie Gard and the limits of parental authority."

3. See "Charlie Gard #charliesfight," GoFundMe.com, accessed June 20, 2021, https://www.gofundme.com/please-help-to-save-charlies-life.

4. Great Ormond Street Hospital (GOSH) position statement, hearing on July 24, 2017.

5. Royal Courts of Justice, Case no. FD17P00103, 2.

6. This is an increasing challenge. The overshoot day has been receding rapidly. In 1985, it was November 5; in 2008, it was September 23. At this rate of consumption, we will need two planets by 2030. See Joignot, "Trop d'humaines pour la planète?" [Too many humans for the planet?].

7. UNICEF, "Diarrhoea remains a leading killer of young children."

8. GOSH position statement, 1.

9. GOSH position statement, 2.

10. GOSH position statement, 3.

11. Royal Courts of Justice, Case no. FD17P00103, 5.

12. Judiciary of England and Wales, "Decision and short reasons to be released to the media in the case of Charlie Gard," 4.

13. Judiciary of England and Wales, 4.

14. Royal Courts of Justice, 2.

15. Royal Courts of Justice, 2, 3; Judiciary of England and Wales, 6.

16. Judiciary of England and Wales, 3.

17. Lantos, "The tragic case of Charlie Gard."

18. Wilkinson and Savulescu, "Hard lessons."

19. Royal Courts of Justice, 5; Judiciary of England and Wales, 5.

20. Judiciary of England and Wales, 3–4.

21. Haidt, "The emotional dog and the relational tail."

22. Bilefsky, "British Court decides Charlie Gard will be moved to a hospice to die."

23. Moschella, "The Charlie Gard case threatens all parents."

24. Bennett, "Charlie Gard: A story of disability bias."

25. Church, "A personal perspective on disability."

26. Wilkinson and Savulescu, "Hard lessons."

27. Hurley, "How a fight for Charlie Gard became a fight against the state."

28. For example, Hammond-Browning, "When doctors and parents don't agree."

29. Wilkinson, "Restoring the balance to 'best interests' disputes in children," 2. Michael Dahnke has correctly pointed out that one of the first cases that disseminated video footage on the web was the case of Terri Schiavo in the early 2000s. This also had enormous public and political impact, but it was limited to the United States. See Dahnke, "Emmanuel Levinas and the face of Terri Schiavo."

30. The notion of "stigmata cases" was introduced by Morgan and Lee, "In the name of the father?" See also Lee and Morgan, "Regulating risk society"; Montgomery, "Law and the demoralization of medicine."

31. Wilkinson and Savulescu, "Hard lessons."

32. Savulescu, "Is it in Charlie Gard's best interest to die?," 1868.

Chapter 2 • *The Establishment of Bioethics*

1. Gaines and Juengst, "Origin myths in bioethics."

2. Jonsen, *The birth of bioethics*, 211–217.

3. Rothman, *Strangers at the bedside*, 5ff., 70ff.

4. Ten Have, *Global bioethics*, 23ff.

5. Stevens, *Bioethics in America*, 40.

6. Evans, *The history and future of bioethics*.

7. "Theologians were the first to appear on the scene" (Jonsen, *The birth of bioethics*, 34).

8. Jonsen, *The birth of bioethics*, 57.

9. Gustafson was University Professor of Theological Ethics at the University of Chicago.

10. Gustafson, *The contributions of theology to medical ethics*, 93–94.

11. Evans, *The history and future of bioethics*, 7.

12. Gracia Guillen, "Bioethics in the Spanish-speaking world."

13. Callahan, "Why America accepted bioethics."

14. Jonsen, *The birth of bioethics*, 83.

15. Reich, "Bioethics in the United States," 100.

16. Jonsen, *The birth of bioethics*, 65.

17. Callahan, "Religion and the secularization of bioethics."

18. Jonsen, *The birth of bioethics*, 333.

19. Thomasma, "Training in medical ethics."

20. Seedhouse, *Ethics: The heart of health care*; Seedhouse and Lovett, *Practical medical ethics*.

21. Cahill, *Theological bioethics*. See also Clark, "Greed is not enough."

22. Cahill, "Bioethics, theology, and social change"; Evans, *The history and future of bioethics*.

23. Savulescu, "Bioethics: Why philosophy is essential for progress."

24. Mills, "Continental philosophy and bioethics."

25. Beauchamp and Childress, *Principles of biomedical ethics*, 2nd ed., ix–x.

26. Ten Have, "Principlism: A Western European appraisal." See also DeMarco and Fox, *New directions in ethics*.

27. Ten Have, "Foundationalism and principles."

28. Gracia, "History of medical ethics."

29. Beauchamp and Childress, *Principles of biomedical ethics*, 1st ed.

30. The full name of the Oviedo Convention is Convention for the Protection of Human Rights and Dignity of the Human Being with regard to the Application of Biology and Medicine. It has been adopted by the members of the Council of Europe and entered into force in December 1999 (see "Details of Treaty No. 164," accessed May 14, 2021, https://www.coe.int/en/web/conventions/full-list/-/conventions/treaty/164).

31. The Universal Declaration on Bioethics and Human Rights was unanimously adopted by member states of UNESCO in October 2005 (http://unesdoc.unesco.org/images/0014/001461/146180E.pdf).

32. Caplan, "Can applied ethics be effective in health care and should it strive to be?" See also Caplan, "Ethical engineers need not apply."

33. Pellegrino, "Bioethics as an interdisciplinary enterprise," 3–4.

34. Ten Have, "Foundationalism and principles," 28.

35. Dewey, *Human nature and conduct*, 239.

36. Williams, "Consequentialism and integrity."

37. Taylor, *Sources of the self*.

38. Jonsen, *The birth of bioethics*, 389ff. These characteristics of the American ethos are reflected in the type of history that Jonsen himself has written; the rise of bioethics is primarily the achievement of personalities.

39. Engelhardt, *The foundations of bioet-hics*.

40. Callahan, "Minimalist ethics."

41. The term "hyper good" is used by Taylor to refer to higher-order goods, "which not only are incomparably more important than others but provide the standpoint from which these must be weighed, judged, decided about." They define what is "moral" in our culture. An example is the principle of universal and equal respect (Taylor, *Sources of the self*, 63).

42. Fox, *The sociology of medicine*, 224–276.

43. Edel, "Ethical theory and moral practice, 323.

44. Callahan, "Why America accepted bioethics," S9.

45. Stevens, *Bioethics in America*, 55.

46. Callahan, "The Hastings Center and the early years of bioethics," 18.

47. Cahill, "Bioethics, theology, and social change," 368–371.

48. Stevens, *Bioethics in America*, 71.

49. Evans, *The history and future of bioethics*, 109.

50. Marshall, "ASBH and moral tolerance."

51. Evans, *The history and future of bioethics,* 108. See also DeBruin, "Ethics on the inside?"

52. Stevens, *Bioethics in America,* 2. Stevens also argues that "bioethics served to transmute potentially hostile impulses of the larger society into an acceptable expertise" (150). However, the disadvantage was co-optation rather than confrontation. Critical questions could no longer be asked.

53. Jacobsen et al., "Feeding the world."

54. Sandler, *Food ethics.*

55. Jonsen, "Why has bioethics become so boring?"

56. Evans, *The history and future of bioethics,* xxi

57. Evans, *The history and future of bioethics,* 77.

58. Evans, *The history and future of bioethics,* 153ff.

59. Illich, *Medical nemesis.*

60. Foucault was not the first to use the concept of biopolitics. See Kakuk, *Bioethics and biopolitics.*

61. McKeown, *The role of medicine.*

62. Ten Have, "The anthropological tradition in the philosophy of medicine."

63. See, for example, Kakuk, *Bioethics and biopolitics.*

64. Gustafson, "Theology confronts technology and the life sciences."

65. Jonsen, *The new medicine and the old ethics,* 128.

66. Gustafson, *The contributions of theology to medical ethics,* 28, 31, 69, 71, 94.

67. Gustafson, *The contributions of theology to medical ethics,* 94.

68. Cahill, *Theological bioethics,* 4.

69. Cahill, *Theological bioethics,* 24–25.

70. Shannon, "Bioethics and religion," 149.

71. Jonsen, *The birth of bioethics,* 399.

72. Cahill, *Theological bioethics,* 374.

73. Caplan, "Can applied ethics be effective in health care and should it strive to be?," 317.

74. Taylor, *Sources of the self,* 3.

75. Blum, *Moral perception and particularity.*

76. Greene and Haidt, "How (and where) does moral judgment work?"

77. Chapman, "In defense of the role of a religiously informed bioethics."

78. Annas, "Reframing the debate on health care reform by replacing our metaphors."

79. Malone, "Policy as product."

80. See, for example, Sherwin, "Foundations, frameworks, lenses"; Sytsma, "The moral authority of symbolic appeals in biomedical ethics."

81. Spencer, *The evolution of the West*; Siedentop, *Inventing the individual.*

82. Verhey, "'Playing God' and invoking a perspective," 353.

83. Ryan, "The new reproductive technologies."

84. Midgley, *The myths we live by*; Midgley, *Science as salvation.* See McElwain, *Mary Midgley.*

85. Stout, *Ethics after Babel.* See also Adams et al., "Social imaginaries in debate."

Chapter 3 • Ghosts

1. Fugh-Berman, "The corporate coauthor."
2. May, "How many people believe in ghosts or dead spirits?"
3. Owens, *The ghost*, 9.
4. Morton, *Ghosts*.
5. Matthews, "Essay mills."
6. Ross et al., "Guest authorship and ghostwriting in publications related to rofecoxib"; Psaty and Kronmal, "Reporting mortality findings in trials of rofecoxib for Alzheimer disease or cognitive impairment."
7. Moffatt and Elliott, "Ghost marketing," 22.
8. Healy and Cattell, "Interface between authorship, industry and science in the domain of therapeutics."
9. Sismondo and Doucet, "Publication ethics and the ghost management of medical publication," 277–278.
10. Sismondo, "Ghost management." It is ironic that to promote the drug Paxill, pharmaceutical company Wyeth called its program CASPPER, alluding to Casper, the friendly ghost from American comic books (Barbour, "How ghost-writing threatened the credibility of medical knowledge and medical journals").
11. Moffatt and Elliott, "Ghost marketing," 20–21.
12. It is generally recognized that clinical trials sponsored and published by pharmaceutical companies are uniformly positive about the product of the sponsor and that they underreport side effects. See Moffatt and Elliott, "Ghost marketing," 24ff.; Sismondo, "Ghost management," 1430; Lexchin, "Those who have the gold make the evidence"; Lexchin et al., "Pharmaceutical industry sponsorship and research outcome and quality"; Kjaergard and Als-Nielsen, "Association between competing interests and authors' conclusions." A recent example is studies about rosiglitazone. This drug significantly increased the risk of myocardial infarction in patients with diabetes. Studies sponsored by the manufacturer did not confirm this risk. Meta-analysis of publications demonstrated that authors with financial relations with the company were three times more positive about the safety of the drug than other authors; 94 percent of all favorable conclusions came from pharma-related authors. See Wang et al., "Association between industry affiliation and position on cardiovascular risk with rosiglitazone."
13. Resnik, "Scientific research and the public trust."
14. Resnik, Peddada, and Brunson, "Research misconduct policies of scientific journals."
15. Lacasse and Leo, "Ghostwriting at elite academic medical centers in the United States."
16. Matheson, "The disposable author."
17. Bekelman, Li, and Gross, "Scope and impact of financial conflicts of interest in biomedical research."
18. Consoli, "Scientific misconduct and science ethics."
19. Judson, *The great betrayal*; Wells and Farthing, *Fraud and misconduct in biomedical research*.

20. Fanelli, "How many scientists fabricate and falsify research?"

21. Raad and Appelbaum, "Relationships between medicine and industry," 467.

22. Sperling, "(Re)disclosing physician financial interests," 180.

23. Choudhry, Stelfox, and Detsky, "Relationships between authors of clinical practice guidelines and the pharmaceutical industry."

24. Raad and Appelbaum, "Relationships between medicine and industry," 466; Moynihan, "Who pays for the pizza? Redefining the relationships between doctors and drug companies."

25. Smith, "Medical journals are an extension of the marketing arm of pharmaceutical companies."

26. Lawrence, "The politics of publication."

27. Van Nuland and Rogers, "Academic nightmares."

28. Bohannon, "Who's afraid of peer review?"

29. Sorokowski et al., "Predatory journals recruit fake editor."

30. Shen and Björk, "'Predatory' open access." The authors found that most predatory publications are in the field of engineering and biomedicine. Many journals (27 percent) are produced in India, followed by North America (17.5 percent). See also Habibzadeh and Simundic, "Predatory journals and their effects on scientific research community."

31. Eriksson and Helgesson, "The false academy"; Ten Have and Gordijn, "Publication ethics."

32. Asadi et al., "Fake/bogus conferences."

33. Steer clear and see Wikipedia's entry for "World Academy of Science, Engineering, and Technology" (https://en.wikipedia.org/wiki/World_Academy_of_Science ,_Engineering_and_Technology, last updated February 10, 2021) and the references therein.

34. Shamseer et al., "Potential predatory and legitimate biomedical journals."

35. Dadkhah, Seno, and Borchardt, "Current and potential cyber attacks on medical journals"; Dadkhah, Maliszewki, and Jazi, "Characteristics of hijacked journals and predatory publishers."

36. Ten Have and Gordijn, "Publication ethics," 159–160.

37. Watson, "Beall's list of predatory open access journals: RIP"; Brezgov, "List of publishers"; Anonymous, *Beall's List of Potential Predatory Journals and Publishers.*

38. Beall, "Predatory journals threaten the quality of published medical research."

39. "We are living in a world where we can no longer trust a scientific journal" (Habibzadeh and Simundic, "Predatory journals and their effects on scientific research community," 271).

40. Gasparyan et al., "The pressure to publish more and the scope of predatory publishing activities"; Vogel, "Researchers may be part of the problem in predatory publishing."

41. Clark and Thompson, "Five (bad) reasons to publish your research in predatory journals."

42. Abbott, "How to choose where to publish your work"; Memon, "Predatory journals spamming for publications." See also Moustafa, "The disaster of the impact factor."

43. Howell and Ford, *The ghost disease, and twelve other stories of detective work in the medical field*, 111–137.

44. Bures, "The strange case of the disappearing penis."

45. Silverstein, "North Korea's nuclear tests are spreading 'ghost disease' causing deformation, defectors say"; Harris, "16 American sickened after attack on embassy staff in Havana"; Entous and Anderson, "The mystery of the Havana syndrome"; Sample, "'Sonick attack' on US embassy in Havana could have been crickets, say scientists."

46. Ihara, "Disease mongering," 924. See Payer, *Disease-mongers: How doctors, drug companies, and insurers are making you feel sick.*

47. Moynihan and Henry, "The fight against disease mongering"; Moynihan, Heath, and Henry, "Selling sickness"; Moynihan and Cassels, *Selling sickness: How the world's biggest pharmaceutical companies are turning us all into patients.*

48. Shankar and Subish, "Disease mongering"; Doran and Henry, "Disease mongering: Expanding the boundaries of treatable disease."

49. Applbaum, "Pharmaceutical marketing and the invention of the medical consumer"; Isaacs, "Disease marketing."

50. Barbour, "How ghost-writing threatened the credibility of medical knowledge and medical journals."

51. Clark, "How halitosis became a medical condition with a 'cure'."

52. Elsner, "The concept of 'fragile skin.'"

53. Lexchin, "Bigger and better: How Pfizer redefined erectile dysfunction"; Tiefer, "Female sexual dysfunction: A case study of disease mongering and activist resistance"; Woloshin and Schwartz, "Giving legs to restless legs: A case study of how the media helps make people sick"; Söderfeldt, Droppe, and Ohnhäuser, "Distress, disease, desire: Perspectives on the medicalization of premature ejaculation."

54. Parens and Johnston, "Facts, values, and Attention-Deficit Hyperactivity Disorder"; Schwarz, "The selling of Attention Deficit Disorder." See also Sadler et al., "Can medicalization be good?"

55. Illich, *Medical nemesis.*

56. Conrad, "The shifting engines of medicalization."

57. Abraham, "Pharmaceuticalization of society in context," 615.

58. Lexchin, "Bigger and better."

59. Parens and Johnston, "Facts, values, and Attention-Deficit Hyperactivity Disorder," 32.

60. Clark, "Medicalization of global health."

61. See Söderfeldt, Droppe, and Ohnhäuser, "Distress, disease, desire."

62. Perls and Handelsman, "Disease mongering of age-associated declines in testosterone and growth hormone levels." See also Mintzes, "Direct to consumer advertising is medicalizing normal human experience."

63. See Moynihan et al., "Coverage by the news media of the benefits and risks of medications"; Zuckerman, "Hype in health reporting"; Schwitzer, "How do US journalists cover treatments, tests, products, and procedures?"; Schwitzer, "Addressing tensions when popular media and evidence-based care collide"; Schwitzer, "Trying to drink from the fire hose: Too much of the wrong kind of health care news."

64. McCoy et al., "Conflicts of interest for patient-advocacy organizations."

65. Herzheimer, "Relationships between the pharmaceutical industry and patients' organizations."

66. For example, adult ADHD is strongly promoted by an advocacy group, receiving its funding from the industry. See Conrad, "The shifting engines of medicalization," 9.

67. Much documentation about ghostwriting has been uncovered due to injury-oriented adversity to the pharmaceutical industry. See Abraham, "Pharmaceuticalization of society in context," 610ff.

68. Barker, "Electronic support groups, patient-consumers, and medicalization: The case of contested illness," 31.

69. Conrad and Stults, "Contestation and medicalization."

70. Fair, "Morgellons: Contested disease, diagnostic compromise and medicalization." Another example is the Gulf War Syndrome. It is an unexplained illness but not recognized as a specific disease (Zavestoski et al., "Patient activism and the struggle for diagnosis").

71. Tungaraza, Talapan-Manikoth, and Jenkins, "Curse of the ghost pills."

72. WHO, "1 in 10 medical products in developing countries is substandard or falsified."

73. It is estimated that in developing countries, 30 percent of the medicine market is affected by falsification, while it is only 1 percent in countries whose distribution channels are regulated and controlled (see Rebiere et al., "Fighting falsified medicines"). WHO estimates that 30 percent of countries have inadequate or no regulation authorities for medicines (see Nayyar et al., "Responding to the pandemic of falsified medicines," 116). Most incidents are reported in China, Peru, Uzbekistan, Russia, and Ukraine (Mackey et al., "Counterfeit drug penetration into global legitimate medicine supply chains," 62). In major cities in India, one in every five medicines sold was fake (see Gautam, Utreja, and Singal, "Spurious and counterfeit drugs: a growing industry in the developing world").

74. Rebiere et al., "Fighting falsified medicines," 295; Campbell et al., "Internet-ordered Viagra (sildenafil citrate) is rarely genuine." The WHO reports that between 19 and 26 million people in the United States now buy medicines over the internet (WHO, *WHO Global Surveillance and Monitoring System for substandard and falsified medical products*, 15).

75. The United States imported 40 percent of drugs and 80 percent of active pharmaceutical ingredients from foreign countries in 2011. See Nayyar et al., "Responding to the pandemic of falsified medicines," 115.

76. Lee et al., "Combating sales of counterfeit and falsified medicines online." See also Cyranoski, "The secret war against counterfeit science."

77. WHO, *WHO Global Surveillance and Monitoring System*, 35.

78. Newton and Timmermann, "Fake penicillin, *The Third Man*, and operation Claptrap."

79. Clark, "Rise in online pharmacies sees counterfeit drugs go global."

80. WHO, *A study on the public health and socioeconomic impact of substandard and falsified medical products*.

81. Nayyar et al. "Responding to the pandemic of falsified medicines," 113–114.

82. Hamilton et al., "Public health interventions to protect against falsified medicines," 1451–1452.

83. El-Jardali et al., "Interventions to combat or prevent drug counterfeiting"; Newton et al., "The primacy of public health considerations in defining poor quality medicines."

84. Ossola, "The fake drug industry is exploding, and we can't do anything about it."

85. 't Hoen and Pascual, "Counterfeit medicines and substandard medicines"; Jack, Williams, and Steen, "Seizure of HIV drugs highlights patent friction."

86. WHO: *WHO Global Surveillance and Monitoring System*, 5.

87. WHO: *WHO Global Surveillance and Monitoring System*, 21. See also Clark, "Rise in online pharmacies sees counterfeit drugs go global"; Tremblay, "Medicines counterfeiting is a complex problem."

88. Sattar, "Bangladesh faces growing trade in fake drugs."

89. Medina, Bel, and Suñé, "Counterfeit medicines in Peru."

90. Havemann, "Sham surgeries, real risks"; Katsnelson, "Why fake it? How 'sham' brain surgery could be killing off valuable therapies for Parkinson's disease."

91. Kim et al., "Science and ethics of sham surgery."

92. Mosely et al., "A controlled trial of arthroscopic surgery for osteoarthritis of the knee." See also Schröder et al., "Sham surgery versus labral repair or biceps tenodesis for type II SLAP lesions of the shoulder"; Wartolowska, Beard, and Carr, "Attitudes and beliefs about placebo surgery among orthopedic shoulder surgeons in the United Kingdom."

93. Dekkers and Boer, "Sham surgery in patients with Parkinson's disease"; Macklin, "The ethical problems with sham surgery in clinical research."

94. AMA, "Use of placebo in clinical practice."

95. AMA, "Use of placebo in clinical practice." See also Brody, "The lie that heals: The ethics of giving placebos."

96. Beecher, "The powerful placebo."

97. Kaptchuk, "Powerful placebo: The dark side of the randomized controlled trial"; De Craen et al., "Placebos and placebo effects in medicine"; Macedo, Farré, and Baños, "Placebo effect and placebos: What are we talking about?"

98. Bok, "The ethics of giving placebos." For the ethical debate, see also Foddy, "A duty to deceive: Placebos in clinical practice."

99. Moerman, "Deconstructing the placebo effect and finding the meaning response."

100. Balint, *The doctor, his patient, and the illness.*

101. Brody, "The lie that heals"; Papakostas and Daras, "Placebos, placebo effect, and the response to the healing situation"; Macedo, Farré, and Baños, "Placebo effect and placebos," 339–340. See also Brody, *Placebos and the philosophy of medicine.*

102. Thomas, "General practice consultations: Is there any point in being positive?"

103. Mill, *Utilitarianism*, 55; Hartzband and Groopman, "Money and the changing culture of medicine."

104. Angell, "Industry-sponsored clinical research: A broken system," 1070.

105. Titus, Wells, and Rhoades, "Repairing research integrity."

106. Lemmens and Freedman, "Ethics review for sale? Conflict of interest and commercial research review boards"; Campbell et al., "Financial relationships between institutional review board members and industry."

107. In the same year there were 1 million doctors in the United States, including

148,000 inactive ones. See Lichter, "Conflict of interest and the integrity of the medical profession."

108. Morris and Taitsman, "The agenda for continuing medical education—limiting industry's influence"; Relman, "Industry support of medical education."

109. Okike et al., "Accuracy of conflict-of-interest disclosures reported by physicians."

110. As suggested by Moynihan, Heath, and Henry, "Selling sickness."

111. Collier, "The price of independence"; Kassirer, "Disclosure's failings: What is the alternative?"

112. The notion of "world-picture" is particularly used by Mary Midgley. See Midgley, *Science as salvation*.

Chapter 4 • Monsters

1. Kaplan, *The science of monsters*.

2. Huet, *Monstrous imagination*.

3. Gilmore, *Monsters*, 46–61. See also Friedman, *The monstrous races in medieval art and thought*.

4. Paré, *Des monstres et prodigies* [Monsters and prodigies]. See also Audeguy, *Les monstres. Si loin et si proches* [Monsters, so far and so near].

5. De Waal Malefijt, "Homo monstrosus."

6. Bates compiled a list of human monstrous births between 1500 and 1700. In most cases, it was possible to provide a retrospective diagnosis on the basis of current genetic and embryological knowledge. See Bates, *Emblematic monsters*. See also Gould and Pyle, *Medical curiosities*.

7. Leroi, *Mutants: On genetic variety and the human body*. See also Hanafi, *The monster in the machine: Magic, medicine, and the marvelous in the time of the Scientific Revolution*.

8. Asma, *On monsters: An unnatural history of our worst fears*.

9. Bondeson, *The two-headed boy, and other medical marvels*.

10. Bondeson, *The two-headed boy*, 64–94.

11. Pastrana was suffering from the syndrome of congenital hypertrichosis and gingival hyperplasia. See Bondeson, *A cabinet of medical curiosities*, 243.

12. Thompson, *Freakery: Cultural spectacles of the extraordinary body*.

13. Holmes, *The Hottentot Venus: The life and death of Saartjie Baartman*.

14. Bondeson, *A cabinet of medical curiosities*, 193–198.

15. Meier, "Skeletons on the shelves." There is now a growing ethical debate about exhibiting human remains. See Gazi, "Exhibition ethics."

16. Clair, *Hubris. La fabrique du monster dans l'art modern* [Hubris: The making of the monster in modern art].

17. Gilmore, *Monsters*, 6ff.; Asma, *On monsters*, 26ff., 39ff., 103ff., 124ff.

18. Coulombe, *Petite philosophie du zombie* [Little zombie philosophy, or how to think through horror].

19. Coulombe argues that zombies are a return to "une animalité première" (a primary animality). This is why they give us a feeling of "l'inquiétante étrangeté" (disturbing strangeness), what Freud has called "Das Unheimliche" (the uncanny). See Coulombe, *Petite philosophie du zombie*, 47, 57.

20. Bondeson, *The two-headed boy*, 160–188.

21. Wiesner-Hanks, *The marvelous hairy girls: The Gonzales sisters and their worlds.*

22. Asma, *On monsters*, 183.

23. Asma, *On monsters*, 204–205.

24. Audeguy, *Les monstres*, 79–95.

25. Arendt, *Eichmann in Jerusalem: A report on the banality of evil.*

26. Huet, *Monstrous imagination*, 108–123. See also Beaune, "Sur la route et les traces des Geoffroy Saint-Hilaire" [On the road and in the footsteps of Geoffroy Saint-Hilaire].

27. Kaplan, *The science of monsters*, 175.

28. Chamayou, *Théorie du drone* [Drone theory].

29. Daston and Park, *Wonders and the order of nature, 1150–1750*; Vallone, *Big & small: A cultural history of extraordinary bodies.*

30. Huet, *Monstrous imagination*, 188–218.

31. Clair, *Hubris*, 1–2.

32. Bates, *Emblematic monsters*, 50.

33. Shildrick, *Embodying the monster.*

34. See Steven, *Bioethics in America*, 120, 127.

35. Hume, *Treatise of human nature*, 235.

36. Midgley, "Biotechnology and monstrosity: Why we should pay attention to the 'yuk factor'."

37. De Dijn, *Taboes, monsters en loterijen* [Taboos, monsters, and lotteries], 12.

38. De Dijn, *Taboes, monsters en loterijen*, 9–10.

39. The argument of sacredness is based on Dworkin, *Life's dominion: An argument about abortion and euthanasia*, 68–101.

40. See Dworkin, *Life's dominion*, 25.

41. This is pointed out by Gilmore, *Monsters*, 19: "represents all that is beyond human control, the uncontrollable and the unruly that threaten the moral order."

42. Goya, *Los Caprichos*, Plate 43: El sueño de la razon produce monstrous [The sleep of reason produces monsters].

43. According to Fiona Sampson, this is the "Frankenstein idea": "the notion that if humans play God with the 'instruments of life' they will produce something monstrous" (*In Search of Mary Shelley*, 4). See also Turney, *Frankenstein's footsteps.*

44. Cohen, *Homo economicus*, 34.

45. Smith, *Culture of death: The assault on medical ethics in America.* See also Nigro, "Bioethics is a monster."

46. Kapur, "Miracle child: Indian baby born with two heads clinging to life."

47. Quinones and Lajka, "'What kind of society do you want to live in?' Inside the country where Down syndrome is disappearing."

48. Culzac, "Boy born with eight limbs doing well after surgery to remove 'parasitic twin,' doctors say."

49. Ager, "The perils of pale."

50. Wright, *Downs: The history of a disability*; Berg and Korossy, "Down syndrome before Down." See also Miller, "Dermatoglyphics and the persistence of 'Mongolism.'"

51. "Humans without some minimum of intelligence or mental capacity are not persons, no matter how many of their organs are active, no matter how spontaneous their living processes are" (Fletcher, *Humanhood: Essays in biomedical ethics*, 135). Also:

"Idiots, however, are another matter. They are not, never were, and never will be in any degree responsible. Idiots, that is to say, are not human" (Fletcher, *Humanhood*, 22).

52. Clunies-Ross, "The development of children with Down's syndrome: Lessons from the past and implications for the future." In the past, Down children were seen as "unfit" residents of institutions, used as subjects of eugenic research, and castrated without their consent. See Lombardo, "Dwarves: Uninformed consent and eugenic research."

53. Thomas, "An elephant in the consultation room? Configuring Down Syndrome in British antenatal care."

54. Thomas and Latimer, "In/exclusion in the clinic: Down's syndrome, dysmorphology and the ethics of everyday medical work."

55. Jacob, Gagnon, and Holmes, "Nursing so-called monsters: On the importance of abjection and fear in forensic psychiatric nursing."

56. Van den Berg, *Medische macht and medische ethiek* [Medical power and medical ethics).

57. Jacob, Gagnon, and Holmes, "Nursing so-called monsters," 155–156.

58. Carroll, *The philosophy of horror; or, Paradoxes of the heart*, 126.

59. International Bioethics Committee, *Report of the IBC on the principle of non-discrimination and non-stigmatization*; Link and Phelan, "Conceptualizing stigma."

60. Jones, *Horror: A biography*. See also Bottig, *Making monstrous: Frankenstein, criticism, theory.*

61. Marcus, "Frankenstein: Myths of scientific and medical knowledge and stories of human relations"; Nagy et al., "The enduring influence of a dangerous narrative: How scientists can mitigate the Frankenstein myth"; Nagy et al., "Why Frankenstein is a stigma among scientists."

62. Annas and Elias, "Thalidomide and the *Titanic*: Reconstructing the technology tragedies of the twentieth century."

63. Marx, "About face: Why is South Korea the world's plastic-surgery capital?" An associated problem is that 80 percent of doctors doing plastic surgery in South Korea are not certified; they are known as "ghost doctors."

64. Radkowska-Walkowicz, "The creation of 'monsters': The discourse of opposition to in vitro fertilization in Poland."

65. Rollin, *The Frankenstein syndrome: Ethical and social issues in the genetic engineering of animals.*

66. Brunk and Hartley, *Designer animals: Mapping the issues in animal biotechnology.*

67. Ten Have, *Wounded planet*, chapter 6, 111ff.

68. For example, Uzogara, "The impact of genetic modification of human foods in the 21st century."

69. See Ten Have, *Wounded planet*, chapter 6, 111ff.

70. Baudrillard, *The vital illusion*, chapter 1, 3–30. For an analysis of argument for and against chimera research, see Sherringham, "Mice, men, and monsters: Opposition to chimera research and the scope of federal regulation."

71. Kass, "The wisdom of repugnance," 18.

72. Bostrom, *Superintelligence*. See also Lin, Abney, and Bekey, *Robot ethics*; Kupferschmidt, "Taming the monsters of tomorrow"; Brundage et al., *The malicious use of artificial intelligence.*

73. Sleasman, "Robots," in *Encyclopedia of global bioethics.*

74. Shapiro, *How to clone a mammoth*; Friese, *Cloning wild life.*

75. Pettit, "First human frozen by cryogenics could be brought back to life 'in just TEN years', claims expert."

76. Van Aken, "Ethics of reconstructing Spanish flu."

77. Marcus, "Frankenstein."

78. Van den Belt, "Frankenstein lives on"; Cohen, "How a horror story haunts science."

79. Carroll, *The philosophy of horror*, 31–32.

80. Carroll calls this the "paradox of fiction" (*The philosophy of horror*, 50ff.): that is, how is it possible that people are frightened and respond so emotionally to what does not exist?

81. Carroll has called this the "paradox of horror" (*The philosophy of horror*, 159ff.).

82. Freud, *The uncanny.*

83. Jones, *Horror*, 263.

84. Midgley, *Heart and mind: The varieties of moral experience.*

85. Fletcher, *Humanhood*, 39.

86. Fletcher, *Humanhood*, 136.

87. Kass, "The wisdom of repugnance," 23.

88. Midgley, *The essential Mary Midgley.*

89. Midgley, *Are you an illusion?* See also Midgley, *The myths we live by.*

90. Midgley, *Are you an illusion?*, 154.

91. Midgley, *Are you an illusion?*, 62.

92. Midgley, *Science as salvation*, 165–169.

93. "The basic idea is this: in contexts where the anomalous or ambiguous character of an object, event, or act seems to threaten disruption of the natural-social order, rather than promising to knit that order together, the object, event, or act will be abominated" (Stout, *Ethics after Babel*, 150).

94. Stout, *Ethics after Babel*, 152.

95. Midgley, *Science as salvation*, 221.

96. De Dijn, "De donkere transcendentie van Prometheus" [The dark transcendence of Prometheus].

97. "The further technical progress advances, the more the social problem of mastering this progress becomes one of an ethical and spiritual kind" (Ellul, "The technological order," 408).

98. Hampshire, "Morality and pessimism," 12.

99. Hampshire, "Morality and pessimism," 9.

Chapter 5 • Pilgrims

1. Woodman, *Patients beyond borders.*

2. Koelbing, *Arzt und Patient in der Antiken Welt* [Doctor and patient in antiquity]; Kasas and Struckmann, *Important medical centres in antiquity: Epidaurus and Corinth.*

3. Tuzunkan, "Wellness tourism: What motivates tourists to participate?"; Voigt, Brown, and Howat, "Wellness tourists: In search of transformation."

4. Hodges, Turner, and Kimball, *Risks and challenges in medical tourism*; Issenberg, *Outpatients*; Cohen, *Patients with passports.*

5. Woodman, *Patients beyond borders.*

6. De la Hoz-Correa, Muñoz-Leiva, and Nakucz, "Past themes and future trends in medical tourism research."

7. Lunt and Carrera, "Medical tourism: Assessing the evidence on treatment abroad."

8. See Issenberg, *Outpatients,* 19–26.

9. Turner, "'First world health care at third world prices'"; Meghani, "The ethics of medical tourism"; Smith, "The problematization of medical tourism." See also Leng, "Medical tourism and the state in Malaysia and Singapore."

10. Burkett, "Medical tourism: Concerns, benefits, and the American legal perspective," 226–227.

11. Cyranoski, "Building a biopolis."

12. Woodman, *Patients beyond borders,* 7.

13. Chen and Flood, "Medical tourism's impact on health care equity and access in low- and middle- income countries," 297.

14. Chen and Flood, "Medical tourism's impact," 294–295.

15. Gupta, "Medical tourism in India."

16. See Gupta, "Medical tourism in India."

17. Woodman, *Patients beyond borders,* 52.

18. Leggat, "Medical tourism."

19. Lunt et al., *Medical tourism: Treatments, markets and health system implications,* 25–26.

20. Lunt and Carrera, "Systematic review of web sites for prospective medical tourists," 57–67; Mason and Wright, "Framing medical tourism: An examination of appeal, risk, convalescence, accreditation, and interactivity in medical tourism web sites."

21. Whittaker, Manderson, and Cartwright, "Patients without borders: Understanding medical travel."

22. Hodges and Kimball, "Unseen travelers: Medical tourism and the spread of infectious diseases."

23. Smith, "The problematization of medical tourism," 1–8; Whittaker, Manderson, and Cartwright, "Patients without borders," 336–343.

24. Chen and Flood, "Medical tourism's impact," 290.

25. WHO, "Density of physicians, 2019."

26. Whittaker, Manderson, and Cartwright, "Patients without borders," 340.

27. Adams et al., "Promoting social responsibility amongst health care users."

28. For example, Pennings, "Ethics of medical tourism."

29. Adams et al., "Promoting social responsibility amongst health care users," 8.

30. This is the conclusion of Turner. See Turner, "Transnational medical travel: Ethical dimensions of global health care," 178.

31. Due to the Affordable Care Act, the number of insured people decreased from 44 million in 2013 to 28 million at the end of 2016. See Cohen, *Long-term trends in health insurance.*

32. Campbell, "193,000 NHS patients a month waiting beyond target time for surgery."

33. Meghani, "A robust, particularist ethical assessment of medical tourism"; Meghani, "The ethics of medical tourism."

34. Meghani, "A robust, particularist ethical assessment of medical tourism," 27.

35. Chen and Flood, "Medical tourism's impact," 291.

36. Chen and Flood, "Medical tourism's impact," 298.

37. Turner, "Transnational medical travel," 170–171.

38. Milstein and Smith, "America's new refugees—seeking affordable surgery offshore."

39. Inhorn, "Globalization and gametes: Reproductive 'tourism,' Islamic bioethics, and Middle Eastern modernity"; Matorras, "Reproductive exile versus reproductive tourism"; Inhorn and Patrizio, "Rethinking reproductive 'tourism' as reproductive 'exile.'"

40. Ackerman, "Plastic paradise: Transforming bodies and selves in Costa Rica's cosmetic surgery tourism industry."

41. Connell, "Medical tourism: Sea, sun, sand and . . . surgery."

42. Lunt et al., *Medical tourism*, 9.

43. Song, "Biotech pilgrims and the transnational quest for stem cell cures."

44. Ackerman, "Plastic paradise," 409, 413.

45. Voigt, Brown, and Howat, "Wellness tourists," 23–24.

46. Bassan and Michaelson, "Honeymoon, medical treatment or big business? An analysis of the meanings of the term 'reproductive tourism' in German and Israeli public media discourses."

47. See, for example, Bennett, *Miracle babies: How babies are made the IVF way.*

48. Blyth, "Fertility patients' experiences of cross-border reproductive care." It is therefore astonishing that the Ethics Committee of the American Society for Reproductive Medicine concludes that "the delivery of CBRC [cross-border reproductive care] does not invoke a duty to inform or warn patients about the potential legal or practical hazards that may accompany such care" ("Cross-border reproductive care: An Ethics Committee opinion," 1632).

49. Merab, "Do you really have to go to India? Set of new rules to curb medical pilgrimages."

50. Walsh, "Transplant tourists flock to Pakistan, where poverty and lack of regulation fuel trade in human organs."

51. Song, "Biotech pilgrims and the transnational quest for stem cell cures."

52. Martin, "Perilous voyages: Travel abroad for organ transplants and stem cell treatments," 139.

53. Cohen, *Patients with passports*, 433; Regensberg et al., "Medicine on the fringe: Stem cell-based interventions in advance of evidence"; Kashihara et al., "Evaluating the quality of website information of private-practice clinics offering cell therapies in Japan."

54. Martin, "Perilous voyages," 141, 152–160; Ryan et al., "Tracking the rise of stem cell tourism"; Cohen and Cohen, "International stem cell tourism and the need for effective regulation; Part I: Stem cell tourism in Russia and India."

55. Jiang and Dong, "Fraudsters operate and officialdom turns a blind eye: A proposal for controlling stem cell therapy in China."

56. Petersen, Seear, and Munsie, "Therapeutic journeys: The hopeful travails of stem cell tourists," 680.

57. Reader, *Pilgrimage*, 100–120.

58. Reader, *Pilgrimage*, 18.

59. Kaufman, *Consuming visions: Mass culture and the Lourdes shrine*; Stausberg, *Religion and tourism*; Reader, *Pilgrimage in the marketplace*.

60. Harpur, *The pilgrim journey: A history of pilgrimage in the Western world*.

61. Eade and Sallnow, *Contesting the sacred: The anthropology of pilgrimage*.

62. Dubisch and Winkelman, *Pilgrimage and healing*.

63. Eade, "Order and power at Lourdes," 62–65.

64. Badone and Roseman, "Approaches to the anthropology of pilgrimage and tourism," 3–4. See also Turner and Turner, *Image and pilgrimage in Christian culture*.

65. Woodman, *Patients beyond borders*, 10.

66. Reader, *Pilgrimage*, 27.

67. Song, *Biomedical odysseys: Fetal cell experiments from cyberspace to China*, 9, 178.

68. Song, "Biotech pilgrims and the transnational quest for stem cell cures," 387.

69. Song, *Biomedical odysseys*, 158ff.

70. Hench et al., "The effect of a hormone of the adrenal cortex (17-hydroxy-11-dehydrocorticosterone: compound E) and of pituitary adrenocorticotrophic hormone on rheumatoid arthritis."

71. Piccart-Gebhart et al., "Trastuzumab after adjuvant chemotherapy in HER2-positive breast cancer."

72. Collier, "Herceptin: Wanting the wonder drug"; Boseley, "The selling of a wonder drug"; Moss, "Hype and Herceptin."

73. Hortobagyi, "Trastuzumab in the treatment of breast cancer," 1735, 1736. The author of this editorial is a paid consultant to Genentech, the distributor of Herceptin in the United States.

74. *Lancet* editors, "Herceptin and early breast cancer."

75. Cameron et al., "11 years' follow-up of trastuzumab after adjuvant chemotherapy in HER2-positive early breast cancer."

76. Abola and Prasad, "The use of superlatives in cancer research."

77. Walsh, "Paralysed man walks again after cell transplant"; Walsh, "The paralysed man who can ride a bike."

78. Einsiedel and Adamson, "Stem cell tourism and future stem cell tourists," 35.

79. Song, *Biomedical odysseys*, 88. See also Einsiedel and Adamson, "Stem cell tourism and future stem cell tourists," 108, 186.

80. Kamenova and Caulfield, "Stem cell hype: Media portrayal of therapy translation."

81. Darrow, Avorn, and Kesselheim, "The FDA breakthrough-drug designation—four years of experience."

82. Russo and Cove, *Genetic engineering: Dreams and nightmares*.

83. White House, Office of the Press Secretary, "Remarks made by the President, Prime Minister Tony Blair of England (via satellite), Dr. Francis Collins, Director of the National Human Genome Research Institute, and Dr. Craig Venter, President and Chief Scientific Officer, Celera Genomics Corporation, on the completion of the first survey of the entire Human Genome Project."

84. Daley, "The promise and perils of stem cell therapeutics."

85. Sreenivasan and Weinberger, "Do you believe in miracles?"; Briggs, "Belief in miracles on the rise"; Theos, "Polls reveal Briton's spiritual side"; Woodward, "What miracles mean." A recent study in the Netherlands indicated that 63 percent of people

believe in miracles (Damen and Van Meur, "Bijna 9 miljoen Nederlanders geloven in wonderen" [Almost 9 million Dutchmen believe in miracles]).

86. Swinburne, *The concept of miracle*. See also Larmer, *Questions of miracle*, 26–39.

87. Stempsey, "Miracles and the limits of medical knowledge," 5.

88. Swinburne, *The concept of miracle*, 2, 9; Larmer, *Questions of miracle*, xi–xiii.

89. Peschel and Peschel, "Medical miracles from a physician-scientist's viewpoint."

90. Larmer, *Questions of miracle*, x; Swinburne, *The concept of miracle*, 61.

91. Mullin, *Miracles and the modern religious imagination*, 6.

92. Duffin, *Medical miracles: Doctors, saints, and healing in the modern world.*

93. Stempsey, "Miracles and the limits of medical knowledge," 7.

94. Dossey, "The possibility of the impossible: Miracles, wonder, and Thomas Jefferson's razor."

95. Peschel and Peschel, "Medical miracles from a physician-scientist's viewpoint," 396–397.

96. Schneiderman, *Embracing our mortality: Hard choices in an age of medical miracles*, 89. See also Duffin, "The doctor was surprised; or, How to diagnose a miracle."

97. The Medical Bureau at Lourdes was established in 1883, 25 years after the apparitions of the Virgin Mary. Dichoso, "Lourdes: A uniquely Catholic approach to medicine"; Harris, *Lourdes: Body and spirit in the secular age*; Gesler, "Lourdes: Healing in a place of pilgrimage"; Diamond, "Miraculous cures."

98. Kuruvilla, "French bishop declares nun's recovery a Lourdes miracle."

99. Duffin, *Medical miracles*, 6.

100. The interesting issue is that the Catholic Church has given a pivotal role to scientific expertise in investigating miraculous events. Religious miracles are associated with the cult of saints, emerging in Europe in late antiquity. Traditionally, saints were recognized and venerated by local communities. They were proclaimed by popular devotion. But the Church increasingly attempted to control popular religion. Around 1588 a formal process of saint-making was initiated. A special congregation was charged with examining asserted miracles. All possible doubt should be explored. Its office of the "devil's advocate" contested the credibility of witnesses, testimonies, and evidence. Saints are nowadays proclaimed by the pope. While there are thousands of saints in popular or folk religion, until 1988 only 285 have been officially recognized. See Brown, *The cult of the saints*. The Holy Congregation of Rites was established in 1587; it is now called the Congregation for the Causes of Saints. See Duffin, *Medical miracles*, 11–15. See also Woodward, *Making saints: How the Catholic Church determines who becomes a saint, who doesn't, and why*. Recent popes have hugely increased the number of official saints. Pope John Paul II canonized 480 saints; Pope Benedict XVI, 45, and until mid-2019 Pope Francis had canonized 892 saints.

101. Mullin, *Miracles and the modern religious imagination*, 11.

102. Research in the United Kingdom indicates that over three-quarters of all adults (77 percent) and three-fifths (61 percent) of nonreligious people believe that "there are things in life that we simply cannot explain through science or any other means." See Hill, *Global church: Reshaping our conversations, renewing our mission, revitalizing our churches*, 131.

103. De Vries, *Kleine filosofie van het wonder* [Little philosophy of the miracle].

104. Cassell, *Expected miracles: Surgeons at work.*

105. Orr, "Responding to patient beliefs in miracles."

106. Kub and Groves, "Miracles and medicine: An annotated bibliography."

107. May, "Claimed contemporary miracles."

108. Clarke, "When they believe in miracles"; Bibler, Shinall, and Stahl, "Responding to those who hope for a miracle: Practices for clinical bioethicists"; Rushton and Russell, "The language of miracles: Ethical challenges."

109. Mansfield, Mitchell, and King, "The doctor as God's mechanic? Beliefs in the Southeastern United States"; Silvestri et al., "Importance of faith on medical decisions regarding cancer care."

110. Sulmasy, "Spiritual issues in the care of dying patients." See also Delisser, "When a miracle is expected: Allowing space to believe."

111. Brett and Jersild, "'Inappropriate' treatment near the end of life: Conflict between religious convictions and clinical judgment."

112. Braithwaite, "The medical miracles delusion"; O'Leary, "Why bioethics should be concerned with medically unexplained symptoms."

113. Labovitz, *Ordinary miracles: True stories about overcoming obstacles and surviving catastrophes.*

114. Nekolaichuk, Jevne, and Maguire, "Structuring the meaning of hope in health and illness."

115. Groopman, *The anatomy of hope: How people prevail in the face of illness*, xiv.

116. Elliott and Olver, "Hope and hoping in the talk of dying cancer patients." Patterson mentions that in the late 1950s a survey among American physicians indicated that 90 percent preferred not to tell cancer patients the truth (*The dread disease: Cancer and modern American culture*, 167).

117. Del Vecchio Good et al., "American oncology and the discourse on hope." See also Patterson, *The dread disease*, 226–227.

118. Brown, "Shifting tenses: Reconnecting regimes of truth and hope."

119. Brown, "Hope against hype—accountability in biopasts, presents and futures."

120. Matthews-King, "Vulnerable would-be parents being sold 'false hopes' by overseas IVF clinics, regulator warns."

121. Examples are xenotransplantation, gene therapy, use of human stem cells, and gene editing. See Brown, "Hope against hype," 6; Charo, "On the road (to a cure?)—stem-cell tourism and lessons for gene editing."

122. Petersen et al., *Stem cell tourism and the political economy of hope*, 59–82.

123. Brown, "Shifting tenses," 335.

124. Brown, "Hope against hype," 4, 5.

125. Brown, "Shifting tenses," 331.

126. Petersen, Tanner, and Munsie, "Between hope and evidence: How community advisors demarcate the boundary between legitimate and illegitimate stem cell treatments."

127. Enserink, "Selling the stem cell dream"; Qiu, "Trading on hope"; Murdoch and Scott, "Stem cell tourism and the power of hope"; Song, "Biotech pilgrims and the transnational quest for stem cell cures."

128. Beste, "Instilling hope and respecting patient autonomy: Reconciling apparently conflicting duties."

129. Koopmeiners et al., "How health care professionals contribute to hope in patients with cancer."

130. Petersen, Seear, and Munsie, "Therapeutic journeys"; Moreira and Palladino, "Between truth and hope: On Parkinson's disease and the production of the 'self.'"

131. Elliott and Olver, "Hope, life, and death: A qualitative analysis of dying cancer patients' talk about hope."

132. Petersen, MacGregor, and Munsie, "Stem cell miracles or Russian roulette? Patients' use of digital media to campaign for access to clinically unproven treatments."

133. Brown, "Hope against hype"; Murdoch and Scott, "Stem cell tourism and the power of hope"; Lau et al., "Stem cell clinics online"; Devereaux and Loring, "Growth of an industry: How US scientists and clinicians have enabled stem cell tourism"; Kamenova and Caulfield, "Stem cell hype."

134. At least the general population in Canada, see Einsiedel and Adamson, "Stem cell tourism and future stem cell tourists."

135. Petersen and Seear, "Technologies of hope: Techniques of the online advertising of stem cell treatments."

136. Brown, "Hope against hype," 6.

137. Petersen et al., *Stem cell tourism and the political economy of hope.*

138. Turner, "US stem cell clinics, patient safety, and the FDA"; Turner and Knoepfler, "Selling stem cells in the USA."

139. Jiang and Dong, "Fraudsters operate and officialdom turns a blind eye," 404.

140. Karlin-Smith, "Libertarians score big victory in 'right-to-try' drug bill"; Karlin-Smith, "House passes right-to-try on second try."

141. White House, "Remarks by President Trump in joint address to Congress." Trump signed the Right to Try Act on May 30, 2018. Trump claimed that hundreds of thousands could be saved as a result of the law.

142. Olsen, *The right to try: How the Federal Government prevents Americans from getting the lifesaving treatments they need*; Sandefur, "Safeguarding the right to try."

143. Bateman-House and Robertson, "The federal Right to Try Act of 2017—a wrong turn for access for investigational drugs and the path forward."

144. Daley, "The promise and perils of stem cell therapeutics."

145. During FY 2010–2014, 20 percent of investigational new drug applications received marketing approval by one year after initial submission; 33 percent was approved by five years. See McKee et al., "How often are drugs made available under the Food and Drug Administration's expanded access process approved?"

146. Pease et al., "Postapproval studies of drugs initially approved by the FDA on the basis of limited evidence."

147. DeTora, "What is safety? Miracles, benefit-risk assessments, and the 'right to try.'"

148. Joffe and Lynch, "Federal right-to-try legislation—threatening the FDA's public health mission."

149. Ghinea, Little, and Lipworth, "Access to high cost cancer medicines through the lens of an Australian senate inquiry." Right-to-try bills were considered in the House Committee on Energy and Commerce.

150. Watson, "A global perspective on compassionate use and expanded access"; Pace et al., "Accelerated access to medicines," 158.

151. Nather and Kaplan, "Public wary of faster approvals of new drugs, STAT-Harvard poll finds."

152. Pace et al., "Accelerated access to medicines," 159.

153. Saluja et al., "Unsafe drugs were prescribed more than one hundred million times in the United States before being recalled."

154. See, for example, Kearns and Bateman-House, "Who stands to benefit? Right to try law provisions and implications."

155. Darrow, Avorn, and Kesselheim, "The FDA breakthrough-drug designation—four years of experience."

156. Jonsen, *The birth of bioethics*, 377–405.

157. Jonsen, *The birth of bioethics*, 397.

158. "Barack Obama's remarks to the Democratic National Convention," *New York Times*, July 27, 2004.

Chapter 6 • Prophets

1. Hippocrates, "Prognostics," 7–9. See also Godderis, *Een arts is vele andere mensen waard* [A physician is as valuable as many other people], 307–355.

2. Rizzi, "Medical prognosis—some fundamentals."

3. Jones, "Prognostication: Medicine's lost art."

4. Hippocrates, "Prognostics," 9.

5. In particular was the case of Eudemus, a famous philosopher at that time in Rome who was seriously ill. While all other doctors prognosticated that he would die, Galen correctly predicted when he would have fevers and when they would end, and the patient healed (Smith, *The Hippocratic tradition*, 82–83).

6. Anschütz, *Ärztliches Handeln* [Medical action], 120–125. See also Rich, "Prognostication in clinical medicine."

7. Wulff, Pedersen, and Rosenberg, *Philosophy of medicine*, 46–60.

8. Wulff, Pedersen, and Rosenberg, *Philosophy of medicine*, 20–21.

9. Kellett, "Prognostication—the lost skill of medicine"; Jones, "Prognostication: Medicine's lost art"; Christakis, "The ellipsis of prognosis in modern medical thought."

10. Wulff, *Rational diagnosis and treatment*, 6–7.

11. De Senneville "Aux Etats-Unis, le boom des tests génétiques" [In the United States, the boom of genetic tests].

12. Coghlan, "Nationwide genetic testing."

13. Bob Abernethy, interview with Dr. Francis Collins, director of the Human Genome Project at the National Institutes of Health, June 16, 2000. See also Davies, *Cracking the genome: Inside the race to unlock human DNA*, 216.

14. Roberts et al., "The predictive capacity of personal genome sequencing."

15. Nebert, "Given the complexity of the human genome, can 'personalised medicine' or 'individualised drug therapy' ever be achieved?," 300.

16. See, for example, Caravagna et al. "Detecting repeated cancer evolution from multi-region tumor sequencing data."

17. Ross and Swetlitz, "IBM pitches its Watson supercomputer as a revolution in cancer care. It's nowhere close."

18. Topol, *The creative destruction of medicine: How the digital revolution will create better health care*. The fear that a new stage of life (Life 3.0) will arrive in the next cen-

tury with the progress of artificial intelligence is the central topic of Max Tegmark's bestseller *Life 3.0: Being human in the age of artificial intelligence.*

19. Meilaender, *Should we live forever? The ethical ambiguities of aging.* Davis, *New Methuselahs: The ethics of life extension.*

20. Temkin, "History and prophecy: Meditations in a medical library."

21. Lewinsohn, *Science, prophecy, and prediction.*

22. The term "genetic prophecy" is used earlier in an editorial by George J. Annas, "Genetic prophecy and genetic privacy—can we prevent the dream from becoming a nightmare?" Nicholas Christakis has related prophecy and prognosis, especially in end-stage medical care. See Christakis, *Death foretold: Prophecy and prognosis in medical care.*

23. Forman, *The story of prophecy: In the life of mankind from early times to the present day.*

24. D. Petersen, *The prophetic literature: An introduction.*

25. Davies, *Prophecy and ethics: Isaiah and the ethical traditions of Israel,* 13–14.

26. Aune, *Prophecy in early Christianity and the ancient Mediterranean world.*

27. Bakan, *Maimonides on prophecy,* 15ff.

28. Davies, *Prophecy and ethics,* 13.

29. Aune, *Prophecy in early Christianity and the ancient Mediterranean world,* 8, 19.

30. Amos 9:11–15.

31. D. Petersen, *The prophetic literature,* 41–42.

32. Aune, *Prophecy in early Christianity and the ancient Mediterranean world,* 108.

33. Bowler, *A history of the future: Prophets of progress from H. G. Wells to Isaac Asimov.*

34. See Silver's contributions to the website Big Think at https://bigthink.com /experts/leesilver.

35. Silver, *Remaking Eden: Cloning and beyond in a brave new world;* Silver, *Remaking Eden: How genetic engineering and cloning will transform the American family.*

36. Harari, *Homo Deus: A brief history of tomorrow.*

37. Topol, *The creative destruction of medicine;* Topol, *The patient will see you now.*

38. Van Leeuwen, *Prophecy in a technocratic era,* 33.

39. PubMed shows that "enhancement" journal publications increased from a few hundred in 1985 to 13,571 in 2017. For "ethics & enhancement" the number of publications grew from 1 in 1985 to 109 in 2017. Using the keyword "predictive medicine" generates 12 publications from 1985 and 13,056 from 2017. For "predictive medicine & ethics" there were publications in 1985 and 64 in 2017. "Life extension" produced 368 publications from 1985 and 2,792 from 2017. For "life extension & ethics" there are 5 publications from 1985 and 35 from 2017.

40. Ferry, *L'homme-Dieu ou le Sens de la Vie* [The man-god, or the meaning of life].

41. Lindeboom, *Descartes and medicine.*

42. Condorcet, *Oeuvres de Condorcet* [Works of Condorcet], vol. 6, 13.

43. Ball, *Unnatural: The heretical idea of making people;* Gordijn, *Medical utopias: Ethical reflections about emerging medical technologies.*

44. Van Dijck, *Imagenation: Popular images of genetics;* Wilkie, *Perilous knowledge: The Human Genome Project and its implications;* Kitcher, *The lives to come: The genetic revolution and human possibilities.*

45. Harris, *Wonderwoman and Superman: The ethics of human biotechnology*, 2.

46. Schwartz, *Life without disease: The pursuit of medical utopia.*

47. Hoedemaekers and ten Have, "Genetic health and disease."

48. Hagerty, "Can your genes make you murder?"

49. Keane, "Survival of the fairest? Evolution and the geneticization of rights."

50. Van Dijck, *Imagenation*; Petersen, "Biofantasies: Genetics and medicine in the print news media"; Condit, *The meanings of the gene: Public debates about human heredity.*

51. "Genetic essentialism reduces the self to a molecular entity, equating human beings, in all their social, historical, and moral complexity, with their genes" (Nelkin and Lindee, *The DNA mystique*, 2; see also 68).

52. Lippman, "Prenatal genetic testing and screening."

53. Lippman, "Prenatal genetic testing and geneticization," 178.

54. Ten Have, "Genetics and culture: The geneticization thesis"; Ten Have, "Genetic advances require comprehensive bioethical debate."

55. Rothman, *Genetic maps and human imaginations.* Steve Jones states, "In the public mind, genetics is no longer a science but a faith; a curse or a salvation" (*The language of the genes: Biology, history and the evolutionary future*, xv).

56. Foucault, *The birth of the clinic.*

57. Arney and Bergen, *Medicine and the management of living.*

58. Clarke et al., "Biomedicalising genetic health, diseases and identities."

59. See Illich, "The medicalization of life"; Zola, "In the name of health and illness."

60. Hoedemaekers and ten Have, "Geneticization: The Cyprus paradigm."

61. Miller, "Geneticization: An interview with Abby Lippman on new genetics."

62. Keller, *The century of the gene*, 3ff. Nelkin and Tancredi define biological reductionism as "the tendency to reduce complex behavior to measurable biological dimensions" (*Dangerous diagnostics: The social power of biological information*, 10).

63. Lippman, "Led (astray) by genetic maps: The cartography of the human genome and health care."

64. Nelkin and Lindee, *The DNA mystique.* See also Nowotny and Testa, *Naked genes: Reinventing the human in the molecular age*, 4.

65. Stempsey, "The geneticization of diagnostics."

66. Torres, "Genetic tools, Kuhnian theoretical shift and the geneticization process"; Petersen, "The genetic conception of health: Is it as radical as claimed?"

67. William Stempsey has pointed at the absurdity of these views. If our genetic constitution determines our identity (genetic essentialism) and if our genetic constitution is causing diseases (genetic reductionism), we *are* our diseases. But also, if genes are constitutive of diseases, in fact all diseases are genetic diseases. See Stempsey, "The geneticization of diagnostics," 198.

68. Keller, *The century of the gene*, 137–138.

69. As Peter Beurton explains, "The more molecular biologists learn about the structure and functioning of the gene, the less they know what a gene really is" ("A unified view of the gene, or how to overcome reductionism," 295).

70. Mukherjee, *The gene: An intimate history*; Carey, *The epigenetic revolution.*

71. Falk, "The gene—a concept in tension."

72. Graham, *Genes: A philosophical inquiry*, 94–96.

73. Rose and Rose, *Genes, cells and brains: The Promethean promises of the new biology*, 52–53, 276–305; Kaplan, *The limits and lies of human genetic research: Dangers for social policy*; Jochemsen, "Reducing people to genetics." See also Lewontin, Rose, and Kamin, *Not in our genes*. For a recent effort to debunk the claims of genetic determinism, see Alexander, *Genes, determinism, and God*. According to Alexander, "the development of human traits is . . . 100 per cent genetic and 100 per cent environmental" (210).

74. Greene and Loscalzo, "Putting the patient back together—social medicine, network medicine, and the limits if reductionism."

75. "The image of genes as clear and distinct causal agents, constituting the basis of all aspects of organismic life, has become so deeply embedded in both popular and scientific thought that it will take far more than good intentions, diligence, or conceptual critique to dislodge it" (Keller, *The century of the gene*, 136). Examining public attitudes toward genetic testing between 2002 and 2010, Henneman et al. concludes that "the public does not share the tempered expectations that are expressed by professionals" ("Public attitudes towards genetic testing revisited," 795).

76. See, for example, Rouvroy, *Human genes and neoliberal governance*; Weiner and Martin, "A genetic future for coronary heart disease?"; Jones, *The language of the genes*; Keane, "Survival of the fairest? Evolution and the geneticization of rights"; Nowotny and Testa, *Naked genes*.

77. Hoedemaekers and ten Have, "Geneticization."

78. Clarke et al., "Biomedicalising genetic health, diseases and identities," 33.

79. Nowotny and Testa, *Naked genes*, 25.

80. See, for example, Hall, "Revolution postponed."

81. See the mission statement on the homepage of the Precision Medicine Initiative website at https://obamawhitehouse.archives.gov/precision-medicine.

82. Keller, *The century of the gene*, 10.

83. Nelkin and Lindee, *The DNA mystique*, 92ff. and 129.

84. Wulff, Pedersen, and Rosenberg, *Philosophy of medicine*, 57.

85. Stent, "That was the molecular biology that was."

86. Spellberg and Taylor-Blake, "On the exoneration of Dr. William H. Stewart: Debunking an urban legend."

87. Christakis, *Death foretold*, 64ff.; Christakis and Lamont, "Extent and determinants of error in doctors' prognoses in terminally ill patients."

88. Hilden and Habbema, "Prognosis in medicine: An analysis of its meaning and roles."

89. Christakis, *Death foretold*, xviii, 179ff.

90. Christakis, *Death foretold*, 33, 137.

91. Tandy-Connor et al., "False-positive results released by direct-to-consumer genetic tests highlight the importance of clinical confirmation testing for appropriate patient care"; United States Government Accountability Office, *Direct-to-consumer genetic tests: Misleading test results are further complicated by deceptive marketing and other questionable practices*.

92. Hamburg and Collins, "The path to personalized medicine," 303.

93. Hamburg and Collins, "The path to personalized medicine," 301.

94. Doz, Marvanne, and Fagot-Largeault, "The person in personalized medicine." It is argued that medicine has always been individualized and personalized. See Fierz,

"Challenge of personalized health care: To what extent is medicine already individualized and what are the future trends?"

95. Tutton, "Personalizing medicine: Futures present and past"; Horwitz et al., "(De)Personalized medicine"; Rose, "Personalized medicine: Promises, problems and perils of a new paradigm for healthcare."

96. Juengst et al., "From 'personalized' to 'precision' medicine: The ethical and social implications of rhetorical reform in genomic medicine"; Schleidgen et al., "What is personalized medicine: Sharpening a vague term based on a systematic literature review."

97. Evans et al., "Deflating the genomic bubble"; Greene and Loscalzo, "Putting the patient back together," 2493.

98. Savard, "Personalised medicine: A critique on the future of health care"; Horne, "The human dimension: Putting the person into personalised medicine"; Chowkwanyun, Bayer, and Galea, " 'Precision' public health—between novelty and hype."

99. Weiss, "Is precision medicine possible?"

100. Chan and Erikainen, "What's in a name? The politics of 'precision medicine.' " This value context is specifically clear in the claims made. For example, the claim that precision medicine reflects a new biomedical cosmology (see Vegter, "Towards precision medicine; a new biomedical cosmology"). Or the claim that it will transform medicine from an art into a science (see Wiesing, "From art to science: A new epistemological status for medicine? On expectations regarding personalized medicine").

101. Harari, *Homo Deus.*

102. Popper, *The poverty of historicism*, vii.

103. Popper, *The poverty of historicism*, 3

104. Popper, *The poverty of historicism*, 51.

105. Popper, *The poverty of historicism*, 43.

106. Popper, *The poverty of historicism*, 64ff.

107. Silver, in the afterword of the new edition of his book, *Remaking Eden* (2007), 294 and 295.

108. Topol, *The creative destruction of medicine*, 243.

109. Topol, *The patient will see you now*, 275ff.

110. Silver, *Remaking Eden* (1997), 11.

111. Future scenarios can be predicted because they emerge from the science and technology that is already available today, according to Silver. See Silver, *Remaking Eden* (1997), 7.

112. Harari, *Homo Deus*, 53.

113. Harari, *Homo Deus*, 65.

114. Harari, *Homo Deus*, 461.

115. Harari, *Homo Deus*, 328.

116. Silver, *Remaking Eden* (1997), 210–211.

117. Silver, *Remaking Eden* (1997), 151.

118. Silver, *Remaking Eden* (1997), 233.

119. Arnold Relman ("A coming medical revolution?") addresses the question of how sure we can be that digitization and virtual communication will indeed replace human contact and redefine personal caring.

120. "we cannot hit the brakes . . . nobody knows where the brakes are" (Harari, *Homo Deus*, 58).

121. Harari, *Homo Deus*, 135.

122. Harari, *Homo Deus*, 220ff.

123. Beresford, "Uncertainty and the shaping of medical decisions."

124. Beresford, "Uncertainty and the shaping of medical decisions," 9.

125. Wray and Loo, "The diagnosis, prognosis, and treatment of medical uncertainty"; Seely, "Embracing the certainty of uncertainty: Implications for health care and research."

126. Nelkin and Tancredi, *Dangerous diagnostics*, 102.

127. Boenink, "Predictive medicine" in *Encyclopedia of global bioethics*.

128. Rizzi, "Medical prognosis—some fundamentals," 369.

129. Karches, "Against the iDoctor: Why artificial intelligence should not replace physician judgment."

130. Hedgecoe, *The politics of personalised medicine: Pharmacogenetics in the clinic*.

131. Ioannidis, "Why most published research findings are false"; Ioannidis, "An epidemic of false claims."

132. Stegenga, *Medical nihilism*.

133. Medicine today, according to Stegenga, should be less aggressive and more gentle: what we need is fewer medical interventions, more lifestyle interventions, and more care (*Medical nihilism*, 186, 190). See also Ehrenreich, *Natural causes: An epidemic of wellness, the certainty of dying, and killing ourselves to live longer*; Wootton, *Bad medicine: Doctors doing harm since Hippocrates*.

134. Silver, *Remaking Eden* (1997), 11, 225.

135. Harris, *Wonderwoman and Superman*, 5.

136. Silver, *Remaking Eden* (1997), 240.

137. Topol, *The patient will see you now*, 13.

138. Slaughter, "The IT revolution reassessed part one: Literature review and key issues."

139. Slaughter, "The IT revolution reassessed part one." See Morozov, *To save everything, click here: The folly of technological solutionism*; Hampson and Jardine, *Look who's watching: Surveillance, treachery, and trust online*.

140. First, "Will big data algorithms dismantle the foundations of liberalism?"

141. Topol, *The patient will see you now*, 275ff.

142. Harari observes, "Wealth and power might become concentrated in the hand of the tiny elite that owns the all-powerful algorithms" (*Homo Deus*, 376). He also writes, "The individual is becoming a tiny chip inside a giant system that nobody really understands" (447). See Gray, "Humanity Mk II: The next great stage of our evolution has begun."

143. Ten Have, "Medical technology assessment and ethics: Ambivalent relations."

144. Cesuroglu et al., "Other side of the coin for personalised medicine and healthcare."

145. Rothman, *Genetic maps and human imaginations*, 14.

146. McAfee, "Neoliberalism on the molecular scale: Economic and genetic reductionism in biotechnology battles."

147. Kalokairinou et al., "Legislation of direct-to-consumer genetic testing in Europe."

148. Silver, *Remaking Eden* (1997), 9, 224.

149. Segal, "When China rules the web: Technology in the service of the state."

150. Kaplan, *The limits and lies of human genetic research*, 170ff.

151. Predictive medicine "serves to expand the potential demand for drugs for people with no symptoms and who feel themselves to be healthy, making patients of those who may or may not have considered themselves as at risk" (Rose and Rose, *Genes, cells and brains*, 185).

152. Otlowski, Taylor, and Bombard. "Genetic discrimination: International perspectives."

153. Smart, Martin, and Parker, "Tailored medicine: Whom will it fit? The ethics of patient and disease stratification."

154. Minari, Brothers, and Morrison, "Tensions in ethics and policy created by National Precision Medicine Programs."

155. "In England, people living in the poorest neighbourhoods will, on average, die seven years earlier than people living in the richest neighbourhoods" (Marmot et al., *Fair society, healthy lives*, 37).

156. Evans et al., "Deflating the genomic bubble"; Maughan, "The promise and the hype of 'personalised medicine.'"

157. Gigerenzer and Garcia-Retamero, "Cassandra's regret: The psychology of not wanting to know."

158. Oster, Shoulson, and Dorsey, "Optimal expectations and limited medical testing: Evidence from Huntington disease."

159. Bishop and Waldholz, *Genome: The story of the most astonishing scientific adventure of our time—the attempt to map all the genes in the human body*, 267–284.

160. McBride et al., "The behavioural response to personalised genetic information"; Marteau et al., "Effects of communicating DNA-based disease risk estimates on risk-reducing behaviours"; Hollands et al., "The impact of communicating genetic risks of disease on risk-reducing health behaviour."

161. Sabaté, *Adherence to long-term therapies: Policy for action.*

162. Ten Have, *Global bioethics*, 67–74.

163. Which is one of the underlying sources for genetic discrimination. See Otlowski, Taylor, and Bombard, "Genetic discrimination"; Taylor et al., "Investigating genetic discrimination in Australia."

164. Ten Have, "The right to know and the right not to know in the ear of neoliberal biopolitics and bioeconomy."

165. Henderson and Petersen, *Consuming health*; Mold, "Repositioning the patient"; Tomes, *Remaking the American patient.*

166. Adams and de Bont, "Information Rx: Prescribing good consumerism and responsible citizenship"; Lee, "Framing choice: The origins and impact of consumer rhetoric in US health care debates."

167. Schneider and Hall, "The patient life: Can consumers direct health care?"

168. As Schneider and Hall comment, "People have better things to do than become health care experts" ("The patient life," 41).

169. Downie, "Patient and consumers"; Hunter, "The case against choice and competition."

170. Levitt, "Empowered by choice?"

171. Juengst et al., "From 'personalized' to 'precision' medicine," 21–33.

172. Juengst et al., "From 'personalized' to 'precision' medicine," 25–26.

173. Davis, *New Methuselahs*, 119–128.

174. For example, Crutchfield, "Compulsory moral bioenhancement should be covert."

175. Pavone and Arias, "Beyond the geneticization thesis: Political economy of PGD/PGS in Spain."

176. Ehrich and Williams, "A 'healthy baby': The double imperative of preimplantation genetic diagnosis."

177. Gefenas et al., "Does the 'new philosophy' in predictive, preventive and personalised medicine require new ethics?"

178. Alexander, *Genes, determinism, and God.*

179. Harari, *Homo Deus*, 327 ("Science does not deal with questions of value") and 329ff.

180. Harari, *Homo Deus*, 177.

181. Damasio, "We must not accept an algorithmic account of human life."

182. Miller, "The composite redesign of humanity's nature"; Sexton, "A reductionist history of humankind."

183. Rose, "Personalized medicine," 349–350.

184. An extreme example is ivacaftor (brand name Kalydeco). This is a drug for children with cystic fibrosis who have a specific mutation that accounts for 4–5 percent of cases. The drug was approved by the FDA in 2012. However, the price of the drug is $300,000 per year. The drug is produced by a private biotech company building on fundamental research in the National Institutes of Health, and with substantial support of the Cystic Fibrosis Foundation. See also Tannock and. Hickman, "Limits to personalized cancer medicine," 1293; Gyawali and Sullivan, "Economics of cancer medicines: For whose benefit?"; Maughan, "The promise and the hype of 'personalized medicine.'"

185. O'Sullivan, Orenstein, and Milla, "Pricing for orphan drugs: Will the market bear what society cannot?"

186. Commission of the European Communities, *Proposal for a council decision adopting a specific research programme in the field of health; Predictive Medicine: Human Genome Analysis (1989–1991)*, 1, 3.

187. Dickson, "Genome Project gets rough ride in Europe"; Blank, "Science and the human genome: Policy issues in genetic intervention in the 1990s"; Durfy and Grotevant, "The Human Genome Project"; McNally, "Eugenics here and now."

188. Condit, *The meanings of the gene*, 5.

189. Condit, *The meanings of the gene*, 214.

190. Rose and Rose, *Genes, cells and brains*, 126. See also Rose, "Eugenics and genetics: The conjoint twins?"

191. "Eugenics as a policy and practice neither originated nor died with the Nazis" (Rose, "Eugenics and genetics," 15).

192. Weindling, "Julian Huxley and the continuity of eugenics in twentieth-century Britain."

193. Koch, "The meaning of eugenics: Reflections on the government of genetic knowledge in the past and the present."

194. Koch, "The meaning of eugenics," 327.

195. According to Rothman, "Eugenics never disappeared from American thinking"

(*Genetic maps and human imaginations*, 111). See also Nelkin and Tancredi, *Dangerous diagnostics*, 13.

196. Commission of the European Communities, *Proposal for a council decision*, 12; McNally, "Eugenics here and now," 69–82.

197. Lederberg, "Orthobiosis: The perception of man."

198. Fletcher, *The ethics of genetic control: Ending reproductive roulette*, 3, 178.

199. Kitcher, *The lives to come*, 187ff.

200. Nelkin and Lindee point out that "eugenics in contemporary culture is less a state ideology than a set of ideals about perfected and 'health' human future" (*The DNA mystique*, 191).

201. Rothman, *Genetic maps and human imaginations*, 177.

202. Rose and Rose, *Genes, cells and brains*, 143; Kitcher, *The lives to come*, 196.

203. Petersen, "Is the new genetics eugenic? Interpreting the past, envisioning the future."

204. Duster, *Backdoor to eugenics*.

205. Taylor et al., "Investigating genetic discrimination in Australia"; Otlowski, Taylor, and Bombard, "Genetic discrimination"; Joly et al., "Comparative approach to genetic discrimination."

206. Miller and Levine, "Avoiding genetic genocide."

207. Comfort, "Can we cure genetic diseases without slipping into eugenics?"

208. People like Ray Kurzweil, Aubrey de Grey, and Max More. See, for example, O'Connell, *To be a machine*; Gollner, *The book of immortality*; Weiner, *Long for this world*; Hall, *Merchants of immortality*.

209. For many, this was a strange movement. Only a few hundred people had their bodies frozen. See Reilly, "Frozen in time."

210. See also "Companies and start-ups working on extending human longevity and life-span" on LongLongLife.org (http://www.longlonglife.org/en/transhumanism-longevity/aging/companies-start-ups-working-extending-human-longevity-life-span).

211. Friend, "Silicon Valley's quest to live forever."

212. Marks and Gottlieb, "Balancing safety and innovation for cell-based regenerative medicine"; Mason and Dunnill, "A brief definition of regenerative medicine."

213. Davis, *New Methuselahs*; Meilaender, *Should we live forever? The ambiguities of aging*.

214. O'Connell, *To be a machine*, 159.

215. Ten Have, *Vulnerability: Challenging bioethics*; May, *A fragile life: Accepting our vulnerability*.

216. Olshansky, "From lifespan to healthspan."

217. See, for example, Dragojlovic, "Canadians' support for radical life extension resulting from advances in regenerative medicine."

218. Popper, *The poverty of historicism*, 56.

219. Tallis, *Enemies of hope: A critique of contemporary pessimism*, 379ff.

220. For a critique of "technological solutionism," see Morozov, *To save everything, click here*, 1–16.

221. Harari, *Homo Deus*, 545.

222. Fukuyama, *Our posthuman future*; Zuboff, *The age of surveillance capitalism*. Countries such as India and China are rapidly developing into surveillance states. In

India, the Aadhaar program, which was launched nine years ago, is building a large biometric database by collecting fingerprints, iris scans, and photos of 1.3 billion people. It is now a gateway to many services, such as food stamps, passports, pensions, or phones. In September 2018, India's Supreme Court ruled that it is constitutional and only prohibited its use by private companies, such as banks and mobile phone service providers.

223. Tallis, *Enemies of hope*, 403. Daniel Callahan agrees but warns that progress often creates its own problems (see Callahan, "Progress: Its glories and pitfalls").

224. Poushter, "Worldwide, people divided on whether life today is better than in the past."

225. Shelley, "A defence of poetry," 293.

226. Bennett, *Cultural pessimism: Narratives of decline in the postmodern world*, 190.

Chapter 7 • Relics

1. Wilkinson, "The saints of Pittsburgh." See also Van Osselaer, de Smaele, and Wils, *Sign or Symptom?*, 116–118.

2. Hunter, "Alder Hey report condemns doctors, management, and coroner"; *The Royal Liverpool Children's Inquiry: Report*. It was also found that Alder Hey has sold organs from living children to pharmaceutical company Aventis Pasteur. Thymus glands were systematically removed during heart surgery; the company paid per organ but of course the parents did not know about this commercial transaction either.

3. Park, "Ten ideas changing the world right now: Biobanks."

4. Orchard-Webb, "10 largest biobanks in the world."

5. Verna, "Museums and the repatriation of indigenous human remains."

6. Knoeff, "Ball pool anatomy: On the public veneration of anatomical relics," 281.

7. Knoeff, "Ball pool anatomy," 284.

8. Woodward, *Making saints: How the Catholic Church determines who becomes a saint, who doesn't, and why*, 404–406.

9. Ten Have, "God en gezondheid" [God and health].

10. Tybjerg, "From bottled babies to biobanks: Medical collections in the twenty-first century," 277.

11. Abraham, *Possessing genius: The bizarre odyssey of Einstein's brain.*

12. Davies, "The unquiet cranium."

13. Cruz, *Relics: What they are and why they matter.*

14. Bartlett, *Why can the dead do such great things? Saints and worshippers from the martyrs to the Reformation*, 7–10.

15. Sumption, *Pilgrimage: An image of medieval religion*, 27.

16. Geary, *Furta sacra: Thefts of relics in the Central Middle Ages.*

17. Bartlett, *Why can the dead do such great things?*, 252; Sumption, *Pilgrimage*, 31–32.

18. There were four ways to procure relics: invention (or "discovery" of relics), gift exchange, purchase, or theft. Stealing relics was illegal but met with general approval. Several justifications were proposed. One is the implicit permission of the saint. It was impossible to steal relics without the cooperation of the saint. Another justification was the security of the relics. Stealing relics from an area of war or occupation, for example in countries conquered by Islamic rulers where veneration is halted, will reinstall

proper veneration. A third argument was that thefts promote the good of the community that receive the relics. Additionally, stolen relics are often more authentic than purchased ones. See Geary, *Furta sacra*, 112ff.

19. The ritual relocation of the relics of a saint was known as *translatio* (Bartlett, *Why can the dead do such great things?*, 282ff.).

20. Sumption, *Pilgrimage*, 23–24.

21. Bartlett, *Why can the dead do such great things?*, 609. Freeman, *Holy bones, holy dust: How relics shaped the history of Medieval Europe*, 13; Brown, *The cult of the saints: Its rise and function in Latin Christianity*.

22. Freeman, *Holy bones, holy dust*, 20.

23. Thomassen, "The uses and meanings of liminality."

24. Woodward, *Making saints*, 404–406.

25. Woodward, *Making saints*, 7.

26. Woodward, *Making saints*, 50.

27. Sumption, *Pilgrimage*, 53; Freeman, *Holy bones, holy dust*, 146–155.

28. Woodward, *Making saints*, 51.

29. Calvin, *A treatise on relics*, 165, 169

30. Wharton, "Relics, Protestants, things."

31. "The living need the dead far more than the dead need the living" (Laqueur, *The work of the dead: A cultural history of mortal remains*, 1).

32. Van der Hoeven, "Levensbericht van Gerardus Vrolik" [Short biography of Gerardus Vrolik].

33. Alberti, "A history of Edinburgh's medical museums"; Knoeff and Zwijnenberg, *The fate of anatomical collections*, 99.

34. Guerrini, "Inside the charnel house: The display of skeletons in Europe, 1500–1800."

35. Knoeff, "Ball pool anatomy," 285.

36. Jones, Gear, and Galvin, "Stored human tissue: An ethical perspective on the fate of anonymous, archival material."

37. Jenkins, *Contesting human remains in museum collections: The crisis of cultural authority*, 3. See also Jenkins, *Keeping their marbles: How the treasures of the past ended up in museums . . . and why they should stay there.*

38. Gazi, "Exhibition ethics—an overview of major issues"; Overholtzer and Argueta, "Letting skeletons out of the closet: The ethics of displaying ancient Mexican human remains."

39. Jenkins, *Contesting human remains in museum collections*, 137.

40. Jenkins, *Contesting human remains in museum collections*, 25–32.

41. Kelly, "Human remains: Objects to study or ancestors to bury?"

42. See, for example, Singer, *A short history of scientific ideas to 1900*; Lindeboom, *Inleiding tot de geschiedenis der geneeskunde* [Introduction to the history of medicine]; Shryock, *The development of modern medicine*; Porter, *The Cambridge history of medicine.*

43. Knoeff, "Touching anatomy: On the handling of preparations in the anatomical cabinets of Frederik Ruysch (1638–1731)."

44. Jones and Whitaker, *Speaking for the dead: The human body in biology and medicine*, 48ff.

45. Knoeff, "Ball pool anatomy," 279–291.

46. Strasser, "The experimenter's museum: GenBank, natural history, and the moral economies of biomedicine"; Strasser, "Collecting nature: Practices, styles, and narratives."

47. Tybjerg, "From bottled babies to biobanks," 277.

48. Nagel, "The afterlife of the reliquary."

49. Eiseman and Haga, *Handbook of human tissue resources*; iSpecimen, "2018 will be a big year for biobanks."

50. Henderson et al., "Characterizing biobank organizations in the US."

51. Vercauteren, "What is a biobank and why is everyone talking about them?," 5.

52. Ogbogu et al., "Newspaper coverage of biobanks."

53. Henderson et al., "Characterizing biobank organizations in the US," 9.

54. Allen, "A failed model."

55. Doucet et al., "Biobank sustainability: Current status and future prospects."

56. Caulfield and Murdoch, "Genes, cells, and biobanks: Yes, there's still a consent problem."

57. Caulfield and Murdoch, "Genes, cells, and biobanks," 3.

58. Nelkin and Lindee, *The DNA mystique: The gene as a cultural icon*, 49ff.

59. Nelkin, "Molecular metaphors: The gene in popular discourse."

60. Rose and Rose, *Genes, cells and brains: The Promethean promises of the new biology*; Fortun, *Promising genomics: Iceland and deCODE Genetics in a world of speculation*; Árnason, "Coding and consent: Moral challenges of the database project in Iceland."

61. Bentham, "Auto-Icon; or, Farther uses of the dead to the living"; Ten Have, *Jeremy Bentham.*

62. Barilan, "Bodyworlds and the ethics of using human remains"; Ameisen and Le Coz, *Avis sur les problèmes éthiques poses par l'utilisation des cadavres à des fins de conservation ou d'exposition muséale* [Opinion on the ethical problems posed by the use of corpses for the purposes of conservation or museum exhibition]; King, Whitaker, and Jones, "I see dead people: Insights from the humanities into the nature of plastinated cadavers"; Jones and Whitaker, *Speaking for the dead*, 87–106; Lantos, *Controversial bodies: Thoughts on the public display of plastinated corpses.*

63. Barilan, "The story of the body and the story of the person: Towards an ethics of representing human bodies and body-parts."

64. Jones and Whitaker, *Speaking for the dead*, 100; Desmond, "Postmortem exhibitions: Taxidermied animals and plastinated corpses in the theaters of the dead."

65. Luyet was a biophysicist but also a Catholic priest in the Salesian Order. See Meryman, "Basile J. Luyet: In Memoriam"; Radin, *Life on ice: A history of new uses for cold blood*, 20–24.

66. Jeuken, *Materie, leven, geest* [Matter, life, spirit], 57–59; Jeuken, "The biological and philosophical definitions of life"; Keilin, "The Leeuwenhoek Lecture: The problem of anabiosis or latent life: History and current concept."

67. "Science: Deep-freeze," *Time*, April 28, 1952, http://content.time.com/time/subscriber/article/0,33009,816378-2,00.html.

68. Ettinger, *The prospect of immortality.*

69. O'Connell, *To be a machine*, 22–41. See also Parry, "Technologies of immortality."

70. Radin, *Life on ice*, 20.

71. See Radin, *Life on ice*, 160ff.

72. Landecker, *Culturing life: How cells became technologies*, 104.

73. See Landecker, "Between beneficence and chattel: The human biological in law and science."

74. Skloot, *The immortal life of Henrietta Lacks*. See also Landecker, *Culturing life*, 140–179.

75. Greely and Cho, "The Henrietta Lacks legacy grows."

76. For most parents, concerns about body wholeness were more important than concerns with informed consent. See Leith, "Consent and nothing but consent? The organ retention scandal."

77. Congregation for the Causes of Saints, *Instruction "Relics in the Church: Authenticity and preservation."*

78. Zika et al., *Biobanks in Europe: Prospects for harmonisation and networking.*

79. Widdows and Cordell, "The ethics of biobanking." See also Master et al., "Biobanks, consent and claims of consensus."

80. Winickoff, "Genome and nation: Iceland's health sector database and its legacy."

81. Hanson, "Ethics and biobanks"; Budimir et al., "Ethical aspects of human biobanks"; Hanson, "Building on relationships of trust in biobank research"; Greely, "The uneasy ethical and legal underpinnings of large-scale genomic biobanks."

82. Andrews, "My body, my property"; De Witte and ten Have, "Ownership of genetic material and information."

83. Campbell, "Body, self, and the property paradigm." See also Kass, "Thinking about the body."

84. Harré, *Physical being*, 11–37, 116–141.

85. Szawarski, "The stick, the eye, and ownership of the body."

86. Ten Have, "Images of man in philosophy of medicine."

87. Buytendijk, "De relatie arts-patient" [The doctor-patient relationship], 2508.

88. Sherkow and Greely, "The history of patenting genetic material."

89. Solbakk, Holm, and Hofmann, *The ethics of research biobanking*, xiii–xviii.

90. See Rajan, *Biocapital: The constitution of postgenomic life.*

91. Szawarski, "The stick, the eye, and ownership of the body," 93; Titmuss, *The gift relationship: From human blood to social policy.*

92. Koplin, "Commodification and human interests," 431. See also Nelkin and Andrews, "*Homo economicus*: Commercialization of body tissue in the age of biotechnology."

93. De Vries et al., "The moral concerns of biobank donors."

94. Melas et al., "Examining the public refusal to consent to DNA biobanking: Empirical data from a Swedish population-based study"; Caulfield et al., "A review of the key issues associated with the commercialization of biobanks"; Nicol et al., "Understanding public reactions to commercialization of biobanks and use of biobank resources."

95. Andrews, "Genes and patent policy: Rethinking intellectual property rights"; Biddle: "Can patents prohibit research?"

96. Zoloth, "The Alexandria plan: Creating libraries for human tissue research and therapeutic use."

97. Chadwick and Berg, "Solidarity and equity: New ethical frameworks for

genetics databases"; Winickoff, "Partnership in UK Biobank: A third way for genomic property?"; Campbell, "Body, self, and the property paradigm," 38–39; Hoedemaekers, Gordijn, and Pijnenburg, "Solidarity and justice as guiding principles in genomic research."

98. Widdows, "Between the individual and the community: The impact of genetics on ethical models"; Cordell, "The biobank as an ethical subject."

99. Nuffield Council on Bioethics, *Human tissue: Ethical and legal issues*, vii, 40.

100. Cantor, *After we die: The life and times of the human cadaver*, 95.

101. Rajan, *Biocapital*, 61.

102. Rajan, *Biocapital*, 182ff.

103. Thomassen, "The uses and meanings of liminality"; Árnason, "Scientific citizenship, benefit, and protection in population-based research." See also Squier, *Liminal lives: Imagining the human at the frontiers of biomedicine*.

104. Guerrini, "Inside the charnel house."

105. Wharton, "Relics, Protestants, things."

106. See Fassin, *Life: A critical user's manual*.

107. Brown, *Human universals*.

108. Van Beers, "TV cannibalism, *Body Worlds* and trade human body parts."

109. Evans, *Playing God*, 3, 135.

110. Evans, *Playing God*, 3.

111. Sandel, *The case against perfection: Ethics in the age of genetic engineering*, 83.

112. Bates, "'Indecent and demoralizing representations': Public anatomy museums in mid-Victorian England."

113. Nuffield Council on Bioethics, *Human tissue*, 40.

114. Cantor, *After we die*, 29.

115. Cantor, *After we die*, 39.

116. Cantor, *After we die*, 4, 43, 254ff.

117. Partridge, "Posthumous interests and posthumous respect," 258.

118. Geary, "Sacred commodities: The circulation of medieval relics."

119. See Congregation for the Causes of Saints, *Instruction "Relics in the Church."*

120. Kane, "Disenchanted America: Accounting for the lack of extraordinary mystical phenomena in Catholic America."

121. Nuffield Council on Bioethics, *Human tissue*, 50ff.

Chapter 8 • Critical Bioethics

1. Vivienne Walt, "France's yellow vests straitjacket Macron"; Diallo, "Why are the 'yellow vests' protesting in France?"; Harding, "Among the *gilets jaunes*." The symbolism of the yellow vest is powerful. By law every automobile needs to be equipped with a yellow vest so in the event of an accident or breakdown, the driver can put on the vest to ensure visibility and to avoid further accidents. Therefore, every car owner has a yellow vest. Government policies have made life much more expensive, especially in rural areas since jobs are usually far away and local transportation has been significantly reduced. People are dependent on cars for shopping, work, and health care more than people in urban areas. They also face lower salaries, pensions, and social security payments.

2. Rovisco, "The indignados social movement and the image of the occupied

square." See also Asara, *Untangling the radical imaginaries of the Indignados' movement*; Asara, "The Indignados as a socio-environmental movement"; Van de Velde, "The 'Indignados': The reasons for outrage"; Ancelovici, Dufour, and Nez, *Street politics in the age of austerity: From the Indignados to Occupy*.

3. Burns, "Charlie Gard's parents want 'Charlie's Law'"; Nicholls, "Three ways the Charlie Gard case could affect future end-of-life cases globally."

4. Marchione, "Chinese researcher claims first gene-edited babies"; Caplan, "He Jiankui's moral mess."

5. Nie and Pickering, "He Jiankui's genetic misadventure, Part 2: How different are Chinese and Western bioethics?" After his university rejected his research proposal, he went to a private hospital. Research results were not published, but He used videos claiming to be the first to make edited babies. The Chinese government was embarrassed and He quickly disappeared, closing his laboratory and office. Up until the scandalous event, he was promoted as a star scientist, and the fallout confirmed some prejudices that in China ethics is only a tool secondary to research and innovation.

6. Batson et al., "Anger at unfairness: Is it moral outrage?"

7. Crockett, "Moral outrage in the digital age."

8. Crockett, "Moral outrage in the digital age," 769.

9. Bastian, Denson, and Haslam, "The roles of dehumanization and moral outrage in retributive justice."

10. Duhigg, "Why are we so angry? The untold story of how we all got so mad at one another."

11. Getz, Dellaire, and Baylis, "Jiankui He: A sorry tale of high-stakes science."

12. Baylis, "First Crispr babies: Where are our ethics?"; Hasson and Darnovsky, "Gene-edited babies: No one has the moral warrant to go it alone"; Hurlbut, Jasanoff, and Saha, "The Chinese gene-editing experiment was an outrage. The scientific community shares the blame"; Yong, "The CRISPR baby scandal gets worse by the day."

13. See, for example, Roth, "Repugnance as a constraint on markets"; Kass, "Organs for sale? Propriety, property, and the price of progress." See also Millns, "Dwarf-throwing and human dignity: A French perspective"; Russell and Giner-Sorolla, "Bodily moral disgust"; Oaten et al., "Moral violations and the experience of disgust and anger."

14. Ten Have, *Vulnerability: Challenging bioethics*.

15. Giroux, "Poisoned city in the age of casino capitalism."

16. Bauman, *Wasted lives: Modernity and its outcasts*.

17. Lorey, *Die Regierung der Prekären* [The government of the precarious].

18. Greig, Hulme, and Turner, *Challenging global inequality*. See also Navarro, "Neoliberalism as a class ideology; or, The political causes of the growth of inequalities."

19. Ostry, Loungani, and Furceri, "Neoliberalism: Oversold?"

20. Riffkin, "In US, 67% dissatisfied with income, wealth distribution." In August 2018, polls show that 51 percent of the young people between 18 and 29 years old prefer socialism over capitalism in the United States, although it is not clear what exactly they understand socialism to be. See Lott, "Americans warming to socialism over capitalism, polls show."

21. Kirby, *Vulnerability and violence: The impact of globalisation*.

22. Public policies often show, in the words of Henry Giroux, a "pathological

disdain for community, public values, and the common good" ("Poisoned city in the age of casino capitalism," 24).

23. Ten Have, *Wounded planet: How declining biodiversity endangers health and how bioethics can help*.

24. This is what Powers and Faden have called "systematic patterns of disadvantage" (*Social justice: The moral foundations of public health and health policy*, 71–79).

25. Turner, "Bioethics needs to rethink its agenda."

26. Hedgecoe, "Critical bioethics: Beyond the social science critique of applied ethics."

27. Paul Farmer and Nicole Campos have made a case for bioethics "from below." See Farmer and Campos, "Rethinking medical ethics: A view from below." See also Benatar, Daar, and Singer, "Global health challenges: The need for an expanded discourse on bioethics."

28. Solomon and Jennings, "Bioethics and populism: How should our field respond?"

29. Solomon and Jennings, "Bioethics and populism," 12, 15.

30. Benatar, "Moral imagination: The missing component in global health."

31. Ornstein and Thomas, "Prominent doctors aren't disclosing their industry ties in medical journal studies. And journals are doing little to enforce their rules."

32. Ten Have, *Global bioethics*, 66–71.

33. Jennings and Dawson, "Solidarity in the moral imagination of bioethics."

34. Jennings and Dawson, "Solidarity in the moral imagination of bioethics," 34.

35. Midgley, *What is philosophy for?*, 73.

36. Midgley, *What is philosophy for?*, 41.

37. Midgley, *What is philosophy for?*, 59.

38. Midgley, *Are you an illusion?*

39. Midgley, *What is philosophy for?*, 85.

40. Midgley, *What is philosophy for?*, 90.

41. Campbell, "Religion and bioethics: Taking symbolism seriously," 24.

42. Stahl, "Patient reflections on the disenchantment of techno-medicine"; Stahl, *Imaging and imagining illness: Becoming whole in a broken body*. See also Shuman, "Re-enchanting the body: Overcoming the melancholy of anatomy."

43. Stuurman, *The invention of humanity: Equality and cultural difference in world history*.

44. Zwiebach, *The common life: Ambiguity, agreement, and the structure of morals*, 83, 125.

45. For example, Robert Reich in his recent book concludes, "The key political and economic institutions of our society—political parties, corporations, and the free market—have abandoned their commitment to the common good" (*The common good*, 90).

46. Tirole, *Economics for the common good*.

47. Ten Have, *Global bioethics*, 113–137.

48. Elliott, "Where ethics comes from and what to do about it," 31.

49. Zwiebach, *The common life*, 179.

50. Commission on the Social Determinants of Health, *Closing the gap in a generation: Health equity through action on the social determinants of health*, 10.

51. Marmot, "Just societies, health equity, and dignified lives: The PAHO Equity Commission."

52. See ten Have, *Wounded planet*, 223ff.

53. World Wildlife Fund, *Living planet report, 2018: Aiming higher*.

54. Watts et al., "The 2018 report of the *Lancet* countdown on health and climate change."

55. In January 2019, PubMed shows that the combination of the keywords "bio-ethics" and "climate change" delivers thirty publications. The same number results from a search with "bioethics" and "biodiversity." The combination of "bioethics" and "environment" results in more publications: 878. Over the past five years, the number of publications focused on bioethics and the environment have increased from 3 percent to 4.6 percent (284 out of a total of 6,115 articles).

56. Davies, Kamradt-Scott, and Rushton, *Disease diplomacy: International norms and global health security*.

57. Ashby and Morrell, "To the barricades or the blackboard: Bioethics activism and the 'stance of neutrality,'" *Journal of Bioethical Inquiry* 15 (2018): 479.

58. Dawson et al., "Bioethics and the myth of neutrality."

59. Solomon and Jennings, "Bioethics and populism," 14–15.

60. Kekes, "Disgust and moral taboos," 443.

61. Horton, "Rediscovering human dignity."

62. Dewey, *A common faith*. See also Adams et al., "Social imaginaries in debate."

Abbott, J. Haxby. "How to choose where to publish your work." *Journal of Orthopaedic & Sports Physical Therapy* 47 (2017): 6–10.

Abernethy, Bob. "Interview with Dr. Francis Collins, director of the Human Genome Project at the National Institutes of Health." *PBS*, June 16, 2000. http://www.pbs.org/wnet/religionandethics/2000/06/16/transcript-bob-abernethys-interview-with-dr-francis-collins-director-of-the-human-genome-project-at-the-national-institutes-of-health/15204.

Abola, Matthew V., and Vinay Prasad. "The use of superlatives in cancer research." *JAMA Oncology* 2, no. 1 (2016): 139–141.

Abraham, Carolyn. *Possessing genius: The bizarre odyssey of Einstein's brain*. Toronto: Penguin Random House, 2002.

Abraham, John. "Pharmaceuticalization of society in context: Theoretical, empirical and health dimensions." *Sociology* 44 (2010): 603–622.

Ackerman, Sara L. "Plastic paradise: Transforming bodies and selves in Costa Rica's cosmetic surgery tourism industry." *Medical Anthropology* 29, no. 4 (2010): 403–423.

Adams, Krystyn, Jeremy Snyder, Valorie A. Crooks, and Rory Johnston. "Promoting social responsibility amongst health care users: Medical tourists' perspectives on an information sheet regarding ethical concerns in medical tourism." *Philosophy, Ethics, and Humanities in Medicine* 8 (2013): 19. https://doi.org/10.1186/1747-5341-8-19.

Adams, Samantha, and Antoinette de Bont. "Information Rx: Prescribing good consumerism and responsible citizenship." *Health Care Analysis* 15 (2007): 273–290.

Adams, Suzi, Paul Blokker, Natalie J. Doyle, John W. M. Krummel, and Jeremy C. A. Smith. "Social imaginaries in debate." *Social Imaginaries* 1, no. 1 (2015): 15–52.

Ager, Susan. "The perils of pale." *National Geographic*, June 2017: 70–91.

Alberti, Samuel. "A history of Edinburgh's medical museums." *Journal of the Royal College of Physicians of Edinburgh* 46 (2016): 187–197.

Alexander, Denis. *Genes, determinism, and God*. Cambridge: Cambridge University Press, 2017.

Allen, Dominic. "A failed model." *Pathologist*, August 18, 2017. https://thepathologist.com/issues/0817/a-failed-model.

AMA (American Medical Association). "Use of placebo in clinical practice." Code of

Medical Ethics Opinion 2.1.4. Accessed on May 14, 2021. https://www.ama-assn.org/delivering-care/ethics/use-placebo-clinical-practice.

Ameisen, Jean-Claude, and Pierre Le Coz. *Avis sur les problèmes éthiques poses par l'utilisation des cadavres à des fins de conservation ou d'exposition muséale* [Opinion of the ethical problems posed by the use of corpses for the purposes of conservation and museum exhibition]. Paris: Comité Consultatif National d'Ethique pour les Sciences de la Vie et de la Santé, January 2010. http://www.ccne-ethique.fr/sites/default/files/publications/avis_111.pdf.

Ancelovici, Marcos, Pascale Dufour, and Héloïse Nez, eds. *Street politics in the age of austerity: From the Indignados to Occupy.* Amsterdam: Amsterdam University Press, 2016.

Andrews, Lori B. "My body, my property." *Hastings Center Report* 16, no. 5 (1986): 28–38.

Andrews, Lori B. "Genes and patent policy: Rethinking intellectual property rights." *Nature Reviews Genetics* 3, no. 10 (2002): 803–808.

Angell, Marcia. "Industry-sponsored clinical research: A broken system." *JAMA* 300 (2008): 1069–1071.

Annas, George J. "Genetic prophecy and genetic privacy—can we prevent the dream from becoming a nightmare?" *American Journal of Public Health* 85, no. 9 (1995): 1196–1197.

Annas, George J. "Reframing the debate on health care reform by replacing our metaphors." *New England Journal of Medicine* 332, no. 11 (1995): 744–747.

Annas, George J., and Sherman Elias. "Thalidomide and the *Titanic*: Reconstructing the technology tragedies of the twentieth century." *American Journal of Public Health* 89 (1999): 98–101.

Anonymous. *Beall's List of Potential Predatory Journals and Publishers.* Last updated March 7, 2021. https://beallslist.net.

Anschütz, Felix. *Ärztliches Handeln: Grundlagen, Möglichkeiten, Grenzen, Widersprüche* [Medical action: Foundations, possibilities, limits, contradictions]. Darmstadt: Wissenschaftliche Buchgesellschaft, 1987.

Applbaum, Kalman. "Pharmaceutical marketing and the invention of the medical consumer." *PLoS Medicine* 3, no. 4 (2006): e189.

Arendt, Hannah. *Eichmann in Jerusalem: A report on the banality of evil.* New York: Penguin Books, 2006. First published 1963 by Viking Press.

Arnason, Vilhjalmur. "Coding and consent: Moral challenges of the database project in Iceland." *Bioethics* 18, no. 1 (2004): 27–49.

Arnason, Vilhjalmur. "Scientific citizenship, benefit, and protection in population-based research." In *The ethics of research biobanking*, edited by Jan Helge Solbakk, Søren Holm, and Bjørn Hofmann, 131–141. Dordrecht: Springer, 2009.

Arney, William Ray, and Bernard J. Bergen. *Medicine and the management of living: Taming the last great beast.* Chicago: University of Chicago Press, 1984.

Asadi, Amin, Nader Rahbar, Mohammad Javad Rezvani, and Fahime Asadi. "Fake/bogus conferences: Their features and some subtle ways to differentiate them from real ones." *Science and Engineering Ethics* 24, no. 2 (2018): 779–784. https://doi.org/10.1007/s11948-017-9906-2.

Asara, Viviana. *Untangling the radical imaginaries of the Indignados' movement:*

Commons, autonomy and ecologism. Vienna: WU Vienna University of Economics and Business, 2018. http://epub.wu.ac.at/6431.

Asara, Viviana. "The Indignados as a socio-environmental movement: Framing the crisis and democracy." *Environmental Policy and Governance* 26 (2016): 527–542.

Ashby, Michael A., and Bronwen Morrell. "To the barricades or the blackboard: Bioethics activism and the 'stance of neutrality.'" *Journal of Bioethical Inquiry* 15 (2018): 479–482.

Asma, Stephen T. *On monsters: An unnatural history of our worst fears*. Oxford: Oxford University Press, 2009.

Audeguy, Stéphane. *Les monstres. Si loin et si proches* [Monsters, so far and so near]. Paris: Gallimard, 2007.

Aune, David E. *Prophecy in early Christianity and the ancient Mediterranean world*. Grand Rapids, MI: William B. Eerdmans Publishing, 1983.

Badone, Ellen, and Sharon R. Roseman. "Approaches to the anthropology of pilgrimage and tourism." In *Intersecting journeys: The anthropology of pilgrimage and tourism*, edited by Ellen Badone and Sharon R. Roseman, 1–23. Urbana: University of Illinois Press, 2004.

Bakan, David. *Maimonides on prophecy*. Northvale, NJ: Jason Aronson, 1991.

Balint, Michael. *The doctor, his patient, and the illness*. London: Pitman Medical Publishing, 1957.

Ball, Philip. *Unnatural: The heretical idea of making people*. London: Bodley Head, 2011.

"Barack Obama's remarks to the Democratic National Convention." *New York Times*, July 27, 2004. https://www.nytimes.com/2004/07/27/politics/campaign/barack-obamas-remarks-to-the-democratic-national.html.

Barbour, Virginia. "How ghost-writing threatened the credibility of medical knowledge and medical journals." *Haematologica* 95 (2010): 1–2.

Barilan, Y. Michael. "The story of the body and the story of the person: Towards an ethics of representing human bodies and body-parts." *Medicine, Health Care and Philosophy* 8 (2005): 193–205.

Barilan, Y. Michael. "Bodyworlds and the ethics of using human remains: A preliminary discussion." *Bioethics* 20, no. 5 (2006): 233–247.

Barker, Kristin K. "Electronic support groups, patient-consumers, and medicalization: The case of contested illness." *Journal of Health and Social Behavior* 49 (2008): 20–36.

Bartlett, Robert. *Why can the dead do such great things? Saints and worshippers from the martyrs to the Reformation*. Princeton, NJ: Princeton University Press, 2013.

Bassan, Sharon, and Merle A. Michaelson. "Honeymoon, medical treatment or big business? An analysis of the meanings of the term 'reproductive tourism' in German and Israeli public media discourses." *Philosophy, Ethics, and Humanities in Medicine* 8 (2013): 1–8. https://doi.org/10.1186/1747-5341-8-9.

Bastian, Brock, Thomas F. Denson, and Nick Haslam. "The roles of dehumanization and moral outrage in retributive justice." *PLoS One* 8, no. 4 (2013): e61842. https://doi.org/10.1371/journal.pone.0061842.

Bateman-House, Alison, and Christopher T. Robertson. "The federal Right to Try Act of 2017—a wrong turn for access for investigational drugs and the path forward." *JAMA Internal Medicine* 178 (2018): 321–322.

Bates, A. W. *Emblematic monsters: Unnatural conceptions and deformed births in early modern Europe.* Amsterdam: Rodopi, 2005.

Bates, A. W. "'Indecent and demoralizing representations': Public anatomy museums in mid-Victorian England." *Medical History* 52 (2008): 1–22.

Batson, C. Daniel, Christopher L. Kennedy, Lesly-Anne Nord, E. L. Stocks, D'Yani A. Fleming, Christian M. Marzette, David A. Lishner, Robin E. Hayes, Leah M. Kolchinsky, and Tricia Zerger. "Anger at unfairness: Is it moral outrage?" *European Journal of Social Psychology* 37 (2007): 1272–1285.

Baudrillard, Jean. *The vital illusion.* New York: Columbia University Press, 2000.

Bauman, Zygmunt. *Wasted lives: Modernity and its outcasts.* Cambridge: Polity Press, 2004.

Baylis, Françoise. "First Crispr babies: Where are our ethics?" *Impact Ethics*, November 26, 2018. https://impactethics.ca/2018/11/26/first-crispr-babies-where-are-our-ethics.

Beall, Jeffrey. "Predatory journals threaten the quality of published medical research." *Journal of Orthopaedic & Sports Physical Therapy* 47 (2017): 3–5.

Beauchamp, Tom L., and James F. Childress. *Principles of biomedical ethics.* 1st ed. Oxford: Oxford University Press, 1979.

Beauchamp, Tom L., and James F. Childress. *Principles of biomedical ethics.* 2nd ed. Oxford: Oxford University Press, 1983.

Beaune, Jean-Claude. "Sur la route et les traces des Geoffroy Saint-Hilaire" [On the road and in the footsteps of Geoffroy Saint-Hilaire]. In *La vie et la mort des monstres* [The life and death of monsters], edited by Jean-Claude Beaune, 11–43. Champ Vallon: Seyssel, 2004.

Beecher, Henry K. "The powerful placebo." *JAMA* 159 (1955): 1602–1606.

Bekelman, Justin E., Yan Li, and Cary P. Gross. "Scope and impact of financial conflicts of interest in biomedical research: A systematic review." *JAMA* 289, no. 4 (2003): 454–465.

Benatar, Solomon R. "Moral imagination: The missing component in global health." *PLoS Medicine* 2, no. 12 (2005): e400.

Benatar, Solomon R., Abdallah S. Daar, and Peter A. Singer. "Global health challenges: The need for an expanded discourse on bioethics." *PLoS Medicine* 2, no. 7 (2005): e143. https://doi.org/10.1371/journal.pmed.0020143.

Bennett, Jana. "Charlie Gard: A story of disability bias." *America*, August 7, 2017. https://www.americamagazine.org/politics-society/2017/07/07/charlie-gard-story-disability-bias.

Bennett, Jennifer. *Miracle babies: How babies are made the IVF way.* Charleston, SC: CreateSpace, 2010.

Bennett, Oliver. *Cultural pessimism: Narratives of decline in the postmodern world.* Edinburgh: Edinburgh University Press, 2001.

Bentham, Jeremy. "Auto-Icon; or, Farther uses of the dead to the living." In *Bentham's Auto-Icon and Related Writings*, edited by James E. Crimmins. Bristol: Thoemmes Press, 2002. First published 1842.

Beresford, Eric B. "Uncertainty and the shaping of medical decisions." *Hastings Center Report* 21, no. 4 (1991): 6–11.

Berg, Joseph M., and Marika Korossy. "Down syndrome before Down: A retrospect." *American Journal of Medical Genetics* 102 (2001): 205–211.

Bernard, Claude. *An introduction to the study of experimental medicine.* New York: Dover, 1957.

Beste, Jennifer. "Instilling hope and respecting patient autonomy: Reconciling apparently conflicting duties." *Bioethics* 19, no. 3 (2005): 215–231.

Beurton, Peter J. "A unified view of the gene, or how to overcome reductionism." In *The concept of the gene in development and evolution: Historical and epistemological perspectives,* edited by Peter J. Beurton, Raphael Falk, and Hans-Jörg Rheinberger, 286–314. Cambridge: Cambridge University Press, 2000.

Bibler, Trevor M., Myrick C. Shinall Jr., and Devan Stahl. "Responding to those who hope for a miracle: Practices for clinical bioethicists." *American Journal of Bioethics* 18 (2018): 40–51.

Biddle, Justin B. "Can patents prohibit research? On the social epistemology of patenting and licensing in science." *Studies in History and Philosophy of Science* 45 (2014): 14–23.

Bilefsky, Dan. "British Court decides Charlie Gard will be moved to a hospice to die." *New York Times,* July 27, 2017. https://www.nytimes.com/2017/07/27/world/europe/uk-charlie-gard-baby-parents.html.

Bishop, Jerry E., and Michael Waldholz. *Genome: The story of the most astonishing scientific adventure of our time—the attempt to map all the genes in the human body.* New York: Simon & Schuster, 1990.

Blank, Robert H. "Science and the human genome: Policy issues in genetic intervention in the 1990s." *Project Appraisal* 4, no. 3 (1989): 122–132. https://doi.org/10.1080/02688867.1989.9726722.

Blum, Lawrence A. *Moral perception and particularity.* Cambridge: Cambridge University Press, 1994.

Blyth, Eric. "Fertility patients' experiences of cross-border reproductive care." *Fertility and Sterility* 94 (2010): e11–e15.

Boenink, Marianne. "Predictive medicine." In *Encyclopedia of global bioethics,* vol. 3, edited by Henk ten Have, 2265–2275. Switzerland: Springer Reference, 2016.

Bohannon, John. "Who's afraid of peer review?" *Science* 342 (2013): 60–65.

Bok, Sissela. "The ethics of giving placebos." *Scientific American* 231 (1974): 17–23.

Bondeson, Jan. *A cabinet of medical curiosities.* New York: W. W. Norton, 1999.

Bondeson, Jan. *The two-headed boy, and other medical marvels.* Ithaca, NY: Cornell University Press, 2000.

Boseley, Sarah. "The selling of a wonder drug." *Guardian,* March 29, 2006. https://www.theguardian.com/science/2006/mar/29/medicineandhealth.health.

Bostrom, Nick. *Superintelligence: Paths, dangers, strategies.* Oxford: Oxford University Press, 2014.

Bottig, Fred. *Making monstrous: Frankenstein, criticism, theory.* Manchester: Manchester University Press, 1991.

Bowler, Peter J. *A history of the future: Prophets of progress from H. G. Wells to Isaac Asimov.* Cambridge: Cambridge University Press, 2017.

Braithwaite, Jeffrey. "The medical miracles delusion." *Journal of the Royal Society of Medicine* 107 (2014): 92–93.

Brett, Allan S., and Paul Jersild. "'Inappropriate' treatment near the end of life: Conflict between religious convictions and clinical judgment." *Archives of Internal Medicine* 163 (2003): 1645–1648.

Brezgov, Stef. "List of publishers." Scholarly Open Access, May 27, 2019. https://scholarlyoa.com/publishers.

Briggs, David. "Belief in miracles on the rise." *Huffington Post*, October 30, 2012. https://www.huffingtonpost.com/david-briggs/belief-in-miracles-on-the-rise-in-the-age-of-oprah_b_2039372.html.

Brody, Howard. *Placebos and the philosophy of medicine: Clinical, conceptual, and ethical issues*. Chicago: University of Chicago Press, 1980.

Brody, Howard. "The lie that heals: The ethics of giving placebos." *Annals of Internal Medicine* 97 (1982): 112–118.

Brown, Donald E. *Human universals*. New York: McGraw-Hill, 1991.

Brown, Nik. "Hope against hype—accountability in biopasts, presents and futures." *Science Studies* 16 (2003): 3–21.

Brown, Nik. "Shifting tenses: Reconnecting regimes of truth and hope." *Configurations* 13, no. 3 (2005): 331–355.

Brown, Peter. *The cult of the saints: Its rise and function in Latin Christianity*. Chicago: University of Chicago Press, 2015. First published 1981.

Brundage, Miles, Shahar Avin, Jack Clark, Helen Toner, Peter Eckersley, Ben Garfinkel, Allan Dafoe, et al. *The malicious use of artificial intelligence: Forecasting, prevention, and mitigation*. Oxford and Cambridge: Future of Humanity Institute / Centre for the Study of Existential Risk, 2018. https://arxiv.org/ftp/arxiv/papers/1802/1802.07228.pdf.

Brunk, Conrad G., and Sarah Hartley, eds. *Designer animals: Mapping the issues in animal biotechnology*. Toronto: University of Toronto Press, 2013.

Budimir, Danijela, Ozren Polašek, Ana Marušić, Ivana Kolčić, Tatijana Zemunik, Vesna Boraska, Ana Jerončić, Mladen Boban, Harry Campbell, and Igor Rudan. "Ethical aspects of human biobanks: A systematic review." *Croatian Medical Journal* 52 (2011): 262–279.

Bures, Frank. "The strange case of the disappearing penis." *Psychology Today* 43, no. 2 (2010): 22.

Burkett, Levi. "Medical tourism: Concerns, benefits, and the American legal perspective." *Journal of Legal Medicine* 28 (2007): 223–245.

Burns, Catherine. "Charlie Gard's parents want 'Charlie's Law.'" *BBC News*, June 20, 2018. https://www.bbc.com/news/health-44334306.

Buytendijk, F. J. J. "De relatie arts-patient" [The doctor-patient relationship]. *Nederlands Tijdschrift voor Geneeskunde* 103 (1959): 2504–2508.

Cahill, Lisa Sowle. "Bioethics, theology, and social change." *Journal of Religious Ethics* 31, no. 3 (2003): 363–398.

Cahill, Lisa Sowle. *Theological bioethics: Participation, justice, and change*. Washington, DC: Georgetown University Press, 2005.

Callahan, Daniel. "The Hastings Center and the early years of bioethics." *Theoretical Medicine and Bioethics* 33, no. 1 (2012): 11–20.

Callahan, Daniel. "Minimalist ethics: On the pacification of morality." In *Ethics in hard times*, edited by Arthur L. Caplan and Daniel Callahan, 261–281. New York: Plenum Press, 1981.

Callahan, Daniel. "Progress: Its glories and pitfalls." *Hastings Center Report* 48, no. 2 (2018): 18–21.

Callahan, Daniel. "Religion and the secularization of bioethics." *Hastings Center Report* 20, no. 4 (1990): S2–S4.

Callahan, Daniel. "Why America accepted bioethics." *Hastings Center Report* 23, no. 6 (1993): S8-S9.

Calvin, John. *A treatise on relics*. Online: Project Guttenberg, 2010. http://www .gutenberg.org/files/32136/32136-pdf.pdf.

Cameron, David, Martine J. Piccart-Gebhart, Richard D. Gelber, Marion Procter, Aron Goldhirsch, Evandro de Azambuja, Gilberto Castro Jr., et al. "11 years' follow-up of trastuzumab after adjuvant chemotherapy in HER2-positive early breast cancer: Final analysis of the HERceptin Adjuvant (HERA) trial." *Lancet* 398 (2017): 1195–1205.

Campbell, Courtney S. "Body, self, and the property paradigm." *Hastings Center Report* 22, no. 5 (1992): 34–42.

Campbell, Courtney S. "Religion and bioethics: Taking symbolism seriously." *Second Opinion* 8 (2001): 4–26.

Campbell, Denis. "193,000 NHS patients a month waiting beyond target time for surgery." *Guardian*, January 13, 2017. https://www.theguardian.com/society/2017/jan /13/193000-nhs-patients-a-month-waiting-beyond-target-for-surgery.

Campbell, Eric C., Joel S. Weissman, Christine Vogeli, Brian R. Clarridge, Melissa Abraham, Jessica E. Marder, and Greg Koski. "Financial relationships between institutional review board members and industry." *New England Journal of Medicine* 355 (2006): 2321–2329.

Campbell, Neil, John P. Clark, Vera J. Stecher, and Irwin Goldstein. "Internet-ordered Viagra (sildenafil citrate) is rarely genuine." *Journal of Sexual Medicine* 9 (2012): 2943–2951.

Cantor, Norman L. *After we die: The life and times of the human cadaver*. Washington, DC: Georgetown University Press, 2010.

Caplan, Arthur L. "Can applied ethics be effective in health care and should it strive to be?" *Ethics* 93, no. 2 (1983): 311–319.

Caplan, Arthur L. "Ethical engineers need not apply: The state of applied ethics today." *Science, Technology & Human Values* 6, no. 33 (1980): 24–32.

Caplan, Arthur. "He Jiankui's moral mess." *PLOS Biologue*, December 3, 2018. https:// blogs.plos.org/biologue/2018/12/03/he-jiankuis-moral-mess.

Caplan, Arthur, and Kelly McBride Folkers. "Charlie Gard and the limits of parental authority." *Hastings Center Report* 47, no. 5 (2017): 15–16.

Caravagna, Giulio, Ylenia Giarrattano, Daniele Ramazzotti, Ian Tomlinson, Trevor A. Graham, Guido Sanguinetti, and Andrea Sottoriva. "Detecting repeated cancer evolution from multi-region tumor sequencing data." *Nature Methods* 15, no. 9 (2018): 707–714.

Carey, Nessa. *The epigenetic revolution: How modern biology is rewriting our understanding of genetics, disease, and inheritance*. New York: Columbia University Press, 2012.

Carroll, Noël. *The philosophy of horror; or, Paradoxes of the heart*. New York and London: Routledge, 1990.

Cassell, Joan. *Expected miracles: Surgeons at work*. Philadelphia: Temple University Press, 1991.

Caulfield, Timothy, Sarah Burningham, Yann Joly, Zubin Master, Mahsa Shabani, Pascal Borry, Allan Becker, et al. "A review of the key issues associated with the commercialization of biobanks." *Journal of Law and the Biosciences* 1, no. 1 (2014): 94–110.

Caulfield, Timothy, and Blake Murdoch. "Genes, cells, and biobanks: Yes, there's still a consent problem." *PLoS Biology* 15, no. 7 (2017): e2002654. https://doi.org/10.1371 /journal.pbio.2002654.

Cesuroglu, Tomris, Elena Syurina, Frans Feron, and Anja Krumeich. "Other side of the coin for personalised medicine and healthcare: content analysis of 'personalised' practices in the literature." *BMJ Open* 6 (2016): e010243. https://doi.org/10:1136 /bmjopen-2015-010243.

Chadwick, Ruth, and Kare Berg, "Solidarity and equity: New ethical frameworks for genetics databases." *Nature Reviews Genetics* 2 (2001): 318–321.

Chamayou, Grégoire. *Théorie du drone* [Drone theory]. Paris: La Fabrique Éditions, 2013.

Chan, Sarah, and Sonja Erikainen. "What's in a name? The politics of 'precision medicine.'" *American Journal of Bioethics* 18, no. 4 (2018): 50–52.

Chapman, Audrey. "In defense of the role of a religiously informed bioethics." *American Journal of Bioethics* 12 (2012): 26–28.

Charo, Rita Alta. "On the road (to a cure?)—Stem-cell tourism and lessons for gene editing." *New England Journal of Medicine* 374 (2016): 901–903.

Chen, Y.Y. Brandon, and Colleen M. Flood. "Medical tourism's impact on health care equity and access in low- and middle-income countries: Making the case for regulation." *Journal of Law, Medicine and Ethics* 41, no. 1 (2013): 286–300.

Choudhry, Niteesh K., Henry Thomas Stelfox, and Allan S. Detsky. "Relationships between authors of clinical practice guidelines and the pharmaceutical industry." *JAMA* 287 (2002): 612–617.

Chowkwanyun, Merlin, Ronald Bayer, and Sandro Galea. "'Precision' public health—between novelty and hype." *New England Journal of Medicine* 379 (2018): 1398–1400. https://doi.org/10.1056/NEJMp1806634.

Christakis, Nicholas A. "The ellipsis of prognosis in modern medical thought." *Social Science & Medicine* 44, no. 3 (1997): 301–315.

Christakis, Nicholas A. *Death foretold: Prophecy and prognosis in medical care*. Chicago: University of Chicago Press, 1999.

Christakis, Nicholas A., and Elizabeth B. Lamont. "Extent and determinants of error in doctors' prognoses in terminally ill patients: Prospective cohort study." *British Medical Journal* 320 (2000): 469–471.

Church, Paige-Terrien. "A personal perspective on disability: Between the words." *JAMA Pediatrics* 171, no. 10 (2017): 939.

Clair, Jean. *Hubris. La fabrique du monster dans l'art modern. Homoncules, géants et acéphales* [Hubris: The making of the monster in modern art; Homunculi, giants, and acephali]. Paris: Gallimard, 2012.

Clark, Charles M.A. "Greed is not enough: Some insights on globalization from Catholic social thought." *Journal of Catholic Social Thought* 2, no. 1 (2005): 23–51.

Clark, Fiona. "Rise in online pharmacies sees counterfeit drugs go global." *Lancet* 386 (2015): 1327–1328.

Clark, Jocalyn. "Medicalization of global health 1: Has the global health agenda become too medicalized?" *Global Health Action* 7 (2014): 23998.

Clark, Laura. "How halitosis became a medical condition with a 'cure.'" *Smithsonian Magazine*, January 29, 2015. https://www.smithsonianmag.com/smart-news /marketing-campaign-invented-halitosis-180954082.

Clark, Alexander M., and David R. Thompson. "Five (bad) reasons to publish your research in predatory journals." *Journal of Advanced Nursing* 73 (2017): 2499–2501.

Clarke, Adele E., Janet K. Shim, Sara Shostak, and Alondra Nelson. "Biomedicalising genetic health, diseases and identities." In *A handbook of genetics and society: Mapping the new genomic era,* edited by Paul Atkinson, Peter Glasner, and Margaret Lock, 21–40. London: Routledge, 2009.

Clarke, Steve. "When they believe in miracles." *Journal of Medical Ethics* 39 (2013): 582–583.

Clunies-Ross, G. G. "The development of children with Down's syndrome: Lessons from the past and implications for the future." *Australian Paediatric Journal* 22 (1986): 167–169.

Coghlan, Andy. "Nationwide genetic testing." *New Scientist* 238, no. 3172 (2018): 8.

Cohen, Cynthia B., and Peter J. Cohen. "International stem cell tourism and the need for effective regulation; Part I: Stem cell tourism in Russia and India: Clinical research, innovative treatment, or unproven hype? *Kennedy Institute of Ethics Journal* 20, no. 1 (2010): 27–49.

Cohen, Daniel. *Homo economicus. Prophète (égaré) des temps nouveaux* [*Homo economicus*: Prophet (lost) of new times]. Paris: Éditions Albin Michel, 2012.

Cohen, I. Glenn. *Patients with passports: Medical tourism, law and ethics.* Oxford: Oxford University Press, 2015.

Cohen, Jon. "How a horror story haunts science." *Science* 359 (2018): 148–150.

Cohen, R. A. *Long-term trends in health insurance: Estimates from the National Health Interview Survey, United States, 1968–2018.* Washington, DC: National Center for Health Statistics, July 2019. https://www.cdc.gov/nchs/data/nhis/health_insurance /TrendHealthInsurance1968_2018.pdf.

Collier, Joe. "*Panorama*: Herceptin: Wanting the wonder drug." *British Medical Journal* 332 (2006): 368.

Collier, Joe. "The price of independence." *British Medical Journal* 332 (2006): 1447–1449.

Comfort, Nathaniel. "Can we cure genetic diseases without slipping into eugenics?" *Nation*, July 16, 2015. https://www.thenation.com/article/can-we-cure-genetic -diseases-without-slipping-into-eugenics.

Commission of the European Communities. *Proposal for a council decision adopting a specific research programme in the field of health; Predictive medicine: Human genome analysis (1989–1991).* COM(88) 424. Brussels: Commission of the European Communities, 1988. http://aei.pitt.edu/33208/1/COM_(88)_424_final.pdf.

Commission on the Social Determinants of Health. *Closing the gap in a generation: Health equity through action on the social determinants of health.* Geneva: World Health Organization, 2008. http://apps.who.int/iris/bitstream/handle/10665 /43943/9789241563703_eng.pdf.

Condit, Celeste Michelle. *The meanings of the gene: Public debates about human heredity.* Madison: University of Wisconsin Press, 1999.

Condorcet. *Oeuvres de Condorcet* [Works of Condorcet]. Vol. 6. Edited by A. O'Connor and M. F. Arago. Paris: Nabu Press, 2012. First published 1793.

Congregation for the Causes of Saints. *Instruction "Relics in the Church: Authenticity and preservation."* Rome: Vatican, 2017. http://www.vatican.va/roman_curia /congregations/csaints/documents/rc_con_csaints_doc_20171208_istruzione -reliquie_en.html.

Connell, John. "Medical tourism: Sea, sun, sand and . . . surgery." *Tourism Management* 27 (2006): 1093–1100.

Conrad, Peter. "The shifting engines of medicalization." *Journal of Health and Social Behavior* 46 (2005): 3–14.

Conrad, Peter, and Cheryl Stults. "Contestation and medicalization." In *Contesting illness: Processes and practices,* edited by Pamela Moss and Katherine Teghtsoonian, 323–335. Toronto: University of Toronto Press, 2008.

Consoli, Luca. "Scientific misconduct and science ethics: A case study based approach." *Science and Engineering Ethics* 12 (2006): 533–541.

Cordell, Sean. "The biobank as an ethical subject." *Health Care Analysis* 19 (2011): 282–294.

Coulombe, Maxime. *Petite philosophie du zombie, ou comment penser par l'horreur* [Little zombie philosophy, or how to think through horror]. Paris: Presses Universitaires de France, 2012.

Crockett, M. J. "Moral outrage in the digital age." *Nature Human Behavior* 1 (2017): 769–771.

Crutchfield, Parker. "Compulsory moral bioenhancement should be covert." *Bioethics* 30, no. 5 (2018): 389–396.

Cruz, Joan Carroll. *Relics: What they are and why they matter.* Charlotte, NC: TAN Books, 2015.

Culzac, Natasha. "Boy born with eight limbs doing well after surgery to remove 'parasitic twin,' doctors say." *Independent,* September 10, 2014. https://www .independent.co.uk/news/world/africa/boy-born-with-eight-limbs-doing-well -after-surgery-to-remove-parasitic-twin-doctors-say-9723409.html.

Cyranoski, David. "Building a biopolis." *Nature* 412 (2001): 370–371.

Cyranoski, David. "The secret war against counterfeit science." *Nature* 545 (2017): 148–150.

Dadkhah, Mehdi, Seyed Amin Hosseine Seno, and Glenn Borchardt. "Current and potential cyber attacks on medical journals: Guidelines for improving security." *European Journal of Internal Medicine* 38 (2017): 25–29.

Dadkhah, Mehdi, Tomasz Maliszewski, and Mohammad Davarpanah Jazi. "Characteristics of hijacked journals and predatory publishers: Our observations in the academic world." *Trends in Pharmacological Sciences* 37 (2016): 415–418.

Dahnke, Michael D. "Emmanuel Levinas and the face of Terri Schiavo: Bioethical and phenomenological reflections on a private tragedy and public spectacle." *Theoretical Medicine and Bioethics* 33 (2012): 405–420.

Daley, George Q. "The promise and perils of stem cell therapeutics." *Cell Stem Cell* 10 (2012): 740–749.

Damasio, Antonio. "We must not accept an algorithmic account of human life." *New Perspectives Quarterly* 33, no. 3 (2016): 59–62.

Damen, Daan, and Van Meur, Lex. "Bijna 9 miljoen Nederlanders geloven in wonderen" [Almost 9 million Dutchmen believe in miracles]. *Motivaction*, February 21, 2020. https://www.motivaction.nl/kennisplatform/nieuws-en-persberichten/bijna-9 -miljoen-nederlanders-geloven-in-wonderen.

Darrow, Jonathan J., Jerry Avorn, and Aaron S. Kesselheim. "The FDA breakthrough-drug designation—four years of experience." *New England Journal of Medicine* 378 (2018): 1444–1453.

Daston, Lorraine, and Katharine Park. *Wonders and the order of nature, 1150–1750*. New York: Zone Books, 2001.

Davies, Damian Walford. "The unquiet cranium." *Times Literary Supplement*, November 8, 2013: 13–15.

Davies, Eryl W. *Prophecy and ethics: Isaiah and the ethical traditions of Israel*. Sheffield: JSOT Press, 1981.

Davies, Kevin. *Cracking the genome: Inside the race to unlock human DNA*. New York: Free Press, 2001.

Davies, Sara E., Adam Kamradt-Scott, and Simon Rushton. *Disease diplomacy: International norms and global health security*. Baltimore: Johns Hopkins University Press, 2015.

Davis, John K. *New Methuselahs: The ethics of life extension*. Cambridge, MA: MIT Press, 2018.

Dawson, Angus, Christopher E. C. Jordens, Paul Macneill, and Deborah Zion. "Bioethics and the myth of neutrality." *Journal of Bioethical Inquiry* 15 (2018): 483–486.

DeBruin, Debra A. "Ethics on the inside?" In *The ethics of bioethics: Mapping the moral landscape*, edited by Lisa A. Eckenwiler and Felicia G. Cohen, 161–169. Baltimore: Johns Hopkins University Press, 2007.

De Craen, Anton J. M., Ted J. Kaptchuk, Jan G. P. Tijssen, and J. Kleijnen. "Placebos and placebo effects in medicine: Historical overview." *Journal of the Royal Society of Medicine* 92 (1999): 511–515.

De Dijn, Herman. "De donkere transcendentie van Prometheus" [The dark transcendence of Prometheus]. *Tijdschrift voor Filosofie* 62, no. 4 (2000): 743–751.

De Dijn, Herman. *Taboes, monsters en loterijen, Over ethiek in de laat-moderne tijd* [Taboos, monsters and lotteries: On ethics in late modern times]. Kapellen: Uitgeverij Pelckmans, 2003.

Dekkers, Wim, and Gerard Boer. "Sham surgery in patients with Parkinson's disease: is it morally acceptable?" *Journal of Medical Ethics* 27 (2001): 151–156.

De la Hoz-Correa, Andrea, Francisco Muñoz-Leiva, and Marta Nakucz. "Past themes and future trends in medical tourism research: A co-word analysis." *Tourism Management* 65 (2018): 200–211.

Delisser, Horace M. "When a miracle is expected: Allowing space to believe." *American Journal of Bioethics* 18 (2018): 52–53.

Del Vecchio Good, Mary-Jo, Byron J. Good, Cynthia Schaffer, and Stuart E. Lind. "American oncology and the discourse on hope." *Culture, Medicine and Psychiatry* 14 (1990): 59–79.

DeMarco, Joseph P., and Richard M. Fox, eds. *New directions in ethics: The challenge of applied ethics*. New York and London: Routledge & Kegan Paul, 1986.

De Senneville, Loup Besmond. "Aux Etats-Unis, le boom des tests génétiques" [In the

United States, the boom of genetic tests]. *La Croix*, August 16, 2018. https://www
.la-croix.com/Monde/Ameriques/Etats-Unis-boom-tests-genetiques-2018-08
-16-1200962066.

Desmond, Jane. "Postmortem exhibitions: Taxidermied animals and plastinated
corpses in the theaters of the dead." *Configurations* 16, no. 3 (2008): 347–378.

DeTora, Lisa M. "What is safety? Miracles, benefit-risk assessments, and the 'right to
try.'" *International Journal of Clinical Practice* 71 (2017): e12966.

Devereaux, Mary, and Jeanne F. Loring. "Growth of an industry: How US scientists and
clinicians have enabled stem cell tourism." *American Journal of Bioethics* 10 (2010):
45–46.

De Vries, Hent. *Kleine filosofie van het wonder* [Little philosophy of the miracle].
Amsterdam: Boom, 2015.

De Vries, Raymond G., Tom Tomlinson, H. Myra Kim, Chris D. Krenz, Kerry A. Ryan,
Nicole Lehpamer, and Scott Y. H. Kim. "The moral concerns of biobank donors: The
effect of non-welfare interests on willingness to donate." *Life Sciences, Society and
Policy* 12 (2016): 3. https://doi.org/10.1186/s40504-016-0036-4.

De Waal Malefijt, Annemarie. "Homo monstrosus." *Scientific American* 219 (1968):
113–118.

Dewey, John. *A common faith*. New Haven: Yale University Press, 1934.

Dewey, John. *Human nature and conduct: An introduction to social psychology*. New
York: Henry Holt, 1922.

De Witte, Joke I., and Henk ten Have. "Ownership of genetic material and information." *Social Science & Medicine* 45, no. 1 (1997): 51–60.

Diallo, Rokhaya. "Why are the 'yellow vests' protesting in France?" *Al Jazeera*,
December 10, 2018. https://www.aljazeera.com/indepth/opinion/yellow-vests
-protesting-france-181206083636240.html.

Diamond, Eugene F. "Miraculous cures." *Linacre Quarterly* 51 (1984): 224–232.

Dichoso, Travis Jon. "Lourdes: A uniquely Catholic approach to medicine." *Linacre
Quarterly* 82 (2015): 8–12.

Dickson, David. "Genome Project gets rough ride in Europe." *Science* 243, no. 4891
(1989): 599.

Doran, Evan, and Donald Henry. "Disease mongering: Expanding the boundaries of
treatable disease." *International Medicine Journal* 38 (2008): 858–861.

Dossey, Larry. "The possibility of the impossible: Miracles, wonder, and Thomas
Jefferson's razor." *Explore* 6, no. 2 (2010): 57–63.

Doucet, Marika, Martin Yuille, Luke Georghiou, and Georges Dagher. "Biobank sustainability: Current status and future prospects." *Journal of Biorepository Science for
Applied Medicine* 5 (2017): 1–7.

Downie, Robin. "Patient and consumers." *Journal of the Royal College of Physicians of
Edinburgh* 47 (2017): 261–265.

Doz, François, Patrice Marvanne, and Anne Fagot-Largeault. "The person in personalized medicine." *European Journal of Cancer* 49 (2013): 1159–1160.

Dragojlovic, Nick. "Canadians' support for radical life extension resulting from advances in regenerative medicine." *Journal of Aging Studies* 27 (2013): 151–158.

Dubisch, Jill, and Michael Winkelman, eds. *Pilgrimage and healing*. Tucson: University
of Arizona Press, 2005.

Duffin, Jacalyn. "The doctor was surprised; or, How to diagnose a miracle." *Bulletin of the History of Medicine* 81 (2007): 699–729.

Duffin, Jacalyn. *Medical miracle: Doctors, saints, and healing in the modern world.* Oxford: Oxford University Press, 2009.

Duhigg, Charles. "Why are we so angry? The untold story of how we all got so mad at one another." *Atlantic*, January/February 2019: 63–75.

Durfy, Sharon J., and Amy E. Grotevant. "The Human Genome Project." *Kennedy Institute of Ethics Journal* 1, no. 4 (1991): 347–362.

Duster, Troy. *Backdoor to eugenics.* New York: Routledge, 2003.

Dworkin, Ronald. *Life's dominion: An argument about abortion and euthanasia.* London: HarperCollins, 1993.

Dyer, Clare. "Law, ethics, and emotion: The Charlie Gard case." *British Medical Journal* 358 (2017): j3152. https://doi.org/10.1136/bmj.j3152.

Eade, John. "Order and power at Lourdes: Lay helpers and the organization of a pilgrimage shrine." In *Contesting the sacred: The anthropology of pilgrimage*, edited by John Eade and Michael J. Sallnow, 51–76. Eugene, OR: Wipf and Stock, 2013.

Eade, John, and Michael J. Sallnow. *Contesting the sacred: The anthropology of pilgrimage.* Eugene, OR: Wipf and Stock, 2013. First published 1991.

Edel, Abraham. "Ethical theory and moral practice: On the term of their relation." In *New directions in ethics: The challenge of applied ethics*, ed. Joseph P. DeMarco and Richard M. Fox, 317–335. New York and London: Routledge & Kegan Paul, 1986.

Ehrenreich, Barbara. *Natural causes: An epidemic of wellness, the certainty of dying, and killing ourselves to live longer.* New York: Twelve; Hachette Book Group, 2018.

Ehrich, Kathryn, and Clare Williams. "A 'healthy baby': The double imperative of preimplantation genetic diagnosis." *Health* 14, no. 1 (2010): 41–56.

Einsiedel, Edna F., and Hannah Adamson. "Stem cell tourism and future stem cell tourists: Policy and ethical implications." *Developing World Bioethics* 12, no. 1 (2012): 35–44.

Eiseman, Elisa, and S. B. Haga, *Handbook of human tissue resources: A national resource of human tissue samples.* Document no. MR-954-OSTP. Santa Monica, CA: RAND Corporation, 1999.

El-Jardali, Fadi, Elie A. Akl, Racha Fadlallah, Sandy Oliver, Nadine Saleh, Lamya El-Bawab, Rana Rizk, Aida Farha, and Rasha Hamra. "Interventions to combat or prevent drug counterfeiting: A systematic review." *BMJ Open* 5 (2015): e006290. https://doi.org/10.1136/bmjopen-2014-006290.

Elliott, Carl. "Where ethics comes from and what to do about it." *Hastings Center Report* 22 (1992): 28–35.

Elliott, Jaklin A., and Ian N. Olver. "Hope and hoping in the talk of dying cancer patients." *Social Science & Medicine* 64 (2007): 138–149.

Elliott, Jaklin A., and Ian N. Olver. "Hope, life, and death: A qualitative analysis of dying cancer patients' talk about hope." *Death Studies* 33 (2009): 609–638.

Ellul, Jacques. "The technological order." *Technology and Culture* 3, no. 4 (1962): 394–421.

Elsner, Peter. "The concept of 'fragile skin': A case of disease mongering in dermatology?" *Journal of the European Academy of Dermatology & Venereology* 31, no. 6 (2017): e281–e282. https://doi.org/10.1111/jdv.14044.

Engelhardt, H. Tristram. *The foundations of bioethics*. Oxford: Oxford University Press, 1986.

Enserink, Martin. "Selling the stem cell dream." *Science* 313 (2006): 160–163.

Entous, Adam, and Jon Lee Anderson. "The mystery of the Havana syndrome." *New Yorker*, November 19 (2018): 34–47.

Eriksson, Stefan, and Gert Helgesson. "The false academy: Predatory publishing in science and bioethics." *Medicine, Health Care and Philosophy* 20 (2017): 163–170.

Esquire editors. "American rage: The *Esquire* / NBC News survey." *Esquire*, January 3, 2016. https://www.esquire.com/news-politics/a40693/american-rage-nbc-survey.

Ethics Committee of the American Society for Reproductive Medicine. "Cross-border reproductive care: An Ethics Committee opinion." *Fertility and Sterility* 106 (2016): 1627–1633.

Ettinger, Robert C. W. *The prospect of immortality*. Garden City, NY: Doubleday, 1964.

Evans, James P., Eric M. Meslin, Theresa M. Marteau and Timothy Caulfield. "Deflating the genomic bubble." *Science* 331, no. 6019 (2011): 861–862.

Evans, John H. *The history and future of bioethics: A sociological view*. Oxford: Oxford University Press, 2012.

Evans, John H. *Playing God*. Chicago: University of Chicago Press, 2002.

Fair, Brian. "Morgellons: Contested disease, diagnostic compromise and medicalisation." *Sociology of Health & Illness* 32 (2010): 597–612.

Falk, Raphael. "The gene—a concept in tension." In *The concept of the gene in development and evolution: Historical and epistemological perspectives*, edited by Peter J. Beurton, Raphael Falk, and Hans-Jörg Rheinberger, 317–348. Cambridge: Cambridge University Press, 2000.

Fanelli, Daniele. "How many scientists fabricate and falsify research? A systematic review and meta-analysis of survey data." *PLoS ONE* 4, no. 5 (2009): e5738. https://doi.org/10.1371/journal.pone.0005738.

Farmer, Paul, and Nicole Gastineau Campos. "Rethinking medical ethics: A view from below." *Developing World Bioethics* 4, no. 1 (2004): 17–41.

Fassin, Didier. *Life: A critical user's manual*. Cambridge: Polity Press, 2018.

Ferry, Luc. *L'homme-Dieu ou le Sens de la Vie* [The man-god, or the meaning of life]. Paris: Grasset et Fasquelle, 1996.

Fierz, Walter. "Challenge of personalized health care: To what extent is medicine already individualized and what are the future trends?" *Medical Science Monitor* 10, no. 5 (2004): 111–123.

First, Daniel. "Will big data algorithms dismantle the foundations of liberalism? *AI & Society* 33 (2018): 545–556. https://doi.org/10.1007/s00146-017-0733-4.

Fletcher, Joseph. *The ethics of genetic control: Ending reproductive roulette*. Garden City, NY: Anchor Press, 1974.

Fletcher, Joseph. *Humanhood: Essays in biomedical ethics*. Buffalo: Prometheus Books, 1979.

Foddy, Bennett. "A duty to deceive: Placebos in clinical practice." *American Journal of Bioethics* 9 (2009): 4–12.

Forman, Henry James. *The story of prophecy: In the life of mankind from early times to the present day*. New York: Farrar & Rinehart, 1936.

Fortun, Mike. *Promising genomics: Iceland and deCODE Genetics in a world of specula-tion.* Berkeley: University of California Press, 2008.

Foucault, Michel. *The birth of the clinic.* London: Routledge, 1993.

Fox, Renée C. *The sociology of medicine: A participant observer's view.* Englewood Cliffs, NJ: Prentice-Hall, 1989.

Freeman, Charles. *Holy bones, holy dust: How relics shaped the history of Medieval Europe.* New Haven: Yale University Press, 2012.

Freud, Sigmund. *The uncanny.* New York: Penguin Books, 2003. First published 1919.

Friedman, John Block. *The monstrous races in medieval art and thought.* Syracuse: Syracuse University Press, 2000.

Friend, Tad. "Silicon Valley's quest to live forever." *New Yorker*, April 3, 2017. https://www.newyorker.com/magazine/2017/04/03/silicon-valleys-quest-to-live-forever.

Friese, Carrie. *Cloning wild life: Zoos, captivity, and the future of endangered animals.* New York: New York University Press, 2013.

Fugh-Berman, Adriana. "The corporate coauthor." *Journal of General Internal Medicine* 20, no. 6 (2005): 546–548.

Fukuyama, Francis. *Our posthuman future: Consequences of the biotechnology revolu-tion.* New York: Picador, 2002.

Gaines, Atwood D., and Eric T. Juengst. "Origin myths in bioethics: Constructing sources, motives and reason in bioethics(s)." *Culture, Medicine and Psychiatry* 32 (2008): 303–327.

Gasparyan, Armen Yuri, Bekaidar Nurmashev, Alexander A. Voronov, Alexey N. Gerasimov, Anna M. Koroleva, and George D. Kitas. "The pressure to publish more and the scope of predatory publishing activities." *Journal of Korean Medical Science* 31 (2016): 1874–1878.

Gautam, C. S., A. Utreja, and G. L. Singal. "Spurious and counterfeit drugs: A growing industry in the developing world." *Postgraduate Medicine Journal* 85 (2009): 251–256.

Gazi, Andromache. "Exhibition ethics—an overview of major issues," *Journal of Conser-vation and Museum Studies* 12, no. 1 (2014): 1–10.

Geary, Patrick J. *Furta sacra: Thefts of relics in the Central Middle Ages.* Princeton, NJ: Princeton University Press, 1990.

Geary, Patrick [J.]. "Sacred commodities: The circulation of medieval relics." In *The social life of things: Commodities in cultural perspective*, edited by Arjun Appadurai, 169–191. Cambridge: Cambridge University Press, 1986.

Gefenas, Eugenijus, Asta Cekanauskaite, Egle Tuzaite, Vilius Dranseika, and Dainius Characiejus. "Does the 'new philosophy' in predictive, preventive and personalised medicine require new ethics?" *EPMA Journal* 2 (2011): 141–147.

Gesler, Wil. "Lourdes: Healing in a place of pilgrimage." *Health & Place* 2 (1996): 95–105.

Getz, Landon J., Graham Dellaire, and Francoise Baylis. "Jiankui He: A sorry tale of high-stakes science." *Hastings Center*, December 10, 2018. https://www.thehastingscenter.org/jiankui-sorry-tale-high-stakes-science.

Ghinea, Narcyz, Miles Little, and Wendy Lipworth. "Access to high cost medicines through the lens of an Australian senate inquiry—defining the 'goods' at stake." *Bioethical Inquiry* 14 (2017): 401–410.

Gigerenzer, Gerd, and Rocio Garcia-Retamero. "Cassandra's regret: The psychology of not wanting to know." *Psychological Review* 124, no. 2 (2017): 179–196.

Gilmore, David M. *Monsters: Evil beings, mythical beasts, and all manner of imaginary terrors.* Philadelphia: University of Pennsylvania Press, 2003.

Giroux, Henry. "Poisoned city in the age of casino capitalism." *Theory in Action* 10, no. 1 (2017): 7–31.

Godderis, Jan. *Een arts is vele andere mensen waard: Inleiding tot de antieke genees-kunde* [A physician is as valuable as many other people: Introduction to ancient medicine]. Leuven, Belgium: Peeters, 1999.

Gollner, Adam Leith. *The book of immortality: The science, belief, and magic behind living forever.* New York: Scribner, 2013.

Gordijn, Bert. *Medical utopias: Ethical reflections about emerging medical technologies.* Leuven, Belgium: Peeters, 2006.

Great Ormond Street Hospital (GOSH). "GOSH's position statement, hearing on July 24, 2017." High Court of Justice, FD17P00103. Accessed on May 14, 2021. https://media.gosh.nhs.uk/documents/Great_Ormond_Street_Hospital_position_statement_at_High_Court_on_24_July_2017.pdf.

Gould, George M., and Walter L. Pyle. *Medical curiosities.* Mount Morris, MI: Hammond Publishing, 1982. First published 1896.

Goya, Francisco. *Los Caprichos* [The caprices]. Plate 43: El sueño de la razon produce monstrous [The sleep of reason produces monsters], 1799.

Gracia, Diego. "History of medical ethics." In *Bioethics in a European perspective*, edited by Henk A. M. J. ten Have and Bert Gordijn, 17–50. Dordrecht: Kluwer Academic, 1999.

Gracia Guillen, Diego. "Bioethics in the Spanish-speaking world." In *Bioethics: A history*, edited by Corrado Viafora, 169–197. San Francisco, CA: International Scholars Publications, 1996.

Graham, Gordon. *Genes: A philosophical inquiry.* London: Routledge, 2002.

Gray, John. "Humanity Mk II: The next great stage of our evolution has begun." *New Statesman* 23–29 September 2016, 68–71.

Greely, Henry T. "The uneasy ethical and legal underpinnings of large-scale genomic biobanks." *Annual Review of Genomics and Human Genetics* 8 (2007): 343–364.

Greely, Henry T., and Mildred K. Cho. "The Henrietta Lacks legacy grows." *EMBO Reports* 14, no. 10 (2013): 849.

Greene, Jeremy A., and Joseph Loscalzo. "Putting the patient back together—social medicine, network medicine, and the limits if reductionism." *New England Journal of Medicine* 377, no. 25 (2017): 2493–2499.

Greene, Joshua, and Jonathan Haidt. "How (and where) does moral judgment work?" *Trends in Cognitive Sciences* 6, no. 12 (2002): 517–523.

Greig, Alastair, David Hulme, and Mark Turner. *Challenging global inequality: Development theory and practice in the 21st century.* Houndmills, UK: Palgrave Macmillan, 2007.

Groopman, Jerome. *The anatomy of hope: How people prevail in the face of illness.* New York: Random House, 2005.

Guerrini, Anita. "Inside the charnel house: The display of skeletons in Europe,

1500–1800." In *The fate of anatomical collections*, edited by Rina Knoeff and Robert Zwijnenberg, 93–109. Farnham: Ashgate, 2015.

Gupta, Amit Sen. "Medical tourism in India: Winners and losers." *Indian Journal of Medical Ethics* 5, no. 1 (2008): 4–5.

Gustafson, James M. *The contributions of theology to medical ethics*. Milwaukee: Marquette University Press, 1975.

Gustafson, James M. "Theology confronts technology and the life sciences." *Commonweal* 105 (1978): 386–392.

Gyawali, Bishal, and Richard Sullivan. "Economics of cancer medicines: For whose benefit?" *New Bioethics* 23, no. 1 (2017): 95–104.

Habibzadeh, Farrokh, and Ana-Maria Simundic. "Predatory journals and their effects on scientific research community." *Biochemia Medica* 27 (2017): 270–272.

Hagerty, Barbara Bradley. "Can your genes make you murder?" *NPR*, July 1, 2010. http://www.npr.org/templates/story/story.php?storyId=128043329.

Haidt, Jonathan. "The emotional dog and the relational tail: A social intuitionist approach to more judgment." *Psychological Review* 198, no. 4 (2001): 814–834

Hall, Stephen S. *Merchants of immortality: Chasing the dream of human life extension*. Boston: Houghton Mifflin, 2003.

Hall, Stephen S. "Revolution postponed." *Scientific American* 303, no. 4 (2010): 60–67.

Hamburg, Margaret A., and Francis S. Collins. "The path to personalized medicine." *New England Journal of Medicine* 363, no. 4 (2010): 301–304.

Hamilton, William L., Cormac Doyle, Mycroft Halliwell-Ewen, and Gabriel Lambert. "Public health interventions to protect against falsified medicines: A systematic review of international, national and local policies." *Health Policy and Planning* 31 (2016): 1448–1466.

Hammond-Browning, Natasha. "When doctors and parents don't agree: The story of Charlie Gard." *Bioethical Inquiry* 14 (2017): 461–468.

Hampshire, Stuart. "Morality and pessimism." In *Public and private morality*, edited by Stuart Hampshire, 1–22. Cambridge: Cambridge University Press, 1978.

Hampson, Fen Osler, and Eric Jardine. *Look who's watching: Surveillance, treachery, and trust online*. Waterloo, Canada: Centre for International Governance Innovation, 2016.

Hanafi, Zakiya. *The monster in the machine: Magic, medicine, and the marvelous in the time of the Scientific Revolution*. Durham, NC: Duke University Press, 2000.

Hanson, M. G. "Building on relationships of trust in biobank research." *Journal of Medical Ethics* 31 (2005): 415–418.

Hanson, M. G. "Ethics and biobanks." *British Journal of Cancer* 100 (2009): 8–12.

Harari, Yuval Noah. *Homo Deus: A brief history of tomorrow*. New York: HarperCollins, 2017.

Harding, Jeremy. "Among the *gilets jaunes*." *London Review of Books* 41, no. 6 (2019): 3–11.

Harpur, James. *The pilgrim journey: A history of pilgrimage in the Western world*. Katonah, NY: BlueBridge, 2016.

Harré, Rom. *Physical being*. Oxford: Blackwell, 1991.

Harris, John. *Wonderwoman and Superman: The ethics of human biotechnology*. Oxford: Oxford University Press, 1992.

Harris, Gardiner. "16 American sickened after attack on embassy staff in Havana," *New York Times*, August 24, 2017. https://www.nytimes.com/2017/08/24/us/politics /health-attack-us-embassy-havana.html.

Harris, Ruth. *Lourdes: Body and spirit in the secular age*. London: Penguin Books, 1999.

Hartzband, Pamela, and Jerome Groopman. "Money and the changing culture of medicine." *New England Journal of Medicine* 360 (2009): 101–103.

Hasson, Katie, and Marcy Darnovsky. "Gene-edited babies: No one has the moral warrant to go it alone." *Guardian*, November 27, 2018. https://www.theguardian.com /science/2018/nov/27/gene-edited-babies-no-one-has-moral-warrant-go-it-alone.

Havemann, Joel. "Sham surgeries, real risks." *Neurology Now* 6, no. 4 (2010): 35–36.

Healy, David, and Dinah Cattell. "Interface between authorship, industry and science in the domain of therapeutics." *British Journal of Psychiatry* 183 (2003): 22–27.

Hedgecoe, Adam M. "Critical bioethics: Beyond the social science critique of applied ethics." *Bioethics* 18, no. 2 (2004): 120–143.

Hedgecoe, Adam [M.]. *The politics of personalised medicine: Pharmacogenetics in the clinic*. Cambridge: Cambridge University Press, 2004.

Hench, Philip S., Edward C. Kendall, Charles H. Slocumb, and Howard F. Polley. "The effects of the adrenal cortical hormone 17-hydroxy-11-dehydrocorticosterone (Compound E) on the acute phase of rheumatic fever; preliminary report." *Mayo Clinical Proceedings* 24, no. 11 (1949): 277–297.

Hench, Philip S., Edward C. Kendall, Charles H. Slocumb, and Howard F. Polley. "The effect of a hormone of the adrenal cortex (17-hydroxy-11-dehydrocorticosterone: compound E) and of pituitary adrenocorticotrophic hormone on rheumatoid arthritis." *Annals of the Rheumatic Diseases* 8, no. 2 (1949): 97–104.

Henderson, Gail E., R. Jean Cadigan, Teresa P. Edwards, Ian Conlon, Anders G. Nelson, James P. Evans, Arlene M. Davis, Catherine Zimmer, and Bryan J. Weiner. "Characterizing biobank organizations in the US: Results from a national survey." *Genome Medicine* 5 (2013): 3.

Henderson, Saras, and Alan Petersen, eds. *Consuming health: The commodification of health care*. London: Routledge, 2002.

Henneman, Lidewij, Eric Vermeulen, Carla G. van El, Liesbeth Claassen, Danielle R. M. Timmermans, and Marina C. Cornel. "Public attitudes towards genetic testing revisited: Comparing opinions between 2002 and 2010." *European Journal of Human Genetics* 21 (2013): 793–799.

Herzheimer, Andrew. "Relationships between the pharmaceutical industry and patients' organizations." *British Medical Journal* 326 (2003): 1208–1210.

Hilden, Jørgen., and J. Dik F. Habbema. "Prognosis in medicine: An analysis of its meaning and roles." *Theoretical Medicine* 8 (1987): 349–365.

Hill, Graham. *Global church: Reshaping our conversations, renewing our mission, revitalizing our churches*. Downers Grove, IL: InterVarsity Press, 2016.

Hippocrates, "Prognostic." In *Hippocrates*, vol. 2, edited by W. H. S. Jones, 1–55. London: Heinemann; Cambridge, MA: Harvard University Press, 1967.

Hodges, Jill R., Leigh Turner, and Ann Marie Kimball, eds. *Risks and challenges in medical tourism: Understanding the global market for health services*. Santa Barbara, CA: Praeger, 2012.

Hodges, Jill R., and Ann Marie Kimball. "Unseen travelers: Medical tourism and the spread of infectious diseases." In *Risks and challenges in medical tourism: Understanding the global market for health services*, edited by Jill R. Hodges, Leigh Turner, and Ann Marie Kimball, 111–137. Santa Barbara, CA: Praeger, 2012.

Hoedemaekers, Rogeer, Bert Gordijn, and Martien Pijnenburg. "Solidarity and justice as guiding principles in genomic research." *Bioethics* 21 (2007): 342–350.

Hoedemaekers, Rogeer, and Henk A. M. J. ten Have. "Genetic health and disease." In *Genes and morality: New essays*, edited by V. Launis, J. Pietarinen, and J. Raikka, 121–143. Amsterdam: Rodopi, 1999.

Hoedemaekers, Rogeer, and Henk A. M. J. ten Have. "Geneticization: The Cyprus paradigm." *Journal of Medicine and Philosophy* 23 (1998): 274–287.

Hollands, Gareth J., David P. French, Simon J Griffin, A. Toby Prevost, Stephen Sutton, Sarah King, and Theresa M. Marteau. "The impact of communicating genetic risks of disease on risk-reducing health behaviour: Systematic review with meta-analysis." *British Medical Journal* 352 (2016): i1102. https://doi.org/10.1136/bmj.i1102.

Holmes, Rachel. *The Hottentot Venus: The life and death of Saartjie Baartman*. London: Bloomsbury, 2007.

Horne, Rob. "The human dimension: Putting the person into personalised medicine." *New Bioethics* 23 (2017): 38–48.

Hortobagyi, Gabriel N. "Trastuzumab in the treatment of breast cancer." *New England Journal of Medicine* 353 (2006): 1734–1737.

Horton, Richard. "Rediscovering human dignity." *Lancet* 364 (2004): 1081–1085.

Horwitz, Ralph I., Mark R. Cullen, Jill Abell, and Jennifer B. Christian. "(De)Personalized medicine." *Science* 339 (2013): 1155–1156.

Howell, Michael, and Peter Ford. *The ghost disease, and twelve other stories of detective work in the medical field*. Harmondsworth: Penguin Books, 1986.

Huet, Marie-Hélène. *Monstrous imagination*. Cambridge, MA: Harvard University Press, 1993.

Hume, David. *Treatise of human nature*. Book 3. Edited by Jonathan Bennett, www.earlymoderntexts.com, 2017. First published 1740. http://www.earlymoderntexts.com/assets/pdfs/hume1740book3.pdf.

Hunter, David J. "The case against choice and competition." *Health Economics, Policy and Law* 4 (2009): 489–501.

Hunter, Mark. "Alder Hey report condemns doctors, management, and coroner." *British Medical Journal* 322 (2001): 255.

Hurlbut, J. Benjamin, Sheila Jasanoff and Krishanu Saha. "The Chinese gene-editing experiment was an outrage. The scientific community shares the blame." *Washington Post*, November 29, 2018. https://www.washingtonpost.com/outlook/2018/11/29/chinese-gene-editing-experiment-was-an-outrage-broader-scientific-community-shares-some-blame.

Hurley, Richard. "How a fight for Charlie Gard became a fight against the state." *British Medical Journal* 358 (2017): j3675. https://doi.org/10.1136/bmj.j3675.

Ihara, Hiroshi. "Disease mongering," in *Encyclopedia of global bioethics*, vol. 1, edited by Henk ten Have, 924–934. Switzerland: Springer Reference, 2016.

Illich, Ivan. "The medicalization of life." *Journal of Medical Ethics* 1 (1975): 73–77.

Illich, Ivan. *Medical nemesis: The expropriation of health*. London: Calder & Boyars, 1975.

Inhorn, Marcia C. "Globalization and gametes: Reproductive 'tourism,' Islamic bioethics, and Middle Eastern modernity." *Anthropology & Medicine* 18 (2011): 87–103.

Inhorn, Marcia C., and Pasquale Patrizio. "Rethinking reproductive 'tourism' as reproductive 'exile.'" *Fertility and Sterility* 92 (2009): 904–906.

International Bioethics Committee (IBC). *Report of the IBC on the principle of non-discrimination and non-stigmatization.* Paris: UNESCO, 2014.

Ioannidis, John P. A. "An epidemic of false claims." *Scientific American* 304, no. 6 (2011): 16–17.

Ioannidis, John P. A. "Why most clinical research is not useful." *PLoS Med* 13, no. 6 (2016): e1002049. https://doi.org/10.1371/journal.pmed.1002049.

Ioannidis, John P. A. "Why most published research findings are false." *PLoS Med* 2, no. 8 (2005): e124.

Isaacs, David. "Disease marketing." *Journal of Paediatrics and Child Health* 53 (2017): 1141–1142.

iSpecimen. "2018 will be a big year for biobanks." December 11, 2017. https://www.ispecimen.com/blog/2018-will-big-year-biobanks.

Issenberg, Sasha. *Outpatients: The astonishing new world of medical tourism.* New York: Columbia Global Reports, 2016.

Jack, Andrew, Frances Williams, and Michael Steen. "Seizure of HIV drugs highlights patent friction." *Financial Times*, March 4, 2009.

Jacob, Jean Daniel, Marilou Gagnon and Dave Holmes. "Nursing so-called monsters: On the importance of abjection and fear in forensic psychiatric nursing." *Journal of Forensic Nursing* 5 (2009): 153–161.

Jacobsen, Sven-Erik, Marten Sörensen, Söre Marcus Pedersen, and Jocon Weiner. "Feeding the world: Genetically modified crops versus agricultural biodiversity." *Agronomy for Sustainable Development* 33 (2013): 651–662.

Jenkins, Tiffany. *Contesting human remains in museum collections: The crisis of cultural authority.* New York: Routledge, 2011.

Jenkins, Tiffany. *Keeping their marbles: How the treasures of the past ended up in museums . . . and why they should stay there.* Oxford: Oxford University Press, 2018.

Jennings, Bruce, and Angus Dawson. "Solidarity in the moral imagination of bioethics." *Hastings Center Report* 45, no. 5 (2015): 31–38.

Jeuken, Marius. *Materie, leven, geest. Een wijsgerige biologie* [Matter, life, spirit: A philosophical biology]. Assen: Van Gorcum, 1979.

Jeuken, Marius. "The biological and philosophical definitions of life." *Acta Biotheoretica* 24, no. 1–2 (1975): 14–21.

Jiang, Li, and Bing He Dong. "Fraudsters operate and officialdom turns a blind eye: A proposal for controlling stem cell therapy in China." *Medicine, Health Care and Philosophy* 19 (2016): 403–410.

Jochemsen, Henk. "Reducing people to genetics." In *Genetic ethics: Do the ends justify the genes?*, edited by John F. Kilner, Rebecca D. Pentz, and Frank E. Young, 75–83. Grand Rapids, MI: William B. Eerdmans Publishing, 1997.

Joffe, Steven, and Holly Fernandez Lynch. "Federal right-to-try legislation—threatening the FDA's public health mission." *New England Journal of Medicine* 378 (2018): 695–697.

Joignot, Frédéric. "Trop d'humaines pour la planète?" [Too many humans for the planet?]. *Le Monde*, January 10, 2009: 16–23.

Joly, Yann, Ida Ngueng Feze, Lingqiao Song, and Bartha M. Knoppers. "Comparative approach to genetic discrimination: Chasing shadows? *Trends in Genetics* 33, no. 5 (2017): 299–302.

Jones, Christopher A. "Prognostication: Medicine's lost art." *Geriatrics for the Practicing Physician* 91, no. 11 (2008): 347–348.

Jones, D. Gareth, R. Gear, and K.A. Galvin. "Stored human tissue: An ethical perspective on the fate of anonymous, archival material." *Journal of Medical Ethics* 29 (2003): 343–347.

Jones, D. Gareth, and Maja I. Whitaker. *Speaking for the dead: The human body in biology and medicine*. Farnham: Ashgate, 2009.

Jones, E. Michael. *Horror: A biography*. Dallas: Spence Publishing, 2002.

Jones, Steve. *The language of the genes: Biology, history and the evolutionary future*. London: Flamingo, 2000.

Jonsen, Albert [R.]. *The birth of bioethics*. Oxford: Oxford University Press, 1998.

Jonsen, Albert R. *The new medicine and the old ethics*. Cambridge, MA: Harvard University Press, 1990.

Jonsen, Albert R. "Why has bioethics become so boring?" *Journal of Medicine and Philosophy* 25, no. 6 (2000): 689–699.

Judiciary of England and Wales. "Decision and short reasons to be released to the media in the case of Charlie Gard." Press release, April 11, 2017. https://www.judiciary.gov .uk/wp-content/uploads/2017/04/gard-press-summary-20170411.pdf.

Judson, Horace Freeland. *The great betrayal: Fraud in science*. Orlando: Harcourt, 2004.

Juengst, Eric, Michelle L. McGowan, Jennifer R. Fishman and Richard A. Settersten. "From 'personalized' to 'precision' medicine: The ethical and social implications of rhetorical reform in genomic medicine." *Hastings Center Report* 46 (2016): 21–33.

Kakuk, Peter, ed. *Bioethics and biopolitics: Theories, applications and connections*. Cham, Switzerland: Springer International, 2017.

Kalokairinou, Louiza, Heidi C. Howard, Santa Slokenberga, E. Fisher, Magdalena Flatscher-Thöni, Mette Hartlev, Rachel van Hellemondt, et al., "Legislation of direct-to-consumer genetic testing in Europe: A fragmented regulatory landscape." *Journal of Community Genetics* 9 (2018): 117–132.

Kamenova, Kalina, and Timothy Caulfield. "Stem cell hype: Media portrayal of therapy translation." *Science Translational Medicine* 7 (2015): 278ps4. https://doi.org/10.1126 /scitranslmed.3010496.

Kane, Paula. "Disenchanted America: Accounting for the lack of extraordinary mystical phenomena in Catholic America." In *Sign or symptom? Exceptional corporeal phenomena in religion and medicine in the nineteenth and twentieth centuries*, edited by Tine van Osselaer, Henk de Smaele, and Kaat Wils, 101–124. Belgium: Leuven University Press, 2017.

Kaplan, Jonathan Michael. *The limits and lies of human genetic research: Dangers for social policy*. New York: Routledge, 2000.

Kaplan, Matt. *The science of monsters*. New York: Scribner, 2012.

Kaptchuk, Ted J. "Powerful placebo: The dark side of the randomized controlled trial." *Lancet* 351 (1998): 1722–1725.

Kapur, Vicky. "Miracle child: Indian baby born with two heads clinging to life." *Kemmannu.com*, March 18, 2014. http://www.kemmannu.com/index.php?action =highlights&type=8744.

Karches, Kyle E. "Against the iDoctor: Why artificial intelligence should not replace physician judgment." *Theoretical Medicine and Bioethics* 39 (2018): 91–110.

Karlin-Smith, Sarah. "Libertarians score big victory in 'right-to-try' drug bill." *Politico*, April 8, 2017. https://www.politico.com/story/2017/08/03/libertarians-score-big -victory-drug-bill-241314.

Karlin-Smith, Sarah. "House passes right-to-try on second try," *Politico*, March 21, 2018. https://www.politico.com/story/2018/03/21/drugs-right-to-try-congress-434677.

Kasas, Savas, and Reinhard Struckmann. *Important medical centres in antiquity: Epidaurus and Corinth*. Athens: Editions Kasas, 1979.

Kashihara, Hidenori, Takeo Nakayama, Taichi Hatta, Naomi Takahashi, and Misao Fujita. "Evaluating the quality of website information of private-practice clinics offering cell therapies in Japan." *Interactive Journal of Medical Research* 5 (2016): e15. htttps://doi.org/10.2196/ijmr.5479.

Kass, Leon R. "Organs for sale? Propriety, property, and the price of progress." *Public Interest* 107 (Spring 1992): 65–86.

Kass, Leon R. "Thinking about the body." *Hastings Center Report* 15, no. 1 (1985): 20–30.

Kass, Leon R. "The wisdom of repugnance." In *The ethics of human cloning*, edited by Leon R. Kass and James Q. Wilson, 3–59. Washington, DC: AEI Press, 1998.

Kassirer, Jerome P. "Disclosure's failings: What is the alternative?" *Academic Medicine* 84 (2009): 1180–1181.

Katsnelson, Alla. "Why fake it? How 'sham' brain surgery could be killing off valuable therapies for Parkinson's disease." *Nature* 476 (2011): 143–144.

Kaufman, Suzanne K. *Consuming visions: Mass culture and the Lourdes shrine*. Ithaca, NY: Cornell University Press, 2005.

Keane, David. "Survival of the fairest? Evolution and the geneticization of rights." *Oxford Journal of Legal Studies* 30 (2010): 467–494.

Kearns, Lisa, and Alison Bateman-House. "Who stands to benefit? Right to try law provisions and implications." *Therapeutic Innovation & Regulatory Science* 51 (2017): 170–176.

Keilin, David. "The Leeuwenhoek Lecture: The problem of anabiosis or latent life: History and current concept." *Proceedings of the Royal Society of London, Series B, Biological Sciences* 150, no. 939 (1959): 149–191.

Kekes, John. "Disgust and moral taboos." *Philosophy* 67, no. 262 (1992): 431–446.

Keller, Evelyn Fox. *The century of the gene*. Cambridge, MA: Harvard University Press, 2000.

Kellett, John. "Prognostication—the lost skill of medicine." *European Journal of Internal Medicine* 19 (2008): 155–164.

Kelly, Bryony. "Human remains: Objects to study or ancestors to bury?" *Clinical Medicine* 4 (2004): 465–467.

Kim, Scott Y., Samuel Frank, Robert Holloway, Carol Zimmerman, Renee Wilson, and Karl Kieburtz. "Science and ethics of sham surgery: A survey of Parkinson disease clinical researchers." *Archives of Neurology* 62, no. 9 (2005): 1357–1360.

King, Mike R., Maja L. Whitaker, and D. Gareth Jones. "I see dead people: Insights from the humanities into the nature of plastinated cadavers." *Journal of Medical Humanities* 35 (2014): 361–376.

Kirby, Peadar. *Vulnerability and violence: The impact of globalisation*. London: Pluto Press, 2006.

Kirschner, Kristin L. "Rethinking and advocacy in bioethics." *American Journal of Bioethics* 1, no. 3 (2001): 60–62.

Kitcher, Philip. *The lives to come: The genetic revolution and human possibilities*. London: Penguin Books, 1996.

Kjaergard, Lise L., and Bodil Als-Nielsen. "Association between competing interests and authors' conclusions: Epidemiological study of randomized clinical trials published in the BMJ." *British Medical Journal* 325 (2002): 249.

Knoeff, Rina. "Ball pool anatomy: On the public veneration of anatomical relics." In *The fate of anatomical collections*, edited by Rina Knoeff and Robert Zwijnenberg, 279–291. Farnham: Ashgate, 2015.

Knoeff, Rina. "Touching anatomy: On the handling of preparations in the anatomical cabinets of Frederik Ruysch (1638–1731)." *Studies in the History and Philosophy of Biological and Biomedical Sciences* 49 (2015): 32–44.

Knoeff, Rina, and Robert Zwijnenberg, eds. *The fate of anatomical collections*. Farnham: Ashgate, 2015.

Koch, Lene. "The meaning of eugenics: Reflections on the government of genetic knowledge in the past and the present." *Science in Context* 17, no. 3 (2004): 315–331.

Koelbing, Huldrych M. *Arzt und Patient in der Antiken Welt* [Doctor and patient in antiquity]. Zürich & München: Artemis Verlag, 1977.

Koopmeiners, Louanne, Janice Post-White, Sarah Gutknecht, Carolyn Ceronsky, Kay Nichelson, Debra Drew, Karen Watrud Mackey, and Mary Jo Kreitzer. "How healthcare professionals contribute to hope in patients with cancer." *Oncology Nursing Forum* 24 (1997): 1507–1513.

Koplin, Julian J. "Commodification and human interests." *Bioethical Inquiry* 15 (2018): 429–440.

Kub, Joan, and Sara Groves. "Miracles and medicine: An annotated bibliography." *Southern Medical Journal* 100, no. 12 (2007): 1273–1276.

Kupferschmidt, Kai. "Taming the monsters of tomorrow." *Science* 359, no. 6372 (2018): 152–155.

Kuruvilla, Carol. "French bishop declares nun's recovery a Lourdes miracle." *Huffington Post*, February 12, 2018. https://www.huffingtonpost.com/entry/french-nun -lourdes-miracle_us_5a81b39ae4b044b3821fab9d.

Labovitz, Deborah R., ed. *Ordinary miracles: True stories about overcoming obstacles and surviving catastrophes*. Thorofare, NJ: SLACK, 2003.

Lacasse, Jeffrey R., and Jonathan Leo. "Ghostwriting at elite academic medical centers in the United States." *PLoS Medicine* 7 (2010): 1–4.

Lakoff, George, and Mark Johnson. *Metaphors we live by*. Chicago: University of Chicago Press, 2003.

Lancet editors. "Herceptin and early breast cancer: A moment for caution." Editorial. *Lancet* 366 (2005): 1673.

Landecker, Hannah. "Between beneficence and chattel: The human biological in law and science." *Science in Context* 12 (1999): 203–225.

Landecker, Hannah. *Culturing life: How cells became technologies.* Cambridge, MA: Harvard University Press, 2007.

Lantos, John D. ed. *Controversial bodies: Thoughts on the public display of plastinated corpses.* Baltimore: Johns Hopkins University Press, 2011.

Lantos, John D. "The tragic case of Charlie Gard." *JAMA Pediatrics* 171, no. 10 (2017): 935–936.

Laqueur, Thomas W. *The work of the dead: A cultural history of mortal remains.* Princeton, NJ: Princeton University Press, 2015.

Larmer, Robert A., ed. *Questions of miracle.* Montreal: McGill-Queen's University Press, 1996.

Lau, Darren, Ubaka Ogbogu, Benjamin Taylor, Tania Stafinski, Devidas Menon, and Timothy Caulfield. "Stem cell clinics online: The direct-to-consumer portrayal of stem cell medicine." *Cell Stem Cell* 3 (2008): 591–594.

Lawrence, Peter A. "The politics of publication: Authors, reviewers and editors must act to protect the quality of research." *Nature* 422 (2003): 259–261.

Lederberg, Joshua. "Orthobiosis: The perception of man." In *The place of value in a world of facts,* edited by Sam Nilsson and Arno Tiselius, 29–58. New York: John Wiley & Sons, 1970.

Lee, Kah Seng, Siew Mei Yee, Syed Tabish Razi Zaidi, Rahul P. Patel, Quan Yang, Yaser Mohammed Al-Worafi, and Long Chiau Ming. "Combating sales of counterfeit and falsified medicines online: A losing battle." *Frontiers in Pharmacology* 8 (2017): 268. https://doi.org/10.3389/phar.2017.00268.

Lee, Nancy S. "Framing choice: The origins and impact of consumer rhetoric in US health care debates." *Social Science & Medicine* 138 (2015): 136–143.

Lee, Robert G., and Derek Morgan. "Regulating risk society: Stigmata cases, scientific citizenship & biomedical diplomacy." *Sydney Law Review* 23, no. 3 (2001): 297–318.

Leggat, Peter. "Medical tourism." *Australian Family Physician* 44 (2015): 16–21.

Leith, Valerie M. Sheach. "Consent and nothing but consent? The organ retention scandal." *Sociology of Health and Illness* 29 (2007): 1023–1042.

Lemmens, Trudo, and Benjamin Freedman. "Ethics review for sale? Conflict of interest and commercial research review boards." *Milbank Quarterly* 78 (2000): 547–584.

Leng, Chee Heng. "Medical tourism and the state in Malaysia and Singapore." *Global Social Policy* 10 (2010): 336–357.

Leroi, Armand Marie. *Mutants: On genetic variety and the human body.* New York: Penguin Books, 2003.

Levitt, Mairi. "Empowered by choice?" In *The right to know and the right not to know: Genetic privacy and responsibility,* edited by Ruth Chadwick, Mairi Levitt, and Darren Shickle, 85–99. Cambridge: Cambridge University Press, 2014.

Lewinsohn, Richard. *Science, prophecy, and prediction.* New York: Fawcett, 1962.

Lewontin, Richard C., Steven Rose, and Leon J. Kamin. *Not in our genes.* Chicago: Haymarket Books, 2017. First published 1984 by Pantheon.

Lexchin, Joel. "Bigger and better: How Pfizer redefined erectile dysfunction." *PLoS Medicine* 3, no. 4 (2006): e132. https://doi.org/10.1371/journal.pmed.0030132.

Lexchin, Joel. "Those who have the gold make the evidence: How the pharmaceutical industry biases the outcomes of clinical trials of medications." *Science and Engineering Ethics* 18 (2012): 247–261.

Lexchin, Joel, Lisa A. Bero, Benjamin Djulbegovic, and Otavio Clark. "Pharmaceutical industry sponsorship and research outcome and quality: Systematic review." *British Medical Journal* 326 (2003): 1167–1176.

Lichter, Allen S. "Conflict of interest and the integrity of the medical profession." *JAMA* 317 (2017): 1725–1726.

Lin, Patrick, Keith Abney, and George A. Bekey, eds. *Robot ethics: The ethical and social implications of robotics.* Cambridge, MA: MIT Press, 2012.

Lindeboom, Gerrit A. *Descartes and medicine.* Amsterdam: Rodopi, 1978.

Lindeboom, Gerrit A. *Inleiding tot de geschiedenis der geneeskunde* [Introduction to the history of medicine]. Haarlem: De Erven F. Bohn, 1971.

Link, Bruce G., and Jo C. Phelan. "Conceptualizing stigma." *Annual Review of Sociology* 27 (2001): 363–385.

Lippman, Abby. "Prenatal genetic testing and screening: Constructing needs and reinforcing inequities." *American Journal of Law and Medicine* 17 (1991): 15–50.

Lippman, Abby. "Led (astray) by genetic maps: The cartography of the human genome and health care." *Social Science & Medicine* 35 (1992): 1469–1476.

Lippman, Abby. "Prenatal genetic testing and geneticization: Mother matters for all." *Fetal Diagnosis and Therapy* 1 (1993): 175–188.

Lombardo, Paul A. "Tracking chromosomes, castrating dwarves: Uninformed consent and eugenic research." *Ethics & Medicine* 25, no. 3 (2009): 149–164.

Lorey, Isabell. *Die Regierung der Prekären* [The government of the precarious]. Turia + Kant: Berlin/Vienna, 2012.

Lott, Maxim. "Americans warming to socialism over capitalism, polls show." *Fox News,* January 4, 2019. https://www.foxnews.com/politics/americans-warming-to -socialism-over-capitalism-polls-show.

Lunt, Neil, and Percivil Carrera. "Medical tourism: Assessing the evidence on treatment abroad." *Maturitas* 66 (2010): 27–32.

Lunt, Neil, and Percivil Carrera. "Systematic review of web sites for prospective medical tourists." *Tourism Review* 66 (2011): 57–67.

Lunt, Neil, Richard Smith, Mark Exworthy, Stephen T. Green, Daniel Horsfall, and Russell Mannion. *Medical tourism: Treatments, markets and health system implications: A scoping review.* Paris: OECD, Directorate for Employment, Labour and Social Affairs, 2011.

Macedo, Ana, Magi Farré, and Josep-E. Baños. "Placebo effect and placebos: What are we talking about? Some conceptual and historical considerations." *European Journal of Clinical Pharmacology* 59 (2003): 337–342.

Mackey, Tim K., Bryan A. Liang, Peter York, and Thomas Kubic. "Counterfeit drug penetration into global legitimate medicine supply chains: A global assessment." *American Journal of Tropical Medicine and Hygiene* 92, no. S6 (2015): 59–67.

Macklin, Ruth. "The ethical problems with sham surgery in clinical research." *New England Journal of Medicine* 341 (1999): 992–996.

Malone, Ruth E. "Policy as product: Morality and metaphor in health policy discourse." *Hastings Center Report* 29, no. 3 (1999): 16–22.

Mansfield, Christopher J., Jim Mitchell, and Dana E. King. "The doctor as God's mechanic? Beliefs in the Southeastern United States." *Social Science & Medicine* 54 (2002): 399–409.

Marchione, Marilynn. "Chinese researcher claims first gene-edited babies." *AP News*, November 26, 2018. https://www.apnews.com/4997bb7aa36c45449b488e19 ac83e86d.

Marcus, Steven. "Frankenstein: Myths of scientific and medical knowledge and stories of human relations." *Southern Review* 38, no. 1 (2002): 188–201.

Marks, Peter, and Scott Gottlieb. "Balancing safety and innovation for cell-based regenerative medicine." *New England Journal of Medicine* 378 (2018): 954–959.

Marmot, Michael. "Just societies, health equity, and dignified lives: The PAHO Equity Commission." *Lancet* 392 (2018): 2247–2250.

Marmot, Michael, Jessica Allen, Peter Goldblatt, Tammy Boyce, Di McNeish, Mike Grady, and Ilaria Geddes. *Fair society, healthy lives: The Marmot Review.* N.p.: Strategic Review of Health Inequalities in England Post-2010, 2010. http://www .instituteofhealthequity.org/resources-reports/fair-society-healthy-lives-the -marmot-review/fair-society-healthy-lives-full-report-pdf.pdf.

Marshall, Mary Faith. "ASBH and moral tolerance." In *The ethics of bioethics: Mapping the moral landscape*, ed. Lisa A. Eckenwiler and Felicia G. Cohen, 134–143. Baltimore: Johns Hopkins University Press, 2007.

Marteau, Theresa M., David P. French, Simon J. Griffin, A. T. Prevost, Stephen Sutton, Clare Watkinson, Sophie Attwood, and Gareth J. Hollands. "Effects of communicating DNA-based disease risk estimates on risk-reducing behaviours. *Cochrane Database of Systematic Reviews* 10 (2010): CD007275. https://doi.org/10.1002 /14651858. CD007275.pub2.

Martin, Dominique. "Perilous voyages: Travel abroad for organ transplants and stem cell treatments." In *Risks and challenges in medical tourism: Understanding the global market for health services*, edited by Jill R. Hodges, Leigh Turner, and Ann Marie Kimball, 138–166. Santa Barbara, CA: Praeger, 2012.

Marx, Patricia. "About face: Why is South Korea the world's plastic-surgery capital?" *New Yorker*, March 23 (2015): 50–55.

Mason, Alicia, and Kevin B. Wright. "Framing medical tourism: An examination of appeal, risk, convalescence, accreditation, and interactivity in medical tourism web sites." *Journal of Health Communication* 16 (2011): 163–177.

Mason, Chris, and Peter Dunnill. "A brief definition of regenerative medicine." *Regenerative Medicine* 3, no. 1 (2008): 1–5.

Master, Zubin, Erin Nelson, Blake Murdoch, and Timothy Caulfield. "Biobanks, consent and claims of consensus." *Nature Methods* 9, no. 9 (2012): 885–888.

Matheson, Alastair. "The disposable author: How pharmaceutical marketing is embraced with medicine's scholarly literature." *Hastings Center Report* 46 (2016): 31–37.

Matorras, Roberto. "Reproductive exile versus reproductive tourism." *Human Reproduction* 20 (2005): 3571.

Matthews, David. "Essay mills: University course work to order." *Times Higher Education*, October 10, 2013. https://www.timeshighereducation.com/features/essay -mills-university-course-work-to-order/2007934.article.

Matthews-King, Alex. "Vulnerable would-be parents being sold 'false hopes' by overseas IVF clinics, regulator warns." *Independent*, April 13, 2018. https://www .independent.co.uk/news/health/ivf-clinics-infertility-parents-pregnancy-embryo -cyprus-uk-prague-spain-czech-republic-a8303721.html.

Maughan, Tim. "The promise and the hype of 'personalized medicine.'" *New Bioethics* 23, no. 1 (2017): 13–20.

May, Ashley. "How many people believe in ghosts or dead spirits?" *USA Today*, October 25, 2017. https://www.usatoday.com/story/news/nation-now/2017/10/25/how -many-people-believe-ghosts-dead-spirits/794215001.

May, Peter. "Claimed contemporary miracles." *Medico-Legal Journal* 71 (2003): 144–158.

May, Todd. *A fragile life: Accepting our vulnerability*. Chicago: University of Chicago Press, 2017.

McAfee, Kathleen. "Neoliberalism on the molecular scale: Economic and genetic reductionism in biotechnology battles." *Geoforum* 34 (2003): 203–219.

McBride, Colleen, Laura M. Koehly, Saskia C. Sanderson, and Kimberly A. Kaphingst. "The behavioural response to personalised genetic information: Will genetic risk profiles motivate individuals and families to choose more healthful behaviors?" *Annual Review of Public Health* 31 (2010): 89–103.

McCoy, Matthew S., Michael Carniol, Katherine Chockley, John W. Urwin, Ezekiel J. Emanuel, and Harald Schmidt. "Conflicts of interest for patient-advocacy organizations." *New England Journal of Medicine* 376 (2017): 880–885.

McElwain, Gregory S. *Mary Midgley: An introduction*. London: Bloomsbury Academic, 2020.

McKee, Amy E., André O. Markon, Kirk M. Chan-Tack, and Peter Lurie. "How often are drugs made available under the Food and Drug Administration's expanded access process approved?" *Journal of Clinical Pharmacology* 57 (2017): S136–S142.

McKeown, Thomas. *The role of medicine: Dream, mirage, or nemesis?* London: Nuffield Provincial Hospitals Trust, 1976.

McNally, Ruth. "Eugenics here and now." In *Genetic imaginations: Ethical, legal and social issues in human genome research*, edited by Peter Glasner and Harry Rothman, 69–82. Aldershot: Ashgate, 1998.

Medina, Edwin, Elvira Bel, and Josep Maria Suñé. "Counterfeit medicines in Peru: A retrospective review (1997–2014)." *BMJ Open* 6 (2016): e010387. https://doi.org /10.1136/bmjopen-2015-010387.

Meghani, Zahra. "The ethics of medical tourism: From the United Kingdom to India seeking medical care." *International Journal of Health Services* 43, no. 4 (2013): 779–800.

Meghani, Zahra. "A robust, particularist ethical assessment of medical tourism." *Developing World Bioethics* 11, no. 1 (2011): 16–29.

Meier, Alison. "Skeletons on the shelves: Museums' slow return of indigenous remains." *Hyperallergic*, October 12, 2015. https://hyperallergic.com/230366/skeletons-on -the-shelves-museums-slow-return-of-indigenous-remains.

Meilaender, Gilbert. *Should we live forever? The ethical ambiguities of aging*. Grand Rapids, MI: William B. Eerdmans Publishing, 2013.

Melas, Philippe A., Louise K. Sjöholm, Tord Forsner, Maigun Edhborg, Niklas Juth, Yvonne Forsell, and Catharina Lavebratt. "Examining the public refusal to consent to DNA biobanking: Empirical data from a Swedish population-based study." *Journal of Medical Ethics* 36, no. 2 (2010): 93–98.

Memon, Aamir Raoof. "Predatory journals spamming for publications: What should researchers do?" *Science and Engineering Ethics* 24, no. 2 (2017): 1617–1639. https:// doi.org/10.1007/s11948-017-9955-6.

Merab, Elizabeth. "Do you really have to go to India? Set of new rules to curb medical pilgrimages." *Daily Nation*, February 20, 2017. Updated on June 29, 2020. https://nation.africa/kenya/healthy-nation/do-you-really-have-to-go-to-india-set-of-new-rules-to-curb-medical-pilgrimages-363468.

Meryman, Harold T. "Basile J. Luyet: In memoriam." *Cryobiology* 12 (1975): 285–292.

Midgley, David, ed. *The essential Mary Midgley*. London and New York: Routledge, 2005.

Midgley, Mary. *Are you an illusion?* London and New York: Routledge, 2014.

Midgley, Mary. "Biotechnology and monstrosity. Why we should pay attention to the 'Yuk Factor.'" *Hastings Center Report* 30, no. 5 (2000): 7–15.

Midgley, Mary. *Heart and mind: The varieties of moral experience*. Rev. ed. London: Routledge, 2003. First published 1981 by Harvester Press.

Midgley, Mary. *The myths we live by*. London and New York: Routledge, 2011.

Midgley, Mary. *Science as salvation: A modern myth and its meaning*. London and New York: Routledge, 1992.

Midgley, Mary. *What is philosophy for?* London: Bloomsbury Academic, 2018.

Mill, John Stuart. *Utilitarianism*. 15th ed. New York: Longmans, Green, 1907.

Miller, Fiona Alice. "Dermatoglyphics and the persistence of 'Mongolism.'" *Social Studies of Science* 33, no. 1 (2003): 75–94.

Miller, Fiona. "Geneticization: An interview with Abby Lippman on new genetics." Canadian Women's Health Network. Accessed on May 14, 2021. http://www.cwhn.ca/fr/node/39708.

Miller, Lantz Fleming. "The composite redesign of humanity's nature: a work in process." *Theoretical Medicine and Bioethics* 39 (2018): 157–164.

Miller, Paul Steven, and Rebecca Leah Levine. "Avoiding genetic genocide: Understanding good intentions and eugenics in the complex dialogue between the medical and disability communities." *Genetics in Medicine* 15, no. 2 (2013): 95–102.

Millns, Susan. "Dwarf-throwing and human dignity: A French perspective." *Journal of Social Welfare & Family Law* 18, no. 3 (1996): 375–380.

Mills, Catherine. "Continental philosophy and bioethics." *Bioethical Inquiry* 7 (2010): 145–148.

Milstein, Arnold, and Mark Smith. "America's new refugees—seeking affordable surgery offshore." *New England Journal of Medicine* 355 (2006): 1637–1640.

Minari, Jusaku, Kyle B. Brothers, and Michael Morrison. "Tensions in ethics and policy created by National Precision Medicine Programs." *Human Genomics* 12 (2018): 22. https://doi.org/10.1186/s40246-018-0151-9.

Mintzes, Barbara. "Direct to consumer advertising is medicalizing normal human experience." *British Medical Journal* 324 (2002): 908–909.

Moerman, Daniel E. "Deconstructing the placebo effect and finding the meaning response." *Annuals of Internal Medicine* 136 (2002): 471–476.

Mold, Alex. "Repositioning the patient: Patient organizations, consumerism, and autonomy in Britain during the 1960s and 1970s." *Bulleting of the History of Medicine* 87 (2013): 225–249.

Montgomery, Jonathan. "Law and the demoralization of medicine." *Legal Studies* 26, no. 2 (2006): 185–210.

Morgan, Derek, and Robert G. Lee. "In the name of the father? *Ex parte Blood*: Dealing with novelty and anomaly." *Modern Law Review* 60, no. 6 (1997): 840–856.

Morris, Lewis, and Julie K. Taitsman. "The agenda for continuing medical education—limiting industry's influence." *New England Journal of Medicine* 361 (2009): 2478–2482.

Moschella, Melissa. "The Charlie Gard case threatens all parents." *USA Today*, July 17, 2017. https://www.usatoday.com/story/opinion/2017/07/16/charlie-gard-parents-right-decide-melissa-moschella-column/478860001.

Mosely, J. Bruce, Kimberly O'Malley, Nancy J. Petersen, Terri J. Menke, Baruch A. Brody, David H. Kuykendall, John C. Hollingsworth, Carol M. Ashton, and Nelda P. Wray. "A controlled trial of arthroscopic surgery for osteoarthritis of the knee." *New England Journal of Medicine* 347 (2002): 81–88.

Moss, Ralph W. "Hype and Herceptin." *New Scientist* 189 (2006): 22.

Moffatt, Barton, and Carl Elliott. "Ghost marketing: Pharmaceutical companies and ghostwritten journal articles." *Perspectives in Biology and Medicine* 50, no. 1 (2007): 18–31.

Moreira, Tiago, and Paolo Palladino. "Between truth and hope: On Parkinson's disease and the production of the 'self.'" *History of the Human Sciences* 18, no. 3 (2005): 55–82.

Morozov, Evgeny. *To save everything, click here: The folly of technological solutionism.* New York: PublicAffairs, 2013.

Morton, Lisa. *Ghosts: A haunted history.* London: Reaktion Books, 2015.

Moustafa, Khaled. "The disaster of the impact factor." *Science and Engineering Ethics* 21 (2015): 139–142.

Moynihan, Ray. "Who pays for the pizza? Redefining the relationships between doctors and drug companies. 1: Entanglement." *British Medical Journal* 326 (2003): 189–192.

Moynihan, Ray, Lisa Bero, Dennis Ross-Degnan, David Henry, Kriby Lee, Judy Watkins, Connie Mah, and Stephen B. Soumerai. "Coverage by the news media of the benefits and risks of medications." *New England Journal of Medicine* 342 (2000): 1645–1650.

Moynihan, Ray, and Alan Cassels. *Selling sickness: How the world's biggest pharmaceutical companies are turning us all into patients.* Vancouver: Greystone Books, 2005.

Moynihan, Ray, Iona Heath, and David Henry. "Selling sickness: The pharmaceutical industry and disease mongering." *British Medical Journal* 324 (2002): 886–890.

Moynihan, Ray, and David Henry. "The fight against disease mongering: Generating knowledge for action." *PLoS Medicine* 3, no. 4 (2006): e191. https://doi.org/10.1371/journal.pmed.0030191.

Mukherjee, Siddhartha. *The gene: An intimate history.* New York: Scribner, 2016.

Mullin, Robert Bruce. *Miracles and the modern religious imagination.* New Haven: Yale University Press, 1996.

Murdoch, Charles E., and Christopher Thomas Scott. "Stem cell tourism and the power of hope." *American Journal of Bioethics* 10 (2010): 16–23.

Nagel, Alexander. "The afterlife of the reliquary." In *Treasures of heaven: Saints, relics, and devotion in Medieval Europe*, edited by Martina Bagnoli, Holger A. Klein, C. Griffith Mann, and James Robinson, 211–222. New Haven: Yale University Press, 2010.

Nagy, Peter, Ruth Wylie, Joey Eschrich, and Ed Finn. "The enduring influence of a dangerous narrative: How scientists can mitigate the Frankenstein myth." *Bioethical Inquiry* 15 (2018): 279–292.

Nagy, Peter, Ruth Wylie, Joey Eschrich, and Ed Finn. "Why Frankenstein is a stigma among scientists." *Science and Engineering Ethics* 24 (2018): 1143–1159.

Nather, David, and Sheila Kaplan. "Public wary of faster approvals of new drugs, STAT-Harvard poll finds." *STAT News*, May 11, 2016. https://www.statnews.com /2016/05/11/stat-harvard-poll-drug-approvals.

Navarro, Vincente. "Neoliberalism as a class ideology; or, The political causes of the growth of inequalities." *International Journal of Health Services* 37, no. 1 (2007): 47–62.

Nayyar, Gaurvika M. L., Amir Attaran, John P. Clark, M. Julia Culzoni, Facundo M. Fernandez, James E. Herrington, Megan Kendall, Paul N. Newton, and Joel G. Breman. "Responding to the pandemic of falsified medicines." *American Journal of Tropical Medicine and Hygiene* 92 (Suppl 6) (2015): 113–118.

Nebert, Daniel W. "Given the complexity of the human genome, can 'personalised medicine' or 'individualised drug therapy' ever be achieved?" *Human Genomics* 3, no. 4 (2009): 299–300.

Nekolaichuk, Cheryl L., Ronna F. Jevne, and Thomas O. Maguire." Structuring the meaning of hope in health and illness." *Social Science & Medicine* 48 (1999): 591–605.

Nelkin, Dorothy. "Molecular metaphors: The gene in popular discourse." *Nature Reviews Genetics* 2 (2001): 555–559.

Nelkin, Dorothy, and Lori Andrews. "*Homo economicus*: Commercialization of body tissue in the age of biotechnology." *Hastings Center Report* 28 (1998): 30–39.

Nelkin, Dorothy, and M. Susan Lindee. *The DNA mystique: The gene as a cultural icon.* New York: W. H. Freeman, 1995.

Nelkin, Dorothy, and Laurence Tancredi. *Dangerous diagnostics: The social power of biological information.* Chicago: University of Chicago Press, 1994.

Newton, Paul N., Abdinasir A. Amin, Chris Bird, Philip Passmore, Graham Dukes, Göran Tomson, Bright Simons, Roger Bate, Philippe J. Gueron, and Nicholas J. White. "The primacy of public health considerations in defining poor quality medicines." *PLoS Medicine* 8, no. 12 (2011): e1001139. https://doi.org/10.1371/journl .pmed.1001139.

Newton, Paul, and Brigitte Timmermann. "Fake penicillin, *The Third Man*, and operation Claptrap." *British Medical Journal* 355 (2016): i6494. https://doi.org /10.1136/bmj.i6494.

Nicholls, Peter. "Three ways the Charlie Gard case could affect future end-of-life cases globally." *Conversation*, July 24, 2017. http://theconversation.com/three-ways-the -charlie-gard-case-could-affect-future-end-of-life-cases-globally-81168.

Nicol, Dianne, Christine Critchley, Rebekah McWhirter, and Tess Whitton. "Under-standing public reactions to commercialization of biobanks and use of biobank resources." *Social Science & Medicine* 162 (2016): 79–87.

Nie, Jing-Bao, and Neil Pickering. "He Jiankui's genetic misadventure, Part 2: How different are Chinese and Western bioethics?" *Hastings Center*, December 13, 2018. https://www.thehastingscenter.org/jiankuis-genetic-misadventure-part-2-different -chinese-western-bioethics.

Nigro, Samuel A. "Bioethics is a monster." *Medical Sentinel* 7, no. 3 (2002): 69–70. http://www.haciendapub.com/medicalsentinel/bioethics-monster.

Nowotny, Helga, and Giuseppe Testa. *Naked genes: Reinventing the human in the molecular age.* Cambridge, MA: MIT Press, 2010.

Nuffield Council on Bioethics. *Human tissue: Ethical and legal issues*. London: Nuffield Council on Bioethics, 1995.

Oaten, Megan, Richard J. Stevenson, Mark A. Williams, Anina N. Rich, Marina Butko, and Trevor I. Case. "Moral violations and the experience of disgust and anger." *Frontiers in Behavioral Neuroscience* 12 (2018): 179. https://doi.org/10.3389/fnbeh.2018.00179.

O'Connell, Mark. *To be a machine: Adventurers among cyborgs, utopians, hackers, and the futurists solving the modern problem of death*. New York: Doubleday, 2017.

Ogbogu, Ubaka, Maeghan Toews, Adam Ollenberger, Pascal Borry, Helene Nobile, Manuala Bergmann, and Timothy Caulfield. "Newspaper coverage of biobanks." *PeerJ* 2 (2014): e500. https://doi.org/10.7717/peerj.500.

Okike, Kanu, Mininder S. Kocher, Erin X. Wei, Charles T. Mehlman, and Mohit Bhandari. "Accuracy of conflict-of-interest disclosures reported by physicians." *New England Journal of Medicine* 361 (2009): 1466–1474.

O'Leary, Diane. "Why bioethics should be concerned with medically unexplained symptoms." *American Journal of Bioethics* 18 (2018): 6–15.

Olshansky, S. Jay. "From lifespan to healthspan." *Journal of the American Medical Association* 320, no. 13 (2018): 1323–1324.

Olsen, Darcy. *The right to try: How the federal government prevents Americans from getting the lifesaving treatments they need*. New York: HarperCollins, 2015.

Orchard-Webb, David. "10 largest biobanks in the world." May 28, 2018. https://www.biobanking.com/10-largest-biobanks-in-the-world.

Ornstein, Charles, and Katie Thomas. "Prominent doctors aren't disclosing their industry ties in medical journal studies. And journals are doing little to enforce their rules." *ProPublica*, December 8, 2018. https://www.propublica.org/article/prominent-doctors-industry-ties-disclosures-medical-journal-studies.

Orr, Robert D. "Responding to patient beliefs in miracles." *Southern Medical Journal* 100 (2007): 1263–1267.

O'Sullivan, Brian, P., David M. Orenstein, and Carlos E. Milla. "Pricing for orphan drugs: Will the market bear what society cannot?" *Journal of the American Medical Association* 310, no. 13 (2013): 1343–1344.

Ossola, Alexandra. "The fake drug industry is exploding, and we can't do anything about it," *Newsweek*, September 17, 2015. http://www.newsweek.com/2015/09/25/fake-drug-industry-exploding-and-we-cant-do-anything-about-it-373088.html.

Oster, Emily, Ira Shoulson, and E. Ray Dorsey. "Optimal expectations and limited medical testing: Evidence from Huntington disease." *American Economic Review* 103, no. 2 (2013): 804–830.

Ostry, Jonathan D., Prakash Loungani, and Davide Furceri. "Neoliberalism: Oversold?" *Finance & Development* 53, no. 2 (2016): 38–41.

Otlowski, Margaret, Sandra Taylor, and Yvonne Bombard. "Genetic discrimination: International perspectives." *Annual Review of Genomics and Human Genetics* 13 (2012): 433–454.

Overholtzer, Lisa, and Juan R. Argueta. "Letting skeletons out of the closet: The ethics of displaying ancient Mexican human remains." *International Journal of Heritage Studies* 24, no. 5 (2018): 508–530.

Owens, Susan. *The ghost: A cultural history*. London: Tate, 2017.

Pace, Jessica, Narcyz Ghinea, Ian Kerridge, and Wendy Lipworth. "Accelerated access to medicines: An ethical analysis." *Therapeutic Innovation & Regulatory Science* 51 (2017): 157–163.

Papakostas, Yiannis G., and Michael D. Daras. "Placebos, placebo effect, and the response to the healing situation: The evolution of a concept." *Epilepsia* 42 (2001): 1614–1625.

Paré, Ambroise. *Des monstres et prodigies* [Monsters and prodigies]. Paris: Éditions L'Œil d'Or, 2003.

Parens, Eric, and Josephine Johnston. "Facts, values, and Attention-Deficit Hyperactivity Disorder (ADHD): An update on the controversies." *Child and Adolescent Psychiatry and Mental Health* 3 (2009): 1. https://doi.org/10.1186/1753-2000-3-1.

Park, Alice. "Ten ideas changing the world right now: Biobanks." *Time*, March 12, 2009. http://content.time.com/time/specials/packages/article/0,28804,1884779 _1884782_1884766,00.html.

Parry, Bronwyn. "Technologies of immortality: The brain on ice." *Studies in History and Philosophy of Biological and Biomedical Sciences* 35 (2004): 391–413.

Partridge, Ernest. "Posthumous interests and posthumous respect." *Ethics* 91, no. 2 (1981): 243–264.

Patterson, James T. *The dread disease: Cancer and modern American culture*. Cambridge, MA: Harvard University Press, 1987.

Pavone, Vincenzo, and Flor Arias. "Beyond the geneticization thesis: Political economy of PGD/PGS in Spain." *Science, Technology, & Human Values* 37, no. 3 (2012): 235–261.

Payer, Lynn. *Disease-mongers: How doctors, drug companies, and insurers are making you feel sick*. New York: John Wiley & Sons, 1992.

Pease, Alison M., Harlan M. Krumholz, Nicholas S. Downing, Jenerius A. Aminawung, Nilay D. Shah, and Joseph S. Ross. "Postapproval studies of drugs initially approved by the FDA on the basis of limited evidence: Systematic review." *British Medical Journal* 357 (2017): j1680.

Pellegrino, Edmund D. "Bioethics as an interdisciplinary enterprise: Where does ethics fit in the mosaic of disciplines?" In *Philosophy of medicine and bioethics: A twenty-year retrospective and critical appraisal*, edited by Ronald A. Carson and Chester R. Burns, 1–23. Dordrecht: Kluwer Academic, 1997.

Pennings, Guido. "Ethics of medical tourism." In *Handbook on medical tourism and patient mobility*, edited by Neil Lunt, Daniel Horsfall, and Johanna Hanefeld, 341–349. Cheltenham, UK: Edward Elgar, 2015.

Perls, Thomas, and David J. Handelsman. "Disease mongering of age-associated declines in testosterone and growth hormone levels." *Journal of the American Geriatrics Society* 63 (2015): 809–811.

Peschel, Richard E., and Enid Rhodes Peschel. "Medical miracles from a physician-scientist's viewpoint." *Perspectives in Biology and Medicine* 31, no. 3 (1988): 391–404.

Petersen, Alan. "Biofantasies: Genetics and medicine in the print news media." *Social Science & Medicine* 52 (2001): 1255–1268.

Petersen, Alan. "The genetic conception of health: Is it as radical as claimed?" *Health* 10 (2006): 481–500.

Petersen, Alan. "Is the new genetics eugenic? Interpreting the past, envisioning the future." *New Formations* 60 (2007): 79–88.

Petersen, Alan, Casimir MacGregor and Megan Munsie. "Stem cell miracles or Russian roulette? Patients' use of digital media to campaign for access to clinically unproven treatments." *Health, Risk & Society* 17 (2015): 592–604.

Petersen, Alan, Megan Munsie, Claire Tanner, Casimir MacGregor, and Jane Brophy. *Stem cell tourism and the political economy of hope.* London: Palgrave Macmillan, 2017.

Petersen, Alan, and Kate Seear. "Technologies of hope: Techniques of the online advertising of stem cell treatments." *New Genetics and Society* 30 (2011): 329–346.

Petersen, Alan, Kate Seear, and Megan Munsie. "Therapeutic journeys: The hopeful travails of stem cell tourists." *Sociology of Health & Illness* 36, no. 5 (2013): 670–685.

Petersen, Alan, Claire Tanner and Megan Munsie. "Between hope and evidence: How community advisors demarcate the boundary between legitimate and illegitimate stem cell treatments." *Health* 19 (2015): 188–206.

Petersen, David L. *The prophetic literature: An introduction.* Louisville: Westminster John Know Press, 2002.

Pettit, Harry. "First human frozen by cryogenics could be brought back to life 'in just TEN years,' claims expert." *Daily Mail,* January 15, 2018. http://www.dailymail.co .uk/sciencetech/article-5270257/Cryogenics-corpses-brought-10-years.html.

Piccart-Gebhart, Martine J., Marion Procter, Brian Leyland-Jones, Aron Glodhirsch, Michael Untch, Ian Smith, Luca Gianni, et al. "Trastuzumab after adjuvant chemo-therapy in HER2-positive breast cancer." *New England Journal of Medicine* 353 (2005): 1659–1672.

Popper, Karl R. *The poverty of historicism.* New York: Harper & Row, 1964.

Porter, Roy, ed. *The Cambridge history of medicine.* Cambridge: Cambridge University Press, 2006.

Poushter, Jacob. "Worldwide, people divided on whether life today is better than in the past." *Pew Research Center,* December 5, 2017. http://www.pewglobal.org/2017/12 /05/worldwide-people-divided-on-whether-life-today-is-better-than-in-the-past.

Powers, Madison, and Ruth Faden. *Social justice: The moral foundations of public health and health policy.* Oxford: Oxford University Press, 2006.

Psaty, Bruce M., and Richard A. Kronmal. "Reporting mortality findings in trials of rofecoxib for Alzheimer disease or cognitive impairment: A case study based on documents from rofecoxib litigation." *JAMA* 299 (2008): 1813–1817.

Qiu, Jane. "Trading on hope." *Nature Biotechnology* 27 (2009): 790–792.

Quinones, Julian, and Arijeta Lajka. "What kind of society do you want to live in? Inside the country where Down syndrome is disappearing." *CBS News,* August 14, 2017. https://www.cbsnews.com/news/down-syndrome-iceland.

Raad, Raymond, and Paul S. Appelbaum. "Relationships between medicine and industry: Approaches to the problem of conflicts of interest." *Annual Review of Medicine* 63 (2012): 465–477.

Radin, Joanna. *Life on ice: A history of new uses for cold blood.* Chicago: University of Chicago Press, 2017.

Radkowska-Walkowicz, Magdalena. "The creation of 'monsters': The discourse of opposition to in vitro fertilization in Poland." *Reproductive Health Matters* 20 (2012): 30–37.

Rajan, Kaushik Sunder. *Biocapital: The constitution of postgenomic life.* Durham, NC: Duke University Press, 2006.

Reader, Ian. *Pilgrimage: A very short introduction*. Oxford: Oxford University Press, 2015.

Reader, Ian. *Pilgrimage in the marketplace*. New York: Routledge, 2014.

Rebiere, Hervé, Pauline Guinot, Denis Chauvey, and Charlotte Brenier. "Fighting falsified medicines: The analytical approach." *Journal of Pharmaceutical and Biomedical Analysis* 142 (2017): 286–306.

Regensberg, Alan C., Lauren A. Hutchinson, Benjamin Schranker, and Debra J. H. Mathews. "Medicine on the fringe: Stem cell-based interventions in advance of evidence." *Stem Cells* 27 (2009): 2312–2319.

Reich, Robert B. *The common good*. New York: Alfred A. Knopf, 2018.

Reich, Warren. "Bioethics in the United States." In *Bioethics: A history*, edited by Corrado Viafora, 83–118. San Francisco, CA: International Scholars Publications, 1996.

Reilly, Claire. "Frozen in time: Inside the facility preserving the dead through cryonics." *CNet*, July 6, 2020. https://www.cnet.com/pictures/frozen-in-time-inside-alcor-life -extension-the-facility-preserving-the-dead-through-cryonics.

Relman, Arnold S. "A coming medical revolution?" *New York Review of Books*, October 25, 2012: 40–42.

Relman, Arnold S. "Industry support of medical education." *JAMA* 300 (2008): 1071–1073.

Resnik, David B. "Scientific research and the public trust." *Science and Engineering Ethics* 17 (2011): 399–409.

Resnik, David B., Shyamal Peddada, and Winnon Brunson. "Research misconduct policies of scientific journals." *Accountability in Research* 16 (2009): 254–267.

Rich, Ben A. "Prognostication in clinical medicine: Prophecy or professional responsibility." *Journal of Legal Medicine* 23 (2002): 297–358.

Riffkin, Rebecca. "In US, 67% dissatisfied with income, wealth distribution." *Gallup News*, January 20, 2014.

Rimmer, Abi. "Charlie Gard's parents end legal fight to keep son alive." *British Medical Journal* 358 (2017): j3589. https://doi.org/10.1136/bmj.j3589.

Rizzi, Dominic A. "Medical prognosis—some fundamentals." *Theoretical Medicine* 14 (1993): 365–375.

Roberts, Nicholas J., Joshua T Vogelstein, Giovanni Parmigiani, Kenneth W Kinzler, Bert Vogelstein, and Victor E Velculescu. "The predictive capacity of personal genome sequencing." *Science Translational Medicine* 4, no. 133 (2012): 133RA58.

Rollin, Bernard E. *The Frankenstein syndrome: Ethical and social issues in the genetic engineering of animals*. Cambridge: Cambridge University Press, 1995.

Rose, Hilary. "Eugenics and genetics: the conjoint twins?" *New Formations* 60 (2007): 13–26.

Rose, Hilary, and Steven Rose. *Genes, cells and brains: The Promethean promises of the new biology*. London: Verso, 2014.

Rose, Nikolas. "Personalized medicine: Promises, problems and perils of a new paradigm for healthcare." *Procedia: Social and Behavioral Sciences* 77 (2013): 341–352.

Ross, Casey, and Ike Swetlitz. "IBM pitches its Watson supercomputer as a revolution in cancer care. It's nowhere close." *STAT News*, September 5, 2017. https://www .statnews.com/2017/09/05/watson-ibm-cancer.

Ross, Joseph S., Kevin P. Hill, David S. Egilman, and Harlan M. Krumholz. "Guest authorship and ghostwriting in publications related to rofecoxib: A case study of industry documents from rofecoxib litigation." *JAMA* 299 (2008): 1800–1812.

Roth, Alvin E. "Repugnance as a constraint on markets." *Journal of Economic Perspectives* 21, no. 3 (2007): 37–58.

Rothman, Barbara Katz. *Genetic maps and human imaginations: The limits of science in understanding who we are.* New York: W. W. Norton, 1998.

Rothman, David J. *Strangers at the bedside: A history of how law and bioethics transformed medical decision making.* New York: Basic Books, 1991.

Rouvroy, A. *Human genes and neoliberal governance: A Foucauldian critique.* New York: Routledge-Cavendish, 2008.

Rovisco, Maria. "The indignados social movement and the image of the occupied square: The making of a global icon." *Visual Communication* 16, no. 3 (2017): 337–359.

Royal Courts of Justice. Case no. FD17P00103. July 24, 2017. https://www.judiciary.uk/wp-content/uploads/2017/07/gosh-v-gard-24072017.pdf

The Royal Liverpool Children's Inquiry: Report. London: Stationery Office, 2001. https://assets.publishing.service.gov.uk/government/uploads/system/uploads/attachment_data/file/250934/0012_ii.pdf.

Rushton, Cindy Hylton, and Kathleen Russell. "The language of miracles: Ethical challenges." *Pediatric Nursing* 22 (1996): 64–67.

Russell, Pascale Sophie, and Roger Giner-Sorolla. "Bodily moral disgust: What it is, how it is different from anger, and why it is an unreasoned emotion." *Psychological Bulletin* 139, no. 2 (2013): 328–351.

Russo, Enzo, and David Cove. *Genetic engineering: Dreams and nightmares.* Oxford: Oxford University Press, 1995.

Ryan, Kirsten A., Amanda N. Sanders, Dong D. Wang, and Aaron D. Levine. "Tracking the rise of stem cell tourism." *Regenerative Medicine* 5, no. 1 (2010): 27–33.

Ryan, Maura Anne. "The new reproductive technologies: Defying God's dominion?" *Journal of Medicine and Philosophy* 20, no. 4 (1995): 419–438.

Sabaté, Eduardo. *Adherence to long-term therapies: Policy for action.* Geneva: World Health Organization, 2001. http://www.who.int/chp/knowledge/publications/adherencerep.pdf.

Sadler, John Z., Fabrice Jotterand, Simon Craddock Lee, and Stephen Inrig. "Can medicalization be good? Situating medicalization within bioethics." *Theoretical Medicine and Bioethics* 30 (2009): 411–425.

Saluja, Sonali, Steffie Woolhandler, David U. Himmelstein, David Bor, and Danny McCormick. "Unsafe drugs were prescribed more than one hundred million times in the United States before being recalled." *International Journal of Health Services* 46 (2016): 523–530.

Sample, Ian. "'Sonick attack' on US embassy in Havana could have been crickets, say scientists." *Guardian*, January 6, 2019. https://www.theguardian.com/world/2019/jan/06/sonic-attack-on-us-embassy-in-havana-could-have-been-crickets-say-scientists.

Sampson, Fiona. *In search of Mary Shelley.* New York: Pegasus Books, 2018.

Sandefur, Christina. "Safeguarding the right to try." *Arizona State Law Journal* 49 (2017): 513–536.

Sandel, Michael J. *The case against perfection: Ethics in the age of genetic engineering.* Cambridge, MA: Belknap Press, 2009.

Sandler, Ronald L. *Food ethics: The basics.* London and New York: Routledge, 2015.

Sattar, Maher. "Bangladesh faces growing trade in fake drugs." *Al Jazeera*, June 24, 2014. https://www.aljazeera.com/news/2014/6/24/bangladesh-faces-growing -trade-in-fake-drugs.

Savard, Jacqueline. "Personalised medicine: A critique on the future of health care." *Bioethical Inquiry* 10 (2010): 197–203.

Savulescu, Julian. "Bioethics: Why philosophy is essential for progress." *Journal of Medical Ethics* 41 (2015): 28–33.

Savulescu, Julian. "Is it in Charlie Gard's best interest to die?" *Lancet* 389 (2017): 1868–1869.

Schleidgen, Sebastian, Corinna Klinger, Teresa Bertram, Wolf H. Rogowski, and Georg Marckmann. "What is personalized medicine: Sharpening a vague term based on a systematic literature review." *BMC Medical Ethics* 14 (2013): 55. https://doi.org /10.1186/1472-6939-14-55.

Schneider, Carl E., and Mark A. Hall. "The patient life: Can consumers direct health care?" *American Journal of Law & Medicine* 35 (2009): 7–65.

Schneiderman, Lawrence J. *Embracing our mortality: Hard choices in an age of medical miracles.* Oxford: Oxford University Press, 2008.

Schröder, Cecilie Piene, Øystein Skare, Olav Reikerås, Peter Mowinckel, and Jens Ivar Brox. "Sham surgery versus labral repair or biceps tenodesis for type II SLAP lesions of the shoulder: A three-armed randomized clinical trial." *British Journal of Sports Medicine* 51 (2017): 1759–1766. https://bjsm.bmj.com/content/51/24/1759.

Schwartz, William B. *Life without disease: The pursuit of medical utopia.* Berkeley: University of California Press, 1998.

Schwarz, Alan. "The selling of Attention Deficit Disorder." *New York Times*, December 14, 2013. http://www.nytimes.com/2013/12/15/health/the-selling-of-attention -deficit-disorder.html.

Schwitzer, Gary. "Addressing tensions when popular media and evidence-based care collide." *BMC Medical Informatics and Decision Making* 13, no. S3 (2013). https:// doi.org/10.1186/1472-6947-13-S3-S3.

Schwitzer, Gary. "How do US journalists cover treatments, tests, products, and pro- cedures? An evaluation of 500 stories." *PLoS Medicine* 5, no. 5 (2008): e95. https:// doi.org/10.1371/journal.pmed.0050095.

Schwitzer, Gary. "Trying to drink from the fire hose: Too much of the wrong kind of health care news." *Trends in Pharmacological Sciences* 36 (2015): 623–627.

Seedhouse, David. *Ethics: The heart of health care.* Chichester, UK: John Wiley & Sons, 1988.

Seedhouse, David, and Lisetta Lovett. *Practical medical ethics.* Chichester, UK: John Wiley & Sons, 1992.

Seely, Andrew J. E. "Embracing the certainty of uncertainty: Implications for health care and research." *Perspectives in Biology and Medicine* 56, no. 1 (2013): 65–77.

Segal, Adam. "When China rules the web: Technology in the service of the state." *Foreign Affairs* 97, no. 5 (2018): 10–18.

Sexton, John. "A reductionist history of humankind," *New Atlantis* 47 (2015): 109–120.

Shah, Seema K., Abby R. Rosenberg, and Douglas S. Diekema. "Charlie Gard and the limits of best interests." *JAMA Pediatrics* 171, no. 10 (2017): 937–938.

Shamseer, Larissa, David Moher, Onyi Maduekwe, Lucy Turner, Virginia Barbour, Rebecca Burch, Jocalyn Clark, James Galipeau, Jason Roberts, and Beverley J. Shea. "Potential predatory and legitimate biomedical journals: Can you tell the difference? A cross-sectional comparison." *BMC Medicine* 15 (2017): 28. https://doi.org/10.1186/s12916-017-0785-9.

Shankar, P. Ravi, and Palaian Subish. "Disease mongering." *Singapore Medical Journal* 48 (2007): 275–280.

Shannon, Thomas A. "Bioethics and religion: A value-added discussion." In *Notes from a narrow ridge: Religion and bioethics*, edited by Dena S. Davis and Laurie Zoloth, 129–150. Hagerstown, MD: University Publishing Group, 1999.

Shapiro, Beth. *How to clone a mammoth: The science of de-extinction*. Princeton, NJ: Princeton University Press, 2015.

Shelley, Percy Bysshe. "A defence of poetry." In *Shelley's Prose; or, The trumpet of a prophecy*, edited by David Lee Clark, 275–297. Albuquerque: University of New Mexico Press, 1954.

Shen, Cenyu, and Bo-Christer Björk. "'Predatory' open access: A longitudinal study of article volumes and market characteristics." *BMC Medicine* 13 (2015): 230. https://doi.org/10.1186/s12916-015-0469-2.

Sherkow, Jacob S., and Henry T. Greely. "The history of patenting genetic material." *Annual Review of Genetics* 49 (2015): 161–182.

Sherringham, Tia. "Mice, men, and monsters: Opposition to chimera research and the scope of federal regulation." *California Law Review* 96, no. 3 (2008): 765–800.

Sherwin, Susan. "Foundations, frameworks, lenses: The role of theories in bioethics." *Bioethics* 13, no. 3–4 (1999): 198–205.

Shildrick, Margrit. *Embodying the monster: Encounters with the vulnerable self*. London: SAGE, 2002.

Shryock, Richard Harrison. *The development of modern medicine*. Madison: University of Wisconsin Press, 1974.

Shuman, Joel James. "Re-enchanting the body: Overcoming the melancholy of anatomy." *Theoretical Medicine and Bioethics* 39 (2018): 473–481.

Siedentop, Larry. *Inventing the individual: The origins of Western liberalism*, London: Allen Lane, 2014.

Silver, Lee M. *Remaking Eden: Cloning and beyond in a brave new world*. New York: Avon Books, 1997.

Silver, Lee M. *Remaking Eden: How genetic engineering and cloning will transform the American family*. New York: HarperCollins, 2007.

Silverstein, Jason. "North Korea's nuclear tests are spreading 'ghost disease' causing deformation, defectors say," *Newsweek*, December 12, 2017. http://www.newsweek.com/north-korea-nuclear-ghost-disease-729693.

Silvestri, Gerard A., Sommer Knittig, James S. Zoller, and Paul J. Nietert. "Importance of faith on medical decisions regarding cancer care." *Journal of Clinical Oncology* 21, no. 7 (2003): 1379–1382.

Singer, Charles. *A short history of scientific ideas to 1900*. Oxford: Oxford University Press, 1959.

Sismondo, Sergio. "Ghost management: How much of the medical literature is shaped behind the scenes by the pharmaceutical industry?" *PLoS Medicine* 4, no. 9 (2007): 1429–1433.

Sismondo, Sergio, and Mathieu Doucet. "Publication ethics and the ghost management of medical publication." *Bioethics* 24, no. 6 (2010): 273–283.

Skloot, Rebecca. *The immortal life of Henrietta Lacks.* New York: Broadway Books, 2010.

Slaughter, Richard A. "The IT revolution reassessed part one: Literature review and key issues." *Futures* 96 (2018): 115–123.

Sleasman, Michael. "Robots." In *Encyclopedia of global bioethics*, vol. 3, edited by Henk ten Have, 2579–2591. Switzerland: Springer Reference, 2016.

Smart, Andrew, Paul Martin, and Michael Parker. "Tailored medicine: Whom will it fit? The ethics of patient and disease stratification." *Bioethics* 18, no. 4 (2004): 322–343.

Smith, Kristen. "The problematization of medical tourism: A critique of neoliberalism." *Developing World Bioethics* 12, no. 1 (2012): 1–8.

Smith, Richard. "Medical journals are an extension of the marketing arm of pharmaceutical companies." *PLoS Medicine* 2 (2005): 364–366.

Smith, Wesley D. *The Hippocratic tradition.* Ithaca, NY: Cornell University Press, 1979.

Smith, Wesley J. *Culture of death: The assault on medical ethics in America.* San Francisco, CA: Encounter Books, 2000.

Söderfeldt, Ylva, Adam Droppe, and Tim Ohnhäuser. "Distress, disease, desire: Perspectives on the medicalization of premature ejaculation." *Journal of Medical Ethics* 43 (December 2017): 865–866. https://doi.org/10.1136/medethics-2015-103248.

Sokol, Daniel. "Charlie Gard case: An ethicist in the courtroom." *British Medical Journal* 358 (2017): j3451. https://doi.org/10.1136/bmj.j3451.

Solbakk, Jan Helge, Søren Holm, and Bjørn Hofmann, eds. *The ethics of research biobanking.* Dordrecht: Springer, 2009.

Solomon, Mildred Z., and Bruce Jennings. "Bioethics and populism: How should our field respond? *Hastings Center Report* 47, no. 2 (2017): 11–16.

Song, Priscilla. *Biomedical odysseys: Fetal cell experiments from cyberspace to China.* Princeton, NJ: Princeton University Press, 2017.

Song, Priscilla. "Biotech pilgrims and the transnational quest for stem cell cures." *Medical Anthropology* 29, no. 4 (2010): 384–402.

Sorokowski, Piotr, Emanuel Kulczycki, Agnieszka Sorokowska, and Katarzyna Pisanski. "Predatory journals recruit fake editor." *Nature* 543, no. 7646 (2017): 481–483.

Spellberg, Brad, and Bonnie Taylor-Blake. "On the exoneration of Dr. William H. Stewart: Debunking an urban legend." *Infectious Diseases of Poverty* 2 (2013): 3. https://doi.org/10.1186/2049-9957-2-3.

Spencer, Nick. *The evolution of the West: How Christianity has shaped our values.* London: SPCK, 2016.

Sperling, Daniel. "(Re)disclosing physician financial interests: Rebuilding trust or making unreasonable burdens on physicians?" *Medicine, Health Care and Philosophy* 20 (2017): 179–186.

Squier, Susan Merrill. *Liminal lives: Imagining the human at the frontiers of biomedicine.* Durham, NC: Duke University Press, 2004.

Sreenivasan, Shoba, and Linda E. Weinberger. "Do you believe in miracles?" *Psychology*

Today, December 15, 2017. https://www.psychologytoday.com/us/blog/emotional-nourishment/201712/do-you-believe-in-miracles.

Stahl, Devan, ed. *Imaging and imagining illness: Becoming whole in a broken body.* Eugene: Cascade Books, 2018.

Stahl, Devan. "Patient reflections on the disenchantment of techno-medicine." *Theoretical Medicine and Bioethics* 39 (2018): 499–513.

Stausberg, Michael. *Religion and tourism: Crossroads, destinations and encounters.* London and New York: Routledge, 2011.

Stegenga, Jacob. *Medical nihilism.* Oxford: Oxford University Press, 2018.

Stempsey, William E. "The geneticization of diagnostics." *Medicine, Health Care and Philosophy* 9 (2006): 193–200.

Stempsey, William E. "Miracles and the limits of medical knowledge." *Medicine, Health Care and Philosophy* 5 (2002): 1–9.

Stent, Gunther S. "That was the molecular biology that was." *Science* 160, no. 3826 (1968): 390–395.

Stevens, M. L. Tina. *Bioethics in America: Origins and cultural politics.* Baltimore: Johns Hopkins University Press, 2000.

Stout, Jeffrey, *Ethics after Babel: The language of morals and their discontents.* Boston: Beacon Press, 1988.

Strasser, Bruno J. "Collecting nature: Practices, styles, and narratives." *Osiris* 27, no. 1 (2012): 303–340.

Strasser, Bruno J. "The experimenter's museum: GenBank, natural history, and the moral economies of biomedicine." *Isis* 102, no. 1 (2011): 60–96.

Stuurman, Siep. *The invention of humanity: Equality and cultural difference in world history.* Cambridge, MA: Harvard University Press, 2017.

Sulmasy, Daniel P. "Spiritual issues in the care of dying patients." *Journal of the American Medical Association* 296 (2006): 1385–1392.

Sumption, Jonathan. *Pilgrimage: An image of medieval religion.* London: Faber and Faber, 1975.

Swinburne, Richard. *The concept of miracle.* London: Palgrave Macmillan, 1970.

Sytsma, Sharon. "The moral authority of symbolic appeals in biomedical ethics." *Cambridge Quarterly of Healthcare Ethics* 13 (2004): 292–301.

Szawarski, Zbigniew. "The stick, the eye, and ownership of the body." In *Ownership of the human body*, edited by H. A. M. J. ten Have and J. V. M. Welie, 81–96. Dordrecht: Kluwer Academic, 1998.

Tallis, Raymond. *Enemies of hope: A critique of contemporary pessimism.* New York: St. Martin's Press, 1997.

Tandy-Connor, Stephany, Jenna Guiltinan, Kate Krempely, Holly LaDuca, Patrick Reineke, Stephanie Gutierrez, Philip Gray, and Brigitte Tippin Davis. "False-positive results released by direct-to-consumer genetic tests highlight the importance of clinical confirmation testing for appropriate patient care." *Genetics in Medicine* 20 (2018): 1515–1521. https://doi.org/10.1038/gim.2018.38.

Tannock, Ian F., and John A. Hickman. "Limits to personalized cancer medicine." *New England Journal of Medicine* 375, no. 13 (2016): 1289–1294.

Taylor, Charles. *Sources of the self: The making of modern identity.* Cambridge: Cambridge University Press, 1989.

Taylor, Sandra, Susan Treloar, Kristine Barlow-Stewart, Mark Stranger, and Margaret Otlowski. "Investigating genetic discrimination in Australia: A large-scale survey of clinical genetics clients." *Clinical Genetics* 74 (2008): 20–30.

Tegmark, Max. *Life 3.0: Being human in the age of artificial intelligence.* New York: Vintage Books, 2017.

Temkin, Owsei. "History and prophecy: Meditations in a medical library." *Bulletin of the History of Medicine* 49 (1975): 305–317.

Ten Have, Henk "The anthropological tradition in the philosophy of medicine." *Theoretical Medicine* 16 (1995): 3–14.

Ten Have, Henk. "Genetic advances require comprehensive bioethical debate." *Croatian Medical Journal* 44, no. 5 (2003): 533–537.

Ten Have, Henk. "Genetics and culture: The geneticization thesis." *Medicine, Health Care and Philosophy* 4 (2001): 295–304.

Ten Have, Henk. *Global bioethics: An introduction.* London and New York: Routledge, 2016.

Ten Have, Henk. "God en gezondheid" [God and health]. In *God en de obsessies van de twintigste eeuw* [God and the obsessions of the twentieth century], edited by Frans Vosman, 133–148. Hilversum: Uitgeverij Gooi en Sticht, 1990.

Ten Have, Henk. "Foundationalism and principles." In *The SAGE handbook of health care ethics: Core and emerging issues,* edited by Ruth Chadwick, Henk ten Have, and Eric M. Meslin, 20–30. Los Angeles: SAGE, 2011.

Ten Have, Henk. "Images of man in philosophy of medicine." In *Critical reflection on medical ethics,* edited by Martyn Evans, 173–193. Stamford, CN: JAI Press, 1998.

Ten Have, Henk. *Jeremy Bentham: Een quantumtheorie van de ethiek* [Jeremy Bentham: A quantum theory of ethics]. Kampen: Kok Agora, 1986.

Ten Have, Henk. "Medical technology assessment and ethics: Ambivalent relations." *Hastings Center Report* 25, no. 5 (1995): 13–19.

Ten Have, Henk. "Principlism: A Western European appraisal." In *A matter of principles? Ferment in U.S. Bioethics,* edited by Edwin R. DuBose, Ronald P. Hamel, and Laurence J. O'Connell, 101–120. Valley Forge, PA: Trinity Press International, 1994.

Ten Have, Henk. "The right to know and the right not to know in the era of neoliberal biopolitics and bioeconomy. In *The right to know and the right not to know: Genetic privacy and responsibility,* edited by Ruth Chadwick, Mairi Levitt, and Darren Shickle, 133–150. Cambridge: Cambridge University Press, 2014.

Ten Have, Henk. *Vulnerability: Challenging bioethics.* London and New York: Routledge, 2016.

Ten Have, Henk. *Wounded planet: How declining biodiversity endangers health and how bioethics can help.* Baltimore: Johns Hopkins University Press, 2019.

Ten Have, Henk, and Bert Gordijn. "Publication ethics: Science versus commerce." *Medicine, Health Care and Philosophy* 20 (2017): 159–161.

Theos. "Polls reveal Briton's spiritual side." October 17, 2013. https://www.theosthink tank.co.uk/in-the-news/2013/10/17/poll-reveals-britons-spiritual-side.

't Hoen, Ellen, and Fernando Pascual. "Counterfeit medicines and substandard medicines: Different problems requiring different solutions." *Journal of Public Health Policy* 36 (2015): 384–389.

Thomas, Gareth M. "An elephant in the consultation room? Configuring Down syndrome in British antenatal care." *Medical Anthropology Quarterly* 30, no. 2 (2016): 238–258.

Thomas, Gareth M., and Joanna Latimer. "In/exclusion in the clinic: Down's syndrome, dysmorphology and the ethics of everyday medical work." *Sociology* 49, no. 5 (2015): 937–954.

Thomas, K. B. "General practice consultations: Is there any point in being positive?" *British Medical Journal* 294 (1987): 1200–1202.

Thomasma, David. "Training in medical ethics: An ethical workup." *Forum in Medicine* 1, no. 9 (1978): 33–36.

Thomassen, Bjorn. "The uses and meanings of liminality." *International Political Anthropology* 2, no. 1 (2009): 5–27.

Thompson, Rosemarie Garland, ed. *Freakery: Cultural spectacles of the extraordinary body*. New York: New York University Press, 1996.

Tiefer, Leonore. "Female sexual dysfunction: A case study of disease mongering and activist resistance." *PLoS Medicine* 3, no. 4 (2006): e178. https://doi.org/10.1371 /journal.pmed.0030178.

Tirole, Jean. *Economics for the common good*. Princeton, NJ: Princeton University Press, 2017.

Titmuss, Richard M. *The gift relationship: From human blood to social policy*. New York: Pantheon Books, 1971.

Titus, Sandra L., James A. Wells, and Lawrence J. Rhoades. "Repairing research integrity." *Nature* 453 (2008): 980–982.

Tomes, Nancy. *Remaking the American patient: How Madison Avenue and modern medicine turned patients into consumers*. Chapel Hill: University of North Carolina Press, 2016.

Topol, Eric. *The creative destruction of medicine: How the digital revolution will create better health care*. New York: Basic Books, 2012.

Topol, Eric. *The patient will see you now: The future of medicine is in your hands*. New York: Basic Books, 2015.

Torres, J. M. "Genetic tools, Kuhnian theoretical shift and the geneticization process." *Medicine, Health Care and Philosophy* 9 (2006): 3–12.

Tremblay, Michael. "Medicines counterfeiting is a complex problem: A review of key challenges across the supply chain." *Current Drug Safety* 8 (2013): 43–55.

Truog, Robert D. "The United Kingdom sets limits on experimental treatment: The case of Charlie Gard." *JAMA* 318, no. 11 (2017): 1001–1002.

Tungaraza, Tongeiji Elifazi, Pravija Talapan-Manikoth, and Rosemary Jenkins. "Curse of the ghost pills: The role of oral controlled-release formulations in the passage of empty intact shells in faeces. Two case reports and a literature review relevant to psychiatry." *Therapeutic Advances in Drug Safety* 4 (2013): 63–71.

Turner, Leigh. "Bioethics needs to rethink its agenda." *British Medical Journal* 328 (2004): 175.

Turner, Leigh. " 'First world health care at third world prices': Globalization, bioethics and medical tourism." *BioSocieties* 2 (2007): 303–325.

Turner, Leigh. "Transnational medical travel: Ethical dimensions of global healthcare." *Cambridge Quarterly of Healthcare Ethics* 22 (2013): 170–180.

Turner, Leigh. "US stem cell clinics, patient safety, and the FDA." *Trends in Molecular Medicine* 21 (2015): 271–273.

Turner, Leigh, and Paul Knoepfler. "Selling stem cells in the USA." *Cell Stem Cell* 19 (2016): 154–157.

Turner, Victor, and Edith Turner. *Image and pilgrimage in Christian culture*. New York: Columbia University Press, 2011. First published 1978.

Turney, Jon. *Frankenstein's footsteps: Science, genetics and popular culture*. New Haven: Yale University Press, 1998.

Tutton, Richard. "Personalizing medicine: Futures present and past." *Social Science & Medicine* 75 (2012): 1721–1728.

Tuzunkan, Demet. "Wellness tourism: What motivates tourists to participate?" *International Journal of Applied Engineering Research* 13, no. 1 (2018): 651–661.

Tybjerg, Karin. "From bottled babies to biobanks: Medical collections in the twenty-first century." In *The fate of anatomical collections*, edited by Rina Knoeff and Robert Zwijnenberg, 263–278. Farnham: Ashgate, 2015.

UNICEF. "Diarrhoea remains a leading killer of young children, despite the availability of a simple treatment solution." Accessed May 14, 2021. https://data.unicef.org/topic /child-health/diarrhoeal-disease.

United States Government Accountability Office (GAO). *Direct-to-consumer genetic tests: Misleading test results are further complicated by deceptive marketing and other questionable practices*. GAO-10-847T. Washington, DC: US Government Accountability Office, July 22, 2010. https://www.gao.gov/assets/gao-10-847t.pdf.

Uzogara, Stella G. "The impact of genetic modification of human foods in the 21st century: A review." *Biotechnology Advances* 18 (2000): 179–206.

Vallone, Lynne. *Big & small: A cultural history of extraordinary bodies*. New Haven: Yale University Press, 2017.

Van Aken, J. "Ethics of reconstructing Spanish flu: Is it wise to reconstruct a deadly virus?" *Heredity* 98 (2007): 1–2.

Van Beers, Britta. "TV cannibalism, *Body Worlds* and trade in human body parts." *Amsterdam Law Forum*, 4, no. 2 (2012): 65–75.

Van den Belt, Henk. "Frankenstein lives on." *Science* 359 (2018): 137.

Van den Berg, J. H. *Medische macht and medische ethiek* [Medical power and medical ethics]. Nijkerk: Uitgeverij G.F. Callenbach, 1970.

Van der Hoeven, Jan. "Levensbericht van Gerardus Vrolik" [Short biography of Gerardus Vrolik]. In *Jaarboek 1859*, 116–134. Amsterdam: Royal Netherlands Academy of Arts and Sciences, 2010. https://www.dwc.knaw.nl/DL/levens berichten/PE00003682.pdf.

Van de Velde, Cécile. "The 'Indignados': The reasons for outrage." *Cités* 3 (2011): 283–287.

Van Dijck, Jose. *Imagenation: Popular images of genetics*. Houndmills, UK: Macmillan Press, 1998.

Van Leeuwen, Arend Theodoor. *Prophecy in a technocratic era*. New York: Scribner's, 1968.

Van Nuland, Sonya E., and Kem A. Rogers. "Academic nightmares: Predatory publishing." *Anatomical Sciences and Education* 10, no. 4 (2017): 392–394.

Van Osselaer, Tine, Henk de Smaele, and Kaat Wils, eds. *Sign or Symptom? Exceptional corporeal phenomena in religion and medicine in the nineteenth and twentieth centuries*. Belgium: Leuven University Press, 2017.

Vegter, M. W. "Towards precision medicine; a new biomedical cosmology." *Medicine, Health Care and Philosophy* 21, no. 4 (2018): 443–456.

Vercauteren, Suzanne. "What is a biobank and why is everyone talking about them?" *University of British Columbia Medical Journal* 9, no. 2 (2018): 5–6.

Verhey, Allen. "'Playing God' and invoking a perspective." *Journal of Medicine and Philosophy* 20, no. 4 (1995): 347–364.

Verna, Mara. "Museums and the repatriation of indigenous human remains." In *Responsibility, fraternity, and sustainability in law: A symposium in honour of Charles D. Gonthier*, May 20–21, 2011, McGill University Faculty of Law. http://www.cisdl .org/wp-content/uploads/2018/05/Conférence-Charles-D-Gonthier-Mara-Verna .pdf.

Vogel, Lauren. "Researchers may be part of the problem in predatory publishing." *CMAJ* 189 (2017): E1324–E1325. https://doi.org/10.1503/cmaj.109-5507.

Voigt, Cornelia, Graham Brown, and Gary Howat. "Wellness tourists: In search of transformation." *Tourism Review* 66 (2011): 16–30.

Walsh, Declan. "Transplant tourists flock to Pakistan, where poverty and lack of regulation fuel trade in human organs." *Guardian*, February 9, 2005. https://www .theguardian.com/world/2005/feb/10/pakistan.declanwalsh.

Walsh, Fergus. "Paralysed man walks again after cell transplant." *BBC News*, October 21, 2014. http://www.bbc.com/news/health-29645760.

Walsh, Fergus. "The paralysed man who can ride a bike." *BBC News*, March 4, 2016. http://www.bbc.com/news/health-35660621.

Walt, Vivienne. "France's yellow vests straitjacket Macron." *Time*, December 17, 2018.

Wang, Amy T., Christopher P. McCoy, Mohammad Hassan Murad, and Victor M. Montori. "Association between industry affiliation and position on cardiovascular risk with rosiglitazone: Cross-sectional systematic review." *British Medical Journal* 340 (2010): c1344.

Wartolowska, Karolina, David J. Beard, and Andrew J. Carr. "Attitudes and beliefs about placebo surgery among orthopedic shoulder surgeons in the United Kingdom." *PLoS One* 9(3) (2014): e91699. https://doi.org/10.1371/journal.pone.0091699.

Watson, Roger. "Beall's list of predatory open access journals: RIP." *Nursing Open* 4, no. 2 (2017): 60.

Watson, Tom. "A global perspective on compassionate use and expanded access." *Therapeutic Innovation & Regulatory Science* 51 (2017): 143–145.

Watts, Nick, Markus Amann, Nigel Arnell , Sonja Ayeb-Karlsson, Kristine Belesova, Helen Berry, Timothy Bouley, et al. "The 2018 report of the *Lancet* countdown on health and climate change: Shaping the health of nations for centuries to come." *Lancet* 392 (2018): 2479–2514.

Weindling, Paul. "Julian Huxley and the continuity of eugenics in twentieth-century Britain." *Journal of Modern European History* 10, no. 4 (2012): 480–499.

Weiner, Jonathan. *Long for this world: The strange science of immortality*. New York: Ecco, 2010.

Weiner, Kate, and Paul Martin. "A genetic future for coronary heart disease?" *Sociology of Health & Illness* 30 (2008): 380–395.

Weiss, Kenneth M. "Is precision medicine possible?" *Issues in Science and Technology* 34, no. 10 (2017): 37–42.

Wells, Frank, and Michael Farthing, eds. *Fraud and misconduct in biomedical research.* 4th ed. London: Royal Society of Medicine Press, 2008.

Wharton, Annabel Jane. "Relics, Protestants, things." *Material Religion* 10, no. 4 (2014): 412–431.

White House, Office of the Press Secretary. "Remarks made by the President, Prime Minister Tony Blair of England (via satellite), Dr. Francis Collins, Director of the National Human Genome Research Institute, and Dr. Craig Venter, President and Chief Scientific Officer, Celera Genomics Corporation, on the completion of the first survey of the entire Human Genome Project." June 26, 2000. https://www.genome.gov/10001356/june-2000-white-house-event.

White House. "Remarks by President Trump in joint address to Congress." February 28, 2017. https://trumpwhitehouse.archives.gov/briefings-statements/remarks-president-trump-joint-address-congress.

Whittaker, Andrea, Lenore Manderson, and Elizabeth Cartwright. "Patients without borders: Understanding medical travel." *Medical Anthropology: Cross-Cultural Studies in Health and Illness* 29, no. 4 (2010): 336–343.

WHO (World Health Organization). "Density of physicians, 2019." Global Health Observatory Statistics Database. http://www.who.int/gho/health_workforce/physicians_density/en.

WHO. "1 in 10 medical products in developing countries is substandard or falsified." News release. November 28, 2017, http://www.who.int/mediacentre/news/releases/2017/substandard-falsified-products/en.

WHO. *A study on the public health and socioeconomic impact of substandard and falsified medical products.* Geneva: World Health Organization, 2017.

WHO. *WHO Global Surveillance and Monitoring System for substandard and falsified medical products.* Geneva: World Health Organization, 2017.

Widdows, Heather. "Between the individual and the community: The impact of genetics on ethical models." *New Genetics & Society* 28, no. 2 (2009): 173–188.

Widdows, Heather, and Sean Cordell. "The ethics of biobanking: Key issues and controversies." *Health Care Analysis* 19 (2011): 207–219.

Wiesing, Urban. "From art to science: A new epistemological status for medicine? On expectations regarding personalized medicine." *Medicine, Health Care and Philosophy* 21, no. 4 (2018): 457–466.

Wiesner-Hanks, Merry. *The marvelous hairy girls: The Gonzales sisters and their worlds.* New Haven: Yale University Press, 2009.

Wilkie, Tom. *Perilous knowledge: The Human Genome Project and its implications.* London and Boston: Faber and Faber, 1993.

Wilkinson, Dominic. "Restoring the balance to 'best interests' disputes in children: Editorial." *British Medical Journal* 358 (2017): j3666. https://doi.org/10.1136/bmj.j3666.

Wilkinson, Dominic, and Julian Savulescu. "After Charlie Gard: Ethically ensuring access to innovative treatment." *Lancet* 390 (2017): 540–542.

Wilkinson, Dominic, and Julian Savulescu. "Hard lessons: Learning from the Charlie Gard case." *Journal of Medical Ethics* 44 (2017): 438–442.

Wilkinson, Rachel. "The saints of Pittsburgh." *Smithsonian Magazine,* July/August 2017.

https://www.smithsonianmag.com/arts-culture/pittsburgh-church-greatest
-collection-relics-outside-vatican-180963680.

Williams, Bernard. "Consequentialism and integrity." In *Consequentialism and its critics*, edited by Samuel Scheffler, 20–50. Oxford: Oxford University Press, 1988.

Winickoff, David E. "Genome and nation: Iceland's health sector database and its legacy." *Innovations: Technology Governance Globalization* 1, no. 2 (2006): 80–105.

Winickoff, David E. "Partnership in UK Biobank: A third way for genomic property? *Journal of Law, Medicine & Ethics* 35, no. 3 (2007): 440–456.

Woloshin, Steven, and Lisa M. Schwartz. "Giving legs to restless legs: A case study of how the media helps make people sick." *PLoS Medicine* 3, no. 4 (2006): e170. https://doi.org/10.1371/journal.pmed.0030170.

Woodman, Josef. *Patients beyond borders: Everybody's guide to affordable, world-class medical travel.* 3rd ed. Chapel Hill, NC: Healthy Travel Media, 2015.

Woodward, Kenneth L. *Making saints: How the Catholic Church determines who becomes a saint, who doesn't, and why.* New York: Simon & Schuster, 1996. First published 1990.

Woodward, Kenneth L. "What miracles mean." *Newsweek*, April 30, 2000. https://www.newsweek.com/what-miracles-mean-157437.

Wootton, David. *Bad medicine: Doctors doing harm since Hippocrates.* Oxford: Oxford University Press, 2007.

World Wildlife Fund (WWF). *Living planet report, 2018: Aiming higher.* Edited by M. Grooten and R. E. A. Almond. Gland, Switzerland: WWF, 2018. https://www.wwf.no/assets/attachments/LPR2018-Full-Report.pdf.

Wray, Charlie M., and Lawrence K. Loo. "The diagnosis, prognosis, and treatment of medical uncertainty." *Journal of Graduate Medical Education* 7, no. 4 (2015): 523–527.

Wright, David. *Downs: The history of a disability.* Oxford: Oxford University Press, 2011.

Wulff, Henrik R. *Rational diagnosis and treatment.* Oxford: Blackwell Scientific, 1976.

Wulff, Henrik R., Stig Andur Pedersen, and Raben Rosenberg. *Philosophy of medicine: An introduction.* Oxford: Blackwell Scientific, 1986.

Yong, Ed. "The CRISPR baby scandal gets worse by the day." *Atlantic*, December 3, 2018. https://www.theatlantic.com/science/archive/2018/12/15-worrying-things
-about-crispr-babies-scandal/577234.

Zavestoski, Stephen, Phil Brown, Sabrina McCormick, Brian Mayer, Maryhelen D'Ottavi, and Jaime C. Lucove. "Patient activism and the struggle for diagnosis: Gulf War illnesses and other medically unexplained physical symptoms in the US." *Social Science & Medicine* 58 (2004): 161–175.

Zika, Eleni, Daniele Paci, Tobias Schulte in den Bäumen, Anette Braun, Sylvie RijKers-Defrasne, Mylène Deschênes, Isabel Fortier, Jens Laage-Hellman, Christian A. Scerri, and Dolores Ibarreta. *Biobanks in Europe: Prospects for harmonisation and networking.* JRC Scientific and Technical Reports. Document no. EUR 24361 EN. Luxembourg: Publication Office of the European Union, 2010.

Zola, Irving K. "In the name of health and illness: On some socio-political consequences of medical influence." *Social Science & Medicine* 9 (1975): 83–87.

Zoloth, Laurie. "The Alexandria plan: Creating libraries for human tissue research and therapeutic use." In *The ethics of research biobanking*, edited by Jan Helge Solbakk, Søren Holm, and Bjørn Hofmann, 173–193. Dordrecht: Springer, 2009.

Zuboff, Shoshana, *The age of surveillance capitalism: The fight for a human future at the new frontier of power*. London: Profile Books, 2019.

Zuckerman, Diana. "Hype in health reporting: 'Checkbook science' buys distortion of medical news." *International Journal of Health Services* 33 (2003): 383–389.

Zwiebach, Burton. *The common life: Ambiguity, agreement, and the structure of morals*. Philadelphia: Temple University Press, 1988.